GNSS Software Receivers

Build and operate multi-GNSS and multi-frequency receivers with state-of-the-art techniques using this up-to-date, thorough and easy-to-follow text. Covering both theory and practice and complemented by MATLAB$^{©}$ code and digital samples with which to test it, this package is a powerful learning tool for students, engineers and researchers everywhere. Suggestions of hardware equipment allow you to get to work straight away and to create your own samples. Concisely but clearly explaining all the fundamental concepts in one place, this is also a perfect resource for readers seeking an introduction to the topic.

Kai Borre was professor of geodesy and founder of the Danish GPS Center at Aalborg University. He was the co-author of several popular books and creator of widely used software within his field. Kai Borre received The Order of Dannebrog and held an honorary Doctorate at Vilnius Technical University.

Ignacio Fernández-Hernández works at the European Commission, DG DEFIS, as responsible for Galileo authentication and high accuracy services. He is also an adjunct professor at KU Leuven, where he teaches satellite navigation, and has been a visiting scholar at Stanford University. He is the recipient of the Institute of Navigation's 2021 Thurlow Award.

José A. López-Salcedo is a professor of electrical engineering at Universitat Autònoma de Barcelona. He has been the technical lead for more than 20 research projects on signal processing techniques for GNSS receivers and a visiting researcher at the University of Illinois Urbana-Champaign, the University of California, Irvine and the European Commission.

M. Zahidul H. Bhuiyan is a research professor at the Department of Navigation and Positioning at the Finnish Geospatial Research Institute (FGI). He has more than 15 years of experience in the GNSS field with expertise in multi-frequency multi-GNSS receiver development. He is actively involved in teaching GNSS related courses in Finnish universities and other international training schools. He has also been working as a technical expert for the EU Agency for the Space Programme (EUSPA) in H2020 project reviewing and proposal evaluation since 2017.

GNSS Software Receivers

Edited by

KAI BORRE
Aalborg University

IGNACIO FERNÁNDEZ-HERNÁNDEZ
European Commission; KU Leuven

JOSÉ A. LÓPEZ-SALCEDO
Universitat Autònoma de Barcelona

M. ZAHIDUL H. BHUIYAN
Finnish Geospatial Research Institute

CAMBRIDGE
UNIVERSITY PRESS

University Printing House, Cambridge CB2 8BS, United Kingdom

One Liberty Plaza, 20th Floor, New York, NY 10006, USA

477 Williamstown Road, Port Melbourne, VIC 3207, Australia

314–321, 3rd Floor, Plot 3, Splendor Forum, Jasola District Centre, New Delhi – 110025, India

103 Penang Road, #05–06/07, Visioncrest Commercial, Singapore 238467

Cambridge University Press is part of the University of Cambridge.

It furthers the University's mission by disseminating knowledge in the pursuit of education, learning, and research at the highest international levels of excellence.

www.cambridge.org
Information on this title: www.cambridge.org/9781108837019
DOI: 10.1017/9781108934176

First published 2023

Printed in the United Kingdom by TJ Books Limited, Padstow Cornwall

A catalogue record for this publication is available from the British Library.

Library of Congress Cataloging-in-Publication Data
Names: Fernández-Hernández, Ignacio, 1977– author. | Borre, K. (Kai), author. | López-Salcedo, José A., 1978– author. | Bhuiyan, M. Zahidul H., 1980– author.
Title: GNSS software receivers / edited by Kai Borre, Aalborg University, Denmark, Ignacio Fernández-Hernández, European Commission, José A. López-Salcedo, Universitat Autònoma de Barcelona, M. Zahidul H. Bhuiyan, Finnish Geospatial Research Institute.
Description: First edition. | Cambridge, United Kingdom ; New York, NY, USA : Cambridge University Press, 2022. | Includes bibliographical references and index.
Identifiers: LCCN 2022016994 | ISBN 9781108837019 (hardback) | ISBN 9781108934176 (ebook)
Subjects: LCSH: GPS receivers. | Signal theory (Telecommunication) | Software radio. | Mobile geographic information systems – Equipment and supplies. | BISAC: TECHNOLOGY & ENGINEERING / Signals & Signal Processing
Classification: LCC TK6565.D5 .F47 2022 |
DDC 621.3841/91–dc23/eng/20220708
LC record available at https://lccn.loc.gov/2022016994

ISBN 978-1-108-83701-9 Hardback

Additional resources for this publication at www.cambridge.org/gnss

Contents

Contributors

M. Zahidul H. Bhuiyan Department of Navigation and Positioning, Finnish Geospatial Research Institute, National Land Survey of Finland, Kirkkonummi, Finland

Padma Bolla DRDO, Telangana, India

Kai Borre Formerly Aalborg University, Aalborg, Denmark

Ignacio Fernández-Hernández Directorate-General Defence Industry and Space, European Commission, Brussels, Belgium. Department of Electrical Engineering, KU Leuven, Leuven, Belgium

Giorgia Ferrara Septentrio, Espoo, Finland

Salomon Honkala Exafore Oy, Tampere, Finland

Martti Kirkko-Jaakkola Department of Navigation and Positioning, Finnish Geospatial Research Institute, National Land Survey of Finland, Kirkkonummi, Finland

Heidi Kuusniemi Digital Economy, University of Vaasa, Vaasa, Finland. Department of Navigation and Positioning, Finnish Geospatial Research Institute, National Land Survey of Finland, Kirkkonummi, Finland

Elena Simona Lohan Communications Engineering Faculty of Information Technology and Communication Sciences Electrical Engineering unit, Tampere University (TAU), Tampere, Finland

José A. López-Salcedo Department of Telecommunications and Systems Engineering, IEEC – CERES, Universitat Autònoma de Barcelona, Barcelona, Spain

Gonzalo Seco-Granados Department of Telecommunications and Systems Engineering, IEEC – CERES, Universitat Autònoma de Barcelona, Barcelona, Spain

Stefan Söderholm Septentrio, Espoo, Finland

Sarang Thombre Department of Navigation and Positioning, Finnish Geospatial Research Institute, National Land Survey of Finland, Kirkkonummi, Finland

Foreword

Software-defined radios (SDRs) brought an unprecedented degree of flexibility to the design of the receivers that support radio navigation. Previously, such design work had been the domain of a handful of hardware manufacturers whose schemes were implemented in hardware. This hardware provided a limited set of outputs and virtually no access to the internal states and data processing. SDRs allowed customization that simply was not possible using mass-produced receiver sets. Implementation specifics that were previously set in stone (or at least silicon) could now be easily tweaked or even changed wholesale. Kai Borre's vision to implement an SDR in MATLAB made the receiver inner workings even more accessible. While nowhere near as practical as a commercial receiver, the code provided by Kai and collaborators was infinitely more interesting and useful as a scientific tool. The ability to easily test new acquisition schemes, different tracking loop designs and obtain wholly new outputs was placed in the hands of anyone who was interested. Equal in importance was the educational opportunity that this tool provided. What better way to understand the signal structure, acquisition and tracking than to use the code to follow the signal processing through from reception to output? Countless scientists and engineers have used this code to better understand how the Global Navigation Satellite System (GNSS) works.

We were very fortunate to gain access to, and assist in the development of early versions of this code in 2002, and even more fortunate to have Kai come to the United States in the Fall of 2005. During this time, he spent two months teaching Stanford students about receiver design and use of the code. This was the first of what became many such visits to Stanford. Kai was the consummate teacher, the enthusiasm he displayed and level of preparation that he put into his lectures was second to none. Our students benefited greatly from his expertise, which he eagerly shared with all. With these tools, we were able to investigate many new aspects related to satellite navigation that simply were not possible with the hardware receivers we had access to at the time. Now we could track scintillation effects to far weaker signal levels than ever before. We could evaluate large numbers of correlator spacings to evaluate the correlation peak distortion to much finer levels of detail. We could perform deep integration with inertial measurements. The list of possibilities is never-ending, and the SDR continues to be one of the most powerful tools for our ongoing research.

Sadly, we lost Kai in the summer of 2017, and we will forever miss the conversations that we had covering his extremely wide range of interests. His sense of humor, his mentorship, and the friendship that he offered were as highly valued as his technical contributions. Fortunately, Ignacio, Jose, and Zahidul have continued his legacy and provided here a complete and thorough update to the design and application of SDRs for GNSS. They greatly expand upon the initial single frequency Global Positioning System (GPS) and early Galileo work to now encompass all four constellations, a novel regional system and the L5 frequency. Further they have significantly expanded on the ever so important MATLAB code base, which continues to truly provide the critical ability to complement the theoretical components detailed in this text with practical, hands-on experience.

This latest text from Kai and his talented co-authors will help to ensure that the knowledge to advance the science of GNSS flourishes, setting the framework for the next generation of satellite navigation leaders. This is something, we feel, the consummate educator Kai Borre had always aspired to do. And in answer to Kai's next page question: yes, this is very exciting!

Dennis M. Akos
Professor, Aerospace Engineering Sciences
University of Colorado at Boulder

Todd Walter
Professor (Research), Aeronautics and
Astronautics Stanford University

Preface

Even if around for decades, SDRs have in the recent years received an enormous recognition and generated widespread interest within the receiver industry. In SDR, a receiver employs an analog-to-digital converter (ADC) that captures all channels of the software radio node and then demodulates the channel waveform using software on a general-purpose processor.

In 1995, I had a vision about coding a GPS receiver in MATLAB. The aim was to create a supplement to existing hardware receivers. It was to serve as an experimental tool for developments. At that time, it was unrealistic to expect a receiver coded in MATLAB to run in real-time. In the spring of 2004, two students at Aalborg University turned the vision into reality. They coded a GPS L1 receiver in MATLAB. In parallel, the idea of writing an accompanying textbook on software-defined GNSS receivers emerged. The textbook was published as *A Software-Defined GPS and Galileo Receiver - A Single-Frequency Approach*, together with the complete MATLAB code. You could additionally purchase a front end and a Universal Serial Bus (USB) driver transferred the digitized data to a laptop. If you had an appropriate antenna, you would be able to change your laptop into a GNSS receiver.

The entire tool became a great success. The technology and applications for software-defined receivers developed over the years. It became obvious that a new text was necessary to gratify the growing demands. Again a group of authors and programmers joined to describe new topics and trends. With the elements provided with this book, readers will be able to construct their own GNSS receiver and compute a position. Is this not exciting?

Kai Borre, December 2016

Our Motivation

First, the GNSS civil community has evolved from one signal, GPS L1 C/A, the focus of our predecessor, to more than a dozen signals from four global constellations, and a similar yet increasing amount of regional systems, adding up to more than 100 satellites. This offers a wide variety of frequencies, signal modulations, coding schemes and message structures. In parallel, receivers have also become very diverse. Our intention is to provide an up-to-date reference that, compared to our predecessor, reflects at least part of this diversity. We focus on three aspects: multisystem (GPS, GLObalnaya NAvigatsionnaya Sputnikova Sistema [GLONASS], Galileo, BeiDou and NavIC), multifrequency (L1 and L5) and multiarchitecture (standard and snapshot).

Second, we have strengthened the link between the book and the MATLAB code provided with it. Most of the figures appearing in the book can be generated with the MATLAB code and available samples. The front end, acquisition, tracking and positioning parameters used are presented and justified. Readers can also change the configurations to obtain different results or record their own samples with very affordable hardware equipment that is also described in the book.

Finally, we hope that the practical orientation of the book makes it a good entry point for engineers and scientists who want to get acquainted with GNSS, but do not know where to start. Our focus is on receiver-specific aspects, and we have made an effort to explain the subtleties that make a GNSS receiver a complicated device in an intuitive way. However, we also include the basic GNSS principles, with emphasis on signal processing, but not neglecting positioning aspects, that make the book self-contained. Throughout the book, the readers are also invited to consult other references that are more comprehensive on GNSS or cover some specific areas, which this book does not intend to replace. There are excellent references on GNSS, and there are also open GNSS software receiver platforms, but an up-to-date combination of both is not so frequent.

The Structure of the Book

The book is divided into 12 chapters. Chapter 1 presents both GNSS signals and receivers. It explains why signals are designed the way they are and presents an overview of the current GNSS signals. This is followed by a detailed signal model, including its different components. Later, the chapter explains signal propagation from the satellite to the receiver and its conditioning in the receiver radio front end. We also describe how receivers can estimate the strength of the signal through the signal-to-noise ratio (SNR) and the carrier-to-noise density ratio (C/N_0). This is followed by the mathematical model for the received signals used in the rest of the book, with emphasis on the down-conversion and sampling process, which leads to shifting our signal representation from continuous time to discrete time. Then, the chapter analyzes how the signals are processed in the receiver. This includes the two main signal processing tasks: acquisition and tracking. We highlight the main acquisition strategies and tracking loops used in GNSS and the engineering trade-offs for their design. The chapter then describes how GNSS navigation data are demodulated in the receiver, including some error detection and correction concepts borrowed from communications. This is followed by the measurement propagation and processing errors, and finalizes with the position computation problem through the least-squares estimator, and some derivations on satellite geometry, measurement residuals, and positioning accuracy. Chapter 1 therefore provides the theoretical foundations on which the SDRs and some receiver design trade-offs, such as sampling frequency, bandwidth, acquisition parameters or tracking loop parameters, are built. The rest of the chapters builds, to a high extent, on this chapter. We have condensed a wealth of information on GNSS signals and receivers in a single chapter, and we hope that readers, ideally but not necessarily with some signal-processing background, can use it as both an entry point to the field of GNSS and a handbook for receiver design aspects.

Chapter 2 introduces GPS and presents the GPS L1 C/A processing chain of the multi-GNSS SDR. After an overview of the system and its signals, the chapter particularizes the general signal model from Chapter 1 to the signal under study and describes the receiver acquisition, tracking, and position computation blocks. This chapter concludes with some experimental results from the MATLAB SDR reproducible by the reader with the samples provided. A reader with a practical, hands-on approach, can go straight to Chapter 2 and consult Chapter 1 for more details on specific matters.

The next four chapters, Chapters 3–6, share the structure and brevity of Chapter 2 and focus on the other satellite navigation systems treated in the book: GLONASS, Galileo, BeiDou, and NavIC. Chapter 3 presents the GLONASS L1 open signal, frequency-division multiple access (FDMA), or L1OF, and the related receiver chain. After a general description of GLONASS and its current signals, the GLONASS L1OF signal is compared, for clarity purposes, with the GPS L1 C/A of the previous chapter. Then, the acquisition, tracking and navigation blocks are described and followed by a presentation of the experimental results. Being GLONASS L1OF the only FDMA signal processed in the book, this chapter presents some particularities of the FDMA signal as well as its processing. Chapter 4 presents the Galileo system, with focus on the E1 signal, E1-B and E1-C components, both processed in the MATLAB receiver. It shows the advantages of combined pilot + data acquisition and pilot tones for signal tracking. Chapters 5 and 6 present the BeiDou B1I and NavIC L5 signals and receiver chain, respectively. The latter is one of the novelties of this book with respect to other books in the GNSS literature.

Chapter 7 describes the integration of all the chains into the single-frequency multi-GNSS receiver. It describes how all functional blocks are implemented in the receiver and then shows some experimental results obtained with the available multi-GNSS sample data. It also describes multi-GNSS positioning, not covered in the previous chapters.

Chapter 8 describes the dual frequency receiver tracking GPS L1 C/A and L5 signals. The dual-frequency signal-processing architecture and algorithms are discussed by taking specific example as GPS L1 and L5 signals. The receiver position uses both L1 and L5 measurements and in particular the ionosphere-free linear combination of carrier-smoothed pseudoranges. This chapter concludes with some results obtained from the second MATLAB GNSS SDR provided with this book: the GPS dual frequency SDR.

Chapter 9 describes snapshot receivers. Snapshot receivers gather just some milliseconds of digital samples to compute a position and time solution. This chapter explains how to maximize the accuracy of the measurements from the acquisition stage and how to compute a position without synchronizing with the GNSS signal and reading its navigation message. We use instantaneous Doppler positioning to initialize the more accurate code-delay positioning stage. We also present some experimental results from the MATLAB GPS L1 C/A snapshot receiver offered with the book.

As a complement to previous chapters, Chapter 10 presents acquisition and tracking schemes with focus on BOC signals. Given the use of Binary Offset Carrier (BOC) signals by most GNSS, and their multipeak autocorrelation function, this chapter describes modern techniques to manage peak ambiguities in signal acquisition and tracking, including advanced multipath mitigation techniques.

Chapter 11 focuses on hardware. In particular, it provides some information on the front ends used for capturing the samples, and front ends that are available on the market and can be purchased by the readers if they want to create and analyze their own samples. It describes the full set up for this purpose. Finally, it presents some trends on GNSS SDR processing platforms, architectures and applications.

Finally, Chapter 12 presents how to download, install and operate the different MATLAB SDRs provided with the book. It includes an overview of the main inputs, outputs, blocks, and functions for each receiver. It also shows how to generate some of the figures in the book that are independent from the SDRs.

The Software

In addition to the textbook, our complete package consists of: a single-frequency multi-system GNSS receiver, a dual frequency GPS L1 C/A and L5 receiver, and a snapshot GPS L1 C/A receiver.

A complete single-frequency multi-GSRx, the Finnish Geospatial Research Institute-GNSS Software Receiver (FGI-GSRx), implemented in MATLAB, was developed by the Department of Navigation and Positioning of the Finnish Geospatial Research Institute. The first version of FGI-GSRx was based on Kai Borre's Soft-GNSS MATLAB GPS receiver. Since this receiver was not originally designed for multi-GNSS operation, it has been significantly modified to support GPS, GLONASS, Galileo, BeiDou and NavIC and to make the receiver more configurable. The experimental results shown in Chapters 2–6 have been obtained by using FGI-GSRx in single-system mode, that is, by processing only one system at the time. The single-frequency multi-GNSS results are presented in Chapter 7. FGI-GSRx provides a unique and easy-to-use platform not only for research and development but also for whoever is interested in learning about GNSS receivers.

Software accompanying this book also provides a dual frequency GPS receiver, supporting Chapter 8. It covers a complete implementation of a GNSS dual-frequency software receiver, dual-frequency GNSS Software Receiver (DF-GSRx), to provide an ionosphere error-free navigation solution. Furthermore, carrier-phase observations are used to enhance the precision of navigation solution. A specific example with real GPS L1/L5 signals demonstrates the performance of DF-GSRx. It was developed by Padma Bolla at Samara National Research University, Russia, also based on SoftGNSS. Interested readers can make use of DF-GSRx software with minor modifications to process dual-frequency signals from other GNSS or regional systems.

Finally, we provide a snapshot of the GPS L1 C/A single frequency MATLAB receiver, supporting Chapter 9. This receiver is based on SoftGNSS, with additions by Oriol Badia-Solé and Tudor Iacobescu, and developed by I. Fernández-Hernández for Aalborg University as part of a PhD thesis. The snapshot receiver has no tracking stage and reads the satellite data from external files, so unlike the other software packages that may take longer to run, the reader can use it to compute an almost-instantaneous position solution.

In addition to the code, we provide the data samples used by the different SDRs to generate the plots in the book, and scripts to other plots from Chapter 10, as part of the package. We have used MATLAB (version R2015b and above) as our coding language because it is the *de facto* programming environment at many universities, it is a flexible language, and it is easy to learn. Additionally, it provides excellent facilities for the presentation of graphical results.

We recall that the software receivers provided with this book are not an operational software. They are not developed according to operational standards, functionalities like exception handling are only partially included, and they have not been exhaustively tested or qualified. Neither are they optimized to yield the best possible performance. Rather, they implement most of the standard signal-processing and position computation algorithms that are described in the book. We believe that this is sufficient for an educational tool, if not better.

One of the challenges of this book and accompanying material is to evolve the original GPS receiver in three dimensions, multi-GNSS, multifrequency and multi-architecture, without exponentially increasing the complexity and work associated. While the three receiver implementations provided cannot cover each and every combination possible, we believe that they provide the necessary building blocks. The reader can reuse the code and build their own receiver at their convenience, for example by combining multi-GNSS with snapshot architectures or by developing multi-GNSS dual frequency receivers. The readers are also invited to implement some of the more modern techniques described in the book (e.g. in Chapter 10 on BOC processing) but not implemented or some developed on their own. The receiver can also be a good starting platform for processing other signals or novel algorithms for anti-jamming, anti-spoofing, multipath detection and mitigation, GNSS authentication, carrier phase high accuracy algorithms, etc. Our work can also be integrated in nowadays' more sophisticated platforms that integrate GNSS with other sensors, signals, and data.

Final Remarks

In the summer of 2017, when this book was at its early stage, Kai Borre left us at the age of 75. Since then, the remaining editors have tried to honor our commitment and put together a book that is worthy of its predecessor. We hope that this book serves, on the one hand, as a tribute to Kai Borre's work in GPS over several decades, which helped so many of us get acquainted with satellite navigation; and on the other hand, we hope it will help engineers build a next generation of receivers that are more accurate, affordable, and safe.

Acknowledgments

We would like to extend our appreciation to the following individuals and organizations for their contributions to this effort: Kai Borre's family members Hanne Lene Jakobsen and Sara Skar for their involvement in the last years; all the chapter authors for their commitment over such a long period of time; Jean-Marie Sleewagen, Francisco Gallardo, David Gómez-Casco, Sergi Locubiche-Serra, Nicola Linty, Guillermo Tobías, David Calle, Irma Rodríguez, Robin Geelen, Jochum Hoes and Thibo Jacqmotte for their reviews and suggestions; Darius Plausinaitis, Oriol Badia-Solé, Tudor Iacobescu, Michele Bavaro and Tyler G. R. Reid for making available their code; Julie Lancashire, Julia Ford, Sarah Strange, Thirumangai Thamizhmani, Jenny van der Meijden and the rest of the Cambridge University Press team for their involvement and patience during the edition process; The National Land Survey of Finland for authorizing the use of their MATLAB code; and Samara State University for their financial support to Professor Borre in the early stages of the book. This book has been written in LaTeX and edited in *Overleaf*.

Ignacio Fernández-Hernández, José A. López-Salcedo, M. Zahidul H. Bhuiyan,
October 2021

Disclaimer: The information and views set out in this book are those of the authors and do not necessarily reflect the official opinion of any organisation.

Abbreviations

ACF	Autocorrelation Function
ADC	Analog-to-Digital Converter
A-GNSS	Assisted GNSS
AltBOC	Alternative Binary Offset Carrier
APME	A-Posteriori Multipath Estimation
ARAIM	Advanced Receiver Autonomous Integrity Monitoring
ARM	Advanced RISC Machine
ARNS	Aeronautical Radio Navigation Service
ASCII	American Standard Code for Information Interchange
ASIC	Application-Specific Integrated Circuit
AWGN	Additive White Gaussian Noise
BCH	Bose-Chaudhuri-Hocquenghem
BDCS	BeiDou Coordinate System
BDS	BeiDou System
BDT	BeiDou Time
BER	Bit Error Rate
B&F	Betz & Fishman
BJ	Bump Jumping
BOC	Binary Offset Carrier
BPSK	Binary Phase-Shift Keying
BW	Band Width
C/A	Coarse Acquisition
CAS	Commercial Authentication Service
CBOC	Composite Binary Offset Carrier
CCART	Correlation Combination Ambiguity Removing Technology
CDMA	Code Division Multiple Access
CDBOC	Complex Double Binary Offset Carrier

CED	Clock and Ephemeris Data
CGCS2000	China Geodetic Coordinate System 2020
C/N_0	Carrier-to-Noise Density Ratio
CNAV	Civil NAVigation (message)
COSPAS	COsmicheskaya Sistyema Poiska Avariynich Sudow - Space System for Search of Distress Vessels
CPU	Central Processing Unit
CRC	Cyclic Redundancy Check
CRMM	Complexity Reduced Multipath Mitigation
CSAC	Chip Scale Atomic Clock
CW	Continuous Wave
dB	Decibels
DBOC	Double-Binary Offset Carrier
DC	Direct Current
DD	Decision Directed
DDPE	Decimation Double Phase Estimator
DE	Double Estimator
DF	Dual Frequency
DET	Double Estimator Technique
DFMC	Dual-Frequency Multiconstellation
DFT	Discrete Fourier Transform
DGNSS	Differential GNSS
DLL	Delay Lock Loop (or Delay-Locked Loop)
DoD	(US) Department of Defense
DOP	Dilution Of Precision
DP	Dot Product
DSP	Digital Signal Processor
DS-SS	Direct Sequence Spread Spectrum
DTFT	Discrete-Time Fourier Transform
DVB-S	Digital Video Broadcasting - Satellite
E_b/N_0	Energy per bit to Noise Density Ratio
ECEF	Earth Centered, Earth Fixed
EIRP	Effective Isotropic Radiated Power
ELS	Early-Late Slope
EM	Expectation Maximization
EML	Early Minus Late

EMLP	Early Minus Late Power
ENU	East-North-Up
ESD	Energy Spectral Density
ESPRIT	Estimation of Signal Parameters via Rotational Invariance Techniques
FDMA	Frequency Division Multiple Access
FEC	Forward Error Correction
FFT	Fast Fourier Transform
FGI	Finnish Geospatial Research Institute
FGI-GSRx	Finnish Geospatial Research Institute - GNSS Software Receiver
FIMLA	Fast Iterative Maximum Likelihood Algorithm
FLL	Frequency Lock Loop (or Frequency-Locked Loop)
FPGA	Field Programmable Gate Array
FSPL	Free Space Path Loss
GDOP	Geometric Dilution of Precision
GEO	Geostationary Earth Orbit
GIVD	Grid Ionospheric Vertical Delay
GIVE	Grid Ionospheric Vertical Error
GLONASS	GLObalnaya NAvigatsionnaya Sputnikova Sistema - Global Navigation Satellite System
GLONASST	GLONASS Time
GNSS	Global Navigation Satellite System
GPP	General Purpose Processor
GPS	Global Positioning System
GPU	Graphical Processing Unit
GPST	GPS Time
GRASS	General Removing Ambiguity via Sidepeak Suppresion
GST	Galileo System Time
GTK	Generalized Teager Kaiser
GTRF	Galileo Terrestrial Reference Frame
HAS	High Accuracy Service
HDOP	Horizontal Dilution Of Precision
HOW	Hand-Over Word
HRC	High Resolution Correlator
Hz	Hertz
IC	Integrated Circuit
ICD	Interface Control Document

IDTFT	Inverse Discrete-Time Fourier Transform
IERS	International Earth Rotation and Reference Systems Service
IF	Intermediate Frequency
IFB	Inter-Frequency Bias
IFFT	Inverse Fast Fourier Transform
IGSO	Inclined Geosynchronous Earth Orbit
IMU	Inertial Measurement Unit
ION	Institute Of Navigation
IoT	Internet of Things
IRNSS	Indian Regional Navigation Satellite System
ISC	Inter Signal Correction
ISPA	Inhibition Side Peak Acquisition
IST	India Standard Time
ITU	International Telecommunications Union
LDPC	Low-Density Parity Check
LFSR	Linear Feedback Shift Register
LHCP	Left-Hand Circularly Polarized
LNA	Low Noise Amplifier
LOS	Line of Sight
LPI	Low Probability of Interception
LS	Least Squares
LTE	Long Term Evolution
MAP	Maximum A Priori
MAT	Mean Acquisition Time
MBOC	Multiplexed Binary Offset Carrier
MCMC	Monte Carlo Markov Chain
MCRW	Modified Correlator Reference Waveform
ME	Multipath Error
MEDLL	Multipath Estimating Delay Lock Loop
MEE	Multipath Error Envelope
MEO	Medium Earth Orbit
MET	Multipath Elimination Technique
MGD	Multiple Gate Delay
M&H	Martin & Heiries
ML	Maximum Likelihood
MMSE	Minimum Mean Square Error

MMT	Multipath Mitigation Technology
MOPS	Minimum Operational Performance Standards
MUSIC	MUltiple SIgnal Classification
NASA	National Aeronautics and Space Administration
NAVIC	NAVigation with Indian Constellation
NC	Narrow Correlator
NCE	Non-Constant Envelope
NCO	Numerically Controlled Oscillator
NDP	Noncoherent Dot Product
NED	North-East-Down
NEML	Normalized Early Minus Late
NEMLP	Normalized Early Minus Late Power
NH	Neumann Hoffman
NLOS	Non-Line of Sight
NMEA	National Marine Electronics Association
NNEML	Normalized Noncoherent Early Minus Late
NWPR	Narrow- and Wideband Power Ratio
OCXO	Oven-Controlled Crystal Oscillator
OEM	Original Equipment Manufacturer
OFDM	Orthogonal Frequency-Division Multiplexing
OS	Open Service
OSNMA	Open Service Navigation Message Authentication
PAC	Pulse Aperture Correlator
PAM	Pulse Amplitude Modulation
PC	Personal Computer
PCS	Parallel Code-phase Search
PDF	Probability Density Function
PDOP	Position Dilution of Precision
PFS	Parallel Frequency Search
PLL	Phase Lock Loop (or Phase-Locked Loop)
POCS	Projection Onto Convex Sets
PPP	Precise Point Positioning
PPS	Precise Positioning Service
PRS	Public Regulated Service
PRN	Pseudo-Random Noise
PSD	Power Spectral Density

PT	Peak Tracking
PVT	Position, Velocity, and Timing
PZ-90	Parametry Zemli 1990 - Earth Parameters 1990
QPSK	Quadrature Phase Shift Keying
QZSS	Quasi-Zenith Satellite System
RAE	Running Average Error
RAIM	Receiver Autonomous Integrity Monitoring
RF	Radio Frequency
RFFE	Radio Frequency Front End
RFI	Radio Frequency Interference
RHCP	Right-Hand Circularly Polarized
RINEX	Receiver INdependent EXchange (format)
RISC	Reduced Instruction Set Computing
RMS	Root Mean Square
RMSE	Root Mean Square Error
RSSML	Reduced Search Space Maximum Likelihood
RTCA	Radio Technical Commission for Aeronautics
RTCM	Radio Technical Commission for Maritime purposes
RTK	Real-Time Kinematics
SAGE	Space-Alternating Generalized Expectation-Maximization
SAM	Slope Asymmetry Metric
SAR	Search And Rescue
SARSAT	Search And Rescue Satellite-Aided Tracking
SBAS	Satellite-Based Augmentation System
SBME	Slope-Based Multipath Estimator
SCM	Side lobe Cancellation Method
SCPC	SubCarrier Phase Cancellation
SDN	Software-Defined Network/Networking
SDR	Software-Defined Radio
SIS	Signal-In-Space
SISA	Signal-In-Space Accuracy
SISE	Signal-In-Space Error
SISRE	Signal-In-Space Range Error
SLC	Side-Lobe Cancellation
SLL	Sub-Carrier Lock Loop
SNPR	Signal-to-Noise Power Ratio

SNR	Signal-to-Noise Ratio
SPS	Standard Positioning Service
SQM	Signal Quality Monitoring
SS	Spread Spectrum
SSB	Single-Side Band
SV	Space Vehicle
TCXO	Temperature-Compensated Crystal Oscillator
TEC	Total Electron Content
TGD	Timing/Total Group Delay
TK	Teager Kaiser
TLM	TeLeMetry word
TMBOC	Time Multiplexed Binary Offset Carrier
ToA	Time of Arrival
TTFF	Time To First Fix
TT&C	Telemetry, Tracking and Control
UAL	Unsuppressed Adjacent Lobes
UAV	Unmanned Aerial Vehicles
UEE	User Equipment Error
UERE	User Equivalent Range Error
URA	User Range Accuracy
URE	User Range Error
USB	Universal Serial Bus
UTC	Universal Time Coordinated
UTM	Universal Transverse Mercator
VC	Vision Correlator
VDOP	Vertical Dilution Of Precision
VLSI	Very Large-Scale Integration
WGS 84	World Geodetic System 1984
WLS	Weigthed Least Squares
ZHD	Zenith Hydrostatic Delay
ZWD	Zenith Wet Delay

Main Constants

K_{boltz}	$1.3806488 \cdot 10^{-23}$	J/K	Boltzmann constant
R_e	$6,378,137$	m	Earth semimajor axis (equatorial radius)
f_e	$1/298.257223563$		Earth flattening factor
μ	$3.986004418 \cdot 10^{14}$	m^3/s^2	Earth gravitational constant
$\dot{\Omega}_e$	$7.2921151467 \cdot 10^{-5}$	rad/s	Earth rotation rate
π	3.1415926535898		Pi
c	$299,792,458$	m/s	Speed of light
G	$6.67408 \cdot 10^{-11}$	$m^3/(kg \cdot s^2)$	Universal gravitational constant

1 GNSS Signals and Receivers

José A. López-Salcedo, Ignacio Fernández-Hernández, M. Zahidul H. Bhuiyan, Elena Simona Lohan and Kai Borre

1.1 Introduction

This chapter introduces the fundamentals of signals and receivers used in Global Navigation Satellite Systems (GNSS). We will start with a general overview of the existing GNSS signals and their main characteristics in Section 1.2, in order to set the grounds for the subsequent sections and chapters. We will continue with Section 1.3, presenting in more detail the structure of GNSS signals and the theoretical models that will be used throughout the book. Section 1.4 will review the GNSS link budget to better understand how GNSS signals are received in the way they are. Then, Section 1.5 will briefly introduce the architecture of GNSS receivers, while their acquisition and tracking modules will be discussed in more detail in Sections 1.6 and 1.7, respectively. Sections 1.8 and 1.9 will cover the navigation message and pseudorange errors, respectively, and finally, Section 1.10 will explain how to calculate a position fix.

1.2 Overview of GNSS Signals

The main purpose of GNSS signals is to provide an accurate ranging measurement to the receiver, while at the same time, to provide the necessary data for the receiver to compute its position. Therefore, well-designed GNSS signals must allow ranging measurements as accurate as possible, as well as data reception without errors, whenever possible. To do so, several constraints and considerations must be borne in mind, as discussed next.

First of all, for both accurate ranging and good data reception, the carrier frequency must be chosen so the signal can propagate well through the atmosphere and be received in all possible weather and visibility conditions. The L band, between 1 and 2 GHz, with a wavelength between 30 and 15 cm is a good candidate for this purpose. In fact, current GNSS signals are transmitted in this band. In their journey from the satellite to the receiver, signals are delayed by the ionosphere, a layer of the atmosphere at around some hundreds of kilometers above the Earth's surface, and then by the troposphere, a lower and thinner layer that determines our weather. Unlike the tropospheric delay, most of the ionospheric delay is difficult to model, but it is related to the carrier frequency of the signal, so having two synchronized frequencies transmitted from the same satellite allows removing it and having more accurate ranging measurements. Transmitting signals in several frequencies also allows more services, increased resilience against interference and precise point positioning (PPP) improvements. Table 1.1 lists the features of GNSS signals used in this book and in the software accompanying it: the American Global Positioning System (GPS), the Russian GLONASS, the European Galileo, the Chinese BeiDou and the Indian Navigation Indian Constellation (NavIC). The first four are global systems, while the latter is a regional system.[1]

In addition to choosing an adequate carrier frequency, the signals need to be powerful and have a broad-enough frequency bandwidth to provide accurate ranging. Unfortunately, generating a powerful signal increases the satellite weight, making it more costly to put it in orbit. GNSS signals are transmitted at a power of some tens of watt (W) and received on Earth at an extremely low power, around 10^{-16} W, which makes the performance of GNSS even more remarkable. Apart from signal power restrictions, the frequency bandwidth is also constrained by satellite and receiver technology. Modern signals such as Multiplexed Binary Offset Carrier (MBOC) used by GPS and Galileo, or Alternative Binary Offset Carrier (AltBOC) used by Galileo and BeiDou, have been designed so that receivers can process a narrower part of it, at a lower complexity, or in its entirety, for a better ranging accuracy. Also, the signal power and bandwidth are constrained by international organizations such as the International Telecommunications Union, which allocates frequencies to services and guarantees that services do not interfere with each other.

For accurate ranging, the signals transmitted by all satellites need also to be synchronized with a common time reference. The satellites have very precise atomic clocks onboard, even though they are not perfect. For a meter-level accuracy, we need to measure time at the level of a few nanoseconds (a nanosecond is 10^{-9} s), since an electromagnetic signal travels 30 cm in 1 ns at the speed of light. The time offset between the individual clocks in the satellites and the common time reference is determined by the continuous tracking of the satellites by stations in the GNSS ground segment. The GNSS ground segment also tracks where the satellites are and sends this information back to the satellites through uplink stations. The satellites, in turn, embed this time offset and their position information into the navigation data that is conveyed by the transmitted GNSS signal. This ensemble of data is called the satellite *ephemeris* (*ephemerides* in plural).[2] The need for transmitting data along with the provision of accurate ranging has driven the design of GNSS signals. However, in order to achieve these same goals, designers prioritized different aspects. Some signals were designed to carry more data than just the satellite ephemerides, other signals were designed to support military signals, some were designed decades later than others, assuming better receiver capabilities, some have been designed to minimize interference with legacy signals, and so on.

In summary, GNSS signals are composed of at least three elements: (i) the data symbols conveying the bits of the navigation message, which contains the satellite

---

[1] Note that the book also indirectly addresses the Japanese Quasi-Zenith Satellite System (QZSS), as it uses GPS L1 C/A-like signals.

[2] Before GPS, this term referred to the positions of celestial bodies used by navigators.

Table 1.1 GNSS signals treated in this book and in the accompanying software.

Signal	L1 C/A	L1OF	E1-B	E1-C	B1I	L5I	L5Q	L5
System	GPS	GLONASS	Galileo	Galileo	BeiDou	GPS	GPS	NavIC
Service	SPS	SPS	OS	OS	OS	SPS	SPS	SPS
Carrier Freq. [MHz]	1575.42	1598.0625–1605.375	1575.42	1575.42	1561.098	1176.45	1176.45	1176.45
Polarization	RHCP	RHCP	RHCP	RHCP	RHCP	RHCP	RHCP	RHCP
Channel Access	CDMA	FDMA	CDMA	CDMA	CDMA	CDMA	CDMA	CDMA
Modulation	BPSK (1)	BPSK (0.5)	MBOC (6,1,1/11)	MBOC (6,1,1/11)	BPSK (2)	BPSK (10)	BPSK (10)	BPSK (1)
Component	Q	Q	I	I	I	I	Q	I
Chip Rate [Mcps]	1.023	0.511	1.023	1.023	2.046	10.23	10.23	1.023
Code Len [chips]	1,023	511	4,092	4,092	2,046	10,230	10,230	1,023
Code Len [ms]	1	1	4	4	1	1	1	1
Code Family	Gold	M-sequence	Memory codes	Memory codes	Gold	2 M-seq	2 M-seq	Gold
Symbol/Bit Rate [sps/bps]	50/50	50/50	250/125	–	50/≈37	100/50	–	50/25
Data Encoding	None	None	FEC 1/2, c:7; interleaving	–	BCH (15,11,1), interleaving	FEC 1/2, c:7	–	FEC 1/2; CL:7; interleaving
Nav Ephemeris	Keplerian	P, V, A	Keplerian	–	Keplerian	Keplerian	–	Keplerian
Nav Iono model	Klobuchar	None	NeQuick	–	Klobuchar	Klobuchar	–	Grid-based; GIVE / GIVD
Time Reference	GPST	UTC	GST	–	BDT	GST	–	IST
Geodetic System	WGS84	PZ-90	GTRF	–	BDCS	WGS84	–	WGS84

ephemerides; (ii) the spreading or pseudo-random noise (PRN) code, which facilitates the distance measurement between the satellite and the receiver and (iii) the carrier wave on which the former two are modulated.[3] The configuration of these three elements for different GNSS signals is shown in Table 1.1. For instance, data symbols are transmitted at a rate of 50 sps in GPS L1 C/A, while for Gailleo E1-B, they are transmitted at 250 sps. Their spreading code length is also different, 1,023 chips for GPS L1 C/A and 4,092 chips for Galileo E1-B. However, both signals share the same carrier frequency at 1575.42 MHz. The impact of using these values will be discussed when specifically addressing each system.

It is interesting to note that all satellites of a given GNSS system transmit similar signals, but their spreading codes are unique and thus each satellite can be univocally identified. Furthermore, since the spreading codes are orthogonal sequences, all satellites can transmit messages using the same carrier frequency and hence simultaneously access the medium without interfering with each other. This leads to a Code Division Multiple Access (CDMA) scheme, as indicated in Table 1.1, for most of the systems. If each of the satellites of the system used a different carrier frequency, we would then have a Frequency Division Multiple Access (FDMA) scheme, as it is the case with GLONASS.

Some of the signals listed in Table 1.1 contain components without navigation data bits, and therefore, their data rate is omitted in the corresponding cell. For instance, this is the case of the E1-C component of the Galileo E1 signal. Unmodulated or *pilot* signal components constitute a new feature of modernized GPS and Galileo signals. They are transmitted with the data components, sometimes orthogonally, and they have their own set of spreading codes including a primary and a secondary code on top of the former. The advantage of this new signal component comes from allocating the two main properties of GNSS signals, i.e., transmitting navigation data and providing ranging information, into two separate channels. The data channel conveys the data for locating the satellites in the constellation. However, this information is encoded in binary $-1, +1$ symbols, whose sign transitions reduce the energy that the receiver can accumulate to obtain a ranging measurement. On the contrary, the pilot channel does not contain unexpected symbol transitions and can be integrated over longer periods, which can drastically enhance the signal-to-noise ratio (SNR) of the received signals.

1.3   Structure of GNSS Signals

1.3.1   Signal Modulation

In Section 1.2, it has been introduced that GNSS signals are composed of three main constituent elements: the navigation data bits, the spreading code and the carrier. The data bits contain binary information that must be converted into signal levels to allow

---

[3] Modulation is the process that allows a signal to be successfully sent through a propagation medium. It will be discussed in Section 1.3.1.

their transmission over the propagation medium. Binary Phase Shift Keying (BPSK) modulation is the simplest way to do so, whereby each bit is converted into a binary symbol with amplitude $\{-1, +1\}$ that modulates a carrier. But before this, the data symbols are multiplied by a spreading code that is unique to each satellite and thus allows identifying each satellite signal at the receiver. The spreading code has a much higher rate than the data symbols, which means that its power content is spread across a much wider bandwidth. The same effect appears when multiplying the data symbols with the spreading code. The result is a signal with a much wider bandwidth than the original one. This effect can be observed in the upper plots of Figure 1.1, which show the frequency representation of the signals at each point of a simplified transmission chain. As can be seen, the spectrum of the signal after multiplication by the spreading code is much wider than that of the original signal.

The process of spreading the power of a signal across a wider range of frequencies is known as *spread spectrum* (SS) modulation. When such spreading is done by multiplying a low-rate signal with a high-rate one, the technique is known as direct-sequence spread spectrum (DS-SS). This is the case in GNSS where the original signal (either containing data or pilots) is multiplied by the spreading code. The ratio between the rate of the spreading code and the rate of the original signal is called the *processing gain*, and it typically runs from 10 to 60 dB. At the GNSS receiver, the original signal can be recovered back by multiplying the received spread signal with the same spreading code signal used at the transmitter. This technique dates back to the 1980s, and it is popular in applications involving radio links in hostile environments. Many radio frequency interference (RFI) signals can easily be rejected because they do not

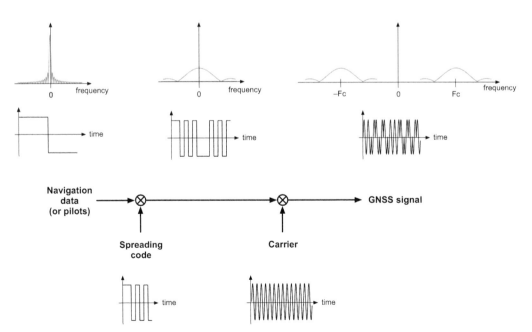

Figure 1.1 Illustration of how GNSS signals are generated, indicating the time and frequency representation of the resulting signal at each point of a simplified transmission chain.

contain the spreading code in their transmitted signal. Therefore, interference is actually spread at the receiver output because the interference signal is only multiplied once by the spreading signal. This feature is the real beauty of SS technology. Only the desired signal, which was generated using the same spreading code, will be seen at the receiver when the despreading operation is applied.

BPSK modulations are generally denoted in the context of GNSS as BPSK(n), where n stands for the rate of the spreading code in multiples of 1.023 MHz.[4] For example, GPS L1 C/A is a BPSK(1) signal. However, most new GNSS signals are modulated in *subcarriers*, which are nothing but a square signal with a rate on the order of the spreading code, or higher, that multiplies the input signal.

Subcarriers were first publicly proposed for GNSS in [1], and they are widely used nowadays in most modernized GNSS signals under the name of BOC signals. They were introduced in order to move the spectral content of the signal away from the carrier frequency, following the well-known modulation principle. By doing so, new signals can share the same carrier frequency as already used by the existing ones, while reducing the spectral overlap. An example comparing the spectrum of BPSK(1) with that of a subcarrier-modulated signal such as BOC(1,1) used in Galileo E1-B is shown in Figure 1.2. As can be seen, both the BPSK(1) and the BOC(1,1) spectra share the same central frequency. However, the spectrum of the BOC(1,1) signal is shifted to both sides of the central frequency so that the overlap with the BPSK(1) is significantly reduced.

The overall transmission and reception scheme of a GNSS signal is therefore the one schematically shown in Figure 1.3, including the possibility that subcarrier

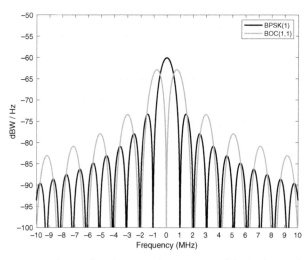

Figure 1.2 Comparison between the spectrum of the BPSK(1) signal used in GPS L1 C/A and that of a subcarrier-modulated signal such as the BOC(1,1) used in Galileo E1-B. The center frequency $f = 0$ corresponds to 1575.42 MHz once the signal is modulated by the carrier.

---

[4] The rate of the spreading code is also known as the *chip rate* because the spreading code is a sequence of binary elements that are referred to as *chips*.

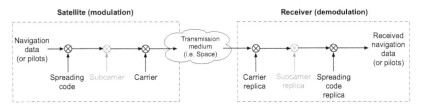

Figure 1.3 Fundamental components in GNSS signal modulation and demodulation.

modulation is present. As previously mentioned, the input signal is composed of data symbols or pilots, depending on the GNSS signal. This low-rate input signal is multiplied by the spreading code so that the resulting spectrum becomes spread in the frequency domain. This operation protects the symbols/pilots against harmful interference, makes the resulting signal exhibit low probability of interception (LPI), allows different satellites to be simultaneously transmitted in the same frequency band and, most importantly, facilitates the use of these signals for ranging. The DS-SS resulting signal is then multiplied by the subcarrier (when present, since not all GNSS signals are subcarrier-modulated) and finally multiplied by the carrier that places the resulting signal at the L band. Once transmitted through the propagation medium, the signal arrives at the receiver and the opposite operations need to be done in order to unbox the data symbols and to obtain the ranging information that is required to position the user.

Bandpass Signal Representation
The modulated signal transmitted by the GNSS satellites is a so-called *bandpass* signal because its frequency representation, or spectral content, is concentrated on a small neighborhood of a high frequency. For example, the BPSK(1) spectrum shown in Figure 1.2 has most of its power concentrated on its main lobe, which spans for approximately 2 MHz bandwidth. This signal is then modulated on a carrier at 1575.42 MHz for the GPS L1 C/A signal, so the spectrum becomes centered at that frequency. This is a clear example of a bandpass signal because most of the power is concentrated in just a 2 MHz bandwidth around a frequency of 1575.42 MHz. Bandpass signals are needed whenever radio transmission is taking place. This is because the size of the antennas needed for transmission and reception is inversely proportional to the central frequency of the signal, so the larger the carrier frequency, the smaller the required antenna. In the communications domain, it is often claimed that the carrier frequency bears no information at all and that is why it is removed at the receiver once its purpose of propagating the radio wave through the medium has been achieved. This is not precisely true in the GNSS domain because part of this carrier, actually, the carrier phase, does bear information on the distance from the satellite to the user's position. So keeping track of the carrier phase is needed at a GNSS receiver, even though this can be done once the carrier frequency has been removed anyway.

In its simplest form, a bandpass signal $s(t)$ can be expressed as follows:

$$s(t) = \sqrt{2P}x(t)\cos(2\pi F_c t), \tag{1.1}$$

where P is the bandpass or radio frequency (RF) transmitted power[5] and $x(t)$ is a unit power and real-valued signal. In our case, $x(t)$ is a train of $\{-1, +1\}$ amplitudes, including the navigation data or pilot symbols, the spreading code and eventually the binary subcarrier as per Figure 1.3. The bandpass nature of the signal s(t) in Eq. (1.1) is obtained after multiplying our signal $x(t)$ with the cosine term at some carrier frequency F_c.

The expression in Eq. (1.1) is the simplest representation of a bandpass signal, but another real-valued data-modulated or pilot signal can be, and is usually, transmitted at the same time using a sine term. This leads to a more general signal model whereby the bandpass signal results in[6]

$$s(t) = \sqrt{2P_I}x_I(t)\cos(2\pi F_c t) - \sqrt{2P_Q}x_Q(t)\sin(2\pi F_c t), \qquad (1.2)$$

which contains two signal components, namely the *in-phase* component $x_I(t)$ and the *quadrature* component $x_Q(t)$, whose bandpass or RF transmitted power is P_I and P_Q, respectively. Note that the term *quadrature* refers to the fact that this component is placed at 90°, that is, in quadraphase, with respect to the in-phase one. It is also interesting to note that $x_I(t)$ and $x_Q(t)$ are two independent real-valued signals that are transmitted at the same carrier frequency. This is possible in virtue of the orthogonality between the cosine and the sine functions, and it is exploited in GNSS, as well as in many other wireless systems, for transmitting two different real-valued signals over a single carrier frequency, thus increasing the spectral efficiency. Being part of the same carrier, these signals are referred to as the signal *components*, while the whole ensemble is just referred to as the GNSS signal. GNSS signals use the in-phase and quadrature components in various ways. Here are some examples based on the signals processed by the software accompanying this book:

(1) The GPS L1 legacy signal conveys data-modulated symbols in both signal components $x_I(t)$ and $x_Q(t)$. $x_I(t)$ is used by the military P(Y) code, out of the scope of this book, and $x_Q(t)$ is the open GPS L1 C/A component that we all use.
(2) GPS L5 includes the two components L5I and L5Q carried by the $x_I(t)$ and $x_Q(t)$ in our model, respectively. L5I includes data-modulated symbols, and L5Q modulates the *pilot* component, used to improve tracking performance. Both components are conceived to be used by civil receivers, for a better performance.
(3) Galileo E1 includes two components in $x_I(t)$: The E1-B, with the data, and the E1-C, with the pilot, which are subtracted from one another. In $x_Q(t)$, the signal modulates the E1-A component, aimed for the Public Regulated Service, encrypted and out of the scope of this book.

---

[5] For a GNSS bandpass signal with amplitude A, its power becomes $P = \frac{1}{T}\int_0^T s^2(t)dt = \frac{1}{T}\int_0^T A^2 x^2(t)\cos^2(2\pi F_c t)dt = \frac{A^2}{2}$, for $T \gg \frac{1}{F_c}$. So we need $A = \sqrt{2P}$ for the bandpass power to be P, as stated in the text.

[6] In some GNSS signals, the two carrier-modulated signal components are explicitly added with a + sign. This involves that the in-phase and quadrature components are sign-reversed one with each other when down converted to the baseband, as it is the case in GPS L1 [2, Section 3.3.1.5.1].

Note that here we are considering two BPSK signal components. If the cosine and sine are processed as a single signal, that is, the signal is processed by looking at the cosine and sine peaks altogether, the signal is defined as QPSK (Quadrature Phase Shift Keying). Higher order PSK signals are possible and, in fact, used by GNSS. For example, the Galileo constant envelope E5 AltBOC signal is processed as an 8-PSK signal [3].

Baseband Signal Representation

The information content of the bandpass signal in Eq. (1.2) lies in the two signal components $x_I(t)$ and $x_Q(t)$. For this reason, it is often convenient to focus on these two signal components only and to remove the cosine and the sine waves that are merely present to make the propagation through the medium possible. If the carrier is perfectly removed, and thus the cosine and sine waves, the resulting signal exhibits two key properties. The first one is that the spectral content is placed back into the base of the frequency axis, in the neighborhood of the zero frequency. So we call the resulting signal a *baseband* signal, denoted herein as $b_s(t)$. The second property is that the baseband signal is composed of two real-valued components, $x_I(t)$ and $x_Q(t)$, which need to be conveniently expressed as an ensemble signal because they are both part of Eq. (1.2). To do so, the following notation is used for the baseband signal:

$$b_s(t) = \sqrt{2P_I}x_I(t) + j\sqrt{2P_Q}x_Q(t), \tag{1.3}$$

where the complex number notation j is used to represent both signal components $x_I(t)$ and $x_Q(t)$, simultaneously. The in-phase component $x_I(t)$ is placed in the real part, while the quadrature component $x_Q(t)$ is placed in the imaginary part, thus being consistent with the fact that the quadrature component is placed at $90°$ with respect to the in-phase one, as in Eq. (1.2).

The baseband signal in Eq. (1.3) contains essentially the same information as the bandpass signal in Eq. (1.2), understanding *information* as the content conveyed by the signal components $x_I(t)$ and $x_Q(t)$. It is for this reason that the baseband signal in Eq. (1.3) is referred to as the baseband *equivalent* signal of Eq. (1.2). It makes use of the well-known Euler's formula (i.e. $e^{jz} = \cos z + j\sin z$), widely used in electrical engineering.[7] In particular, the reader can check that the bandpass signal in Eqs. (1.2) and (1.3) are related as follows:

$$s(t) = \sqrt{2P_I}x_I(t)\cos(2\pi F_c t) - \sqrt{2P_Q}x_Q(t)\sin(2\pi F_c t) \tag{1.4}$$

$$= \text{Re}\left[(\sqrt{2P_I}x_I(t) + j\sqrt{2P_Q}x_Q(t))e^{j2\pi F_c t}\right] \tag{1.5}$$

$$= \text{Re}\left[b_s(t)e^{j2\pi F_c t}\right] \tag{1.6}$$

with Re[·] being the real part operator. The relationship between the baseband and bandpass signals will be further elaborated in Section 1.4.3.

---

[7] We will use complex-plane diagrams to show the correlation results of our signal of interest with a complex replica, modulated in both the in-phase and quadrature components, in order to ensure a proper tracking of the signal. See, for example, the discrete-time scatter plot for GPS tracking in Figure 2.4. This plot is also known as a *constellation* plot.

The baseband signal $b_s(t)$ has its spectral content in the neighborhood of the zero frequency, but an exception is made for GNSS signals whose components are sub-carrier modulated. In this case, we will see that the baseband equivalent signal has its spectral content around the subcarrier frequency, similarly to what happens for a bandpass signal. In fact, a subcarrier-modulated signal can actually be understood as a bandpass signal rather than a baseband one because it is still modulated by either a squared-wave cosine or sine function at a given subcarrier frequency. The plots shown in Figure 1.2 are actually the baseband spectra of the GPS L1 C/A component and the Galileo E1-B component, the latter accounts for the contribution due to the BOC(1,1) modulation only. As can be seen, the main lobes of the BOC(1,1) spectrum are located at ± 1.023 MHz, which is the subcarrier frequency for this modulation.

Tiered Structure of GNSS Signal Components

The present section is intended to set the grounds for understanding the inner struc-ture of GNSS signals and being able to answer questions like how data symbols are conveyed, how different satellites are distinguished or how the frequency representa-tion of GNSS signals looks like. To do so, a general formulation will be introduced to represent either the in-phase or the quadrature components, which may carry either data-modulating symbols or a pilot signal, depending on the GNSS signal. Since we will consider a general model, the $_{I,Q}$ subindex will be dropped for the sake of simplicity and we will simply refer by $x(t)$ to any of the GNSS signal components.

According to Figure 1.1, GNSS signal components consist of the product between a data/pilot signal and a spreading code signal. This is the simplest interpretation, and it often suffices for understanding the underlying nature of GNSS signals, which is nothing but a spread spectrum signal. As such, one can easily understand that the receiver must then implement the inverse operation and multiply the received signal with a local replica of the spreading code. In this way, despreading takes place, and the original data/pilot signal is recovered back. However, this pragmatic interpretation often makes it difficult to analyze the inner structure of the signal and the impact of the different constituent elements and parameters on the signal, for instance, when computing the correlation and power spectral density (PSD) of the resulting signal. It is for this reason that a more in-depth analysis will be derived in this section. To do so, a tiered approach will be followed where GNSS signal components are considered to be composed of layers or tiers of different signals.

With these considerations in mind, it is found that any GNSS signal component can be expressed as[8]

$$x(t) = \sum_{i=-\infty}^{\infty} d[i]g(t - iT_d), \qquad (1.7)$$

where $\{d[i]\}_{i=-\infty}^{\infty}$ is a sequence of data-modulating symbols, each of them transmit-ted through the propagation medium by means of a tiered waveform $g(t)$. The time

---

[8] Note that sequences are indexed with the discrete-time notation [·], while signals are indexed with the continuous-time notation (·) to better emphasize their distinct nature.

duration of this waveform is limited to the symbol period T_d. Note that the signal in Eq. (1.7) is actually a pulse amplitude modulated (PAM) signal because $g(t)$ is a pulse (actually, it is not a simple pulse but a tiered waveform) whose amplitude is modulated by the symbols $d[i]$. PAM signal properties have been widely studied in the context of digital communications, and they will facilitate the understanding of some key features of GNSS signals, such as their correlation and spectral representation. Pilot GNSS signal components transmitting no symbols can be modeled as well with Eq. (1.7) by setting $d[i] = 1$ for all i.

Regarding the tiered symbol waveform $g(t)$, it is composed of N_r concatenated spreading code signals $c(t)$ whose amplitudes are modulated by the binary ± 1 values of the so-called *secondary code* sequence, $\{u[k]\}_{k=0}^{N_r-1}$. That is,

$$g(t) = \sum_{k=0}^{N_r-1} u[k]c(t - kT_{\text{code}}). \tag{1.8}$$

The result is another PAM signal, where the shaping pulse is now given by the spreading code signal $c(t)$, which has a limited time duration equal to the spreading code period, T_{code}.

The spreading code signal $c(t)$ is in turn another tiered waveform. It is composed of N_c concatenated chip pulses $p(t)$ whose amplitudes are modulated by the binary ± 1 values of the spreading code, or so-called *primary code* sequence $\{v[m]\}_{m=0}^{N_c-1}$. That is,

$$c(t) = \sum_{m=0}^{N_c-1} v[m]p(t - mT_c). \tag{1.9}$$

The result is again another PAM signal, where the shaping pulse is now given by the chip pulse $p(t)$, which is a real-valued pulse with a limited time duration equal to the chip period, T_c. Putting together all these terms, we can see that GNSS signal components are formed by encapsulating different layers of pulses and sequences, one inside another like in a Matryoshka doll. A summary of these signals and sequences is schematically illustrated in Figure 1.4 for the sake of clarity.

On the basis of this structure, we can see that the symbol period has a duration $T_d = N_r T_{\text{code}}$, with N_r the length of the secondary code sequence and T_{code} the spreading code period. The symbol rate is then defined as $R_d = 1/T_d$. In turn, the code period is given by $T_{\text{code}} = N_c T_c$, with N_c being the length of the primary code sequence and T_c be the chip period. The chip rate is then defined as $R_c = 1/T_c$. As an example, we

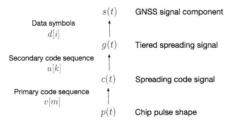

Figure 1.4 Schematic representation of the constituent elements upon which the tiered structure of a GNSS signal component is built.

can consider the GPS L1 C/A signal component whose parameters are indicated in Table 1.1. This component has a symbol rate as low as $R_d = 50$ sps, which means that the symbol period is $T_d = 20$ ms[9]. The chip rate is $R_c = 1.023$ MHz, and the primary code length is $N_c = 1,023$ chips, which means that the primary code period is $T_{code} = N_c/R_c = 1$ ms. There is explicitly no secondary code in GPS L1 C/A, as stated in its interface control document (ICD), so these 1 ms primary codes are repeated one after another within the 20-ms symbol period. But we could understand this repetition as being brought by an all-ones secondary code code with length $N_r = 20$. This interpretation will be useful for determining the PSD of GPS L1 C/A signals.

It is interesting to note that the signal model presented so far is valid for either BPSK- or BOC-modulated signals. The only difference between both is at the chip pulse. For instance, BPSK modulations typically adopt a rectangular chip pulse, occupying the whole chip period. This is the case of GPS L1 and GPS L5 signals that use

$$p_{BPSK}(t) = \Pi\left(\frac{t - T_c/2}{T_c}\right), \tag{1.10}$$

which stands for a rectangular pulse of width T_c, delayed here for convenience by $T_c/2$ to make it causal (i.e. so that the pulse starts at $t = 0$). The resulting chip pulse is shown in Figure 1.5(a). BOC-modulated signals, instead, are multiplied by the sign of a subcarrier waveform. This can be encompassed in the present formulation by having the chip pulse be the subcarrier waveform observed during one chip period. This results into a chip pulse composed of alternating rectangular pulses whose individual duration is a fraction of the chip period. The general formulation is given by

$$p_{BOC}(t) = \sum_{q=0}^{N_{scc}-1} (-1)^q \Pi\left(\frac{t - qT_{sc}/2 - T_{sc}/4}{T_{sc}/2}\right), \tag{1.11}$$

where T_{sc} is the subcarrier period and $N_{scc} = \frac{T_c}{T_{sc}/2}$ is the number of half subcarrier periods within one chip period. The term $T_{sc}/4$ in the numerator is introduced to ensure that the pulse is causal, as the BPSK chip pulse in Eq. (1.10) is. A similar approach to the model used in Eq. (1.11) is adopted for instance in [4, Section 9.10], where the interested reader is referred to for more details. To put an example of Eq. (1.11), let us consider a BOC(1,1) signal such as the one whose spectrum is shown in Figure 1.2. For a BOC(1,1), there is one subcarrier period per chip period so that $T_{sc} = T_c$ and $N_{scc} = 2$. This results into the following chip pulse:

$$p_{BOC(1,1)}(t) = \Pi\left(\frac{t - T_c/4}{T_c/2}\right) - \Pi\left(\frac{t - 3T_c/4}{T_c/2}\right), \tag{1.12}$$

which consists of two consecutive and sign-reversed rectangular pulses, as shown in Figure 1.5(b). It is also worth noting that rectangular pulses in the BOC(1,1) chip pulse are narrower than that in BPSK(1), which intuitively means that BOC(1,1) should provide better accuracy in the time-delay estimation of the received signal. This topic will be discussed when addressing the PSD of both signals.

---

[9] Since GPS L1 C/A is using BPSK modulation, each BPSK symbol conveys one bit. This means that the bit rate and symbol rate are the same, as well as the bit period and the symbol period.

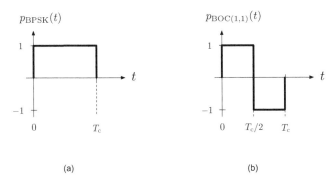

Figure 1.5 Illustration of (a) a BPSK chip pulse and (b) a BOC(1,1) chip pulse.

Merging together all the constituent elements introduced so far, Eq. (1.7) can equivalently be expressed as:

$$x(t) = \sum_{i=-\infty}^{\infty} d[i] \sum_{k=0}^{N_r-1} \sum_{m=0}^{N_c-1} u[k]v[m]p(t - mT_c - kT_{\text{code}} - iT_d) = \sum_{m=-\infty}^{\infty} a[m]p(t - mT_c),$$

$$(1.13)$$

where the right-hand side is expressed as a function of some equivalent symbols $a[m]$ whose values depend on the combination between data/pilot symbols, secondary code and primary code of the GNSS signal. This latter expression corresponds to that of a conventional PAM.

An example of a BPSK-modulated GNSS signal component is shown in Figure 1.6 to illustrate the tiered structure that has just been presented. A generic signal has been assumed for the sake of simplicity. The top plot in the figure shows the primary spreading code $v[m]$, which is a discrete-time sequence composed of $N_c = 15$ chips at a chip rate of $R_c = 1.023$ MHz. The primary spreading code sequence is then shaped with a rectangular pulse of duration $T_c = 1/R_c$. The resulting spreading code signal is shown in the second top plot, corresponding to the continuous-time signal $c(t)$ introduced in Eq. (1.9). It can easily be seen that $c(t)$ is composed of rectangular pulses, each of them having an amplitude given by the sign of the primary spreading code sequence $v[m]$. Next, N_r replicas of the spreading code signal $c(t)$ are concatenated to form the tiered spreading signal $g(t)$ introduced in Eq. (1.8). This is done in this example using a secondary code of length $N_r = 4$, whose values are all equal to 1, that is, $u[k] = 1$ for $k = 0, 1, 2, 3$ in Eq. (1.8). The resulting signal $g(t)$ can be seen in the middle plot of Figure 1.6, where alternate dark and light colors have been used to facilitate the visual identification of each of the replicas of $c(t)$ that form $g(t)$. The resulting signal $g(t)$ becomes the waveform conveying the data-modulated symbols following the expression in Eq. (1.7), and similar to what occurs in a conventional PAM signal. This can be seen in the bottom plot of Figure 1.6, where the data-modulated signal $x(t)$ is shown for an observation period of two symbols. It can be seen how each of these symbols is shaped with the waveform $g(t)$ and then concatenated to form the resulting data-modulated signal $x(t)$. Alternate dark and light

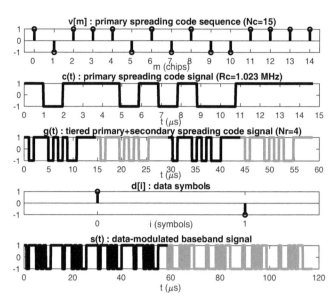

Figure 1.6 Illustration of the tiered structure of a generic GNSS signal. Example assuming a primary code sequence with $N_c = 15$ chips, chip rate $R_c = 1.023$ MHz a rectangular chip pulse with duration $T_c = 1/R_c$, and a secondary code sequence with length $N_r = 4$. The resulting signal is shown for an observation period where two binary data-modulated symbols are present.

colors are used here again to facilitate the visual identification of the two repeated waveforms.

1.3.2   Spreading Codes

The primary code sequence $v[m]$ is the cornerstone of GNSS signals. It is selected in such a way that provides a pseudorandom sequence of ± 1 values with orthogonality properties. At the same time, these antipodal values guarantee that once the sequence is pulse shaped, the resulting continuous-time signal becomes zero-mean regardless of the chip pulse, thus ensuring a power-efficient signal transmission. As for the orthogonality property, this is the key underlying principle behind CDMA. It allows signals from different satellites to coexist in the same frequency band without interfering one with each other. Perfect orthogonality has been shown to be possible for synchronous systems where the spreading sequences of different users are all perfectly aligned. Unfortunately, this is not the case in GNSS, where even if the signals were transmitted synchronously from all satellites, they travel different distances to the receiver and, therefore, they experience different propagation delays. Because of this asynchronous behavior, no perfect orthogonality can be achieved in GNSS, thus sparking the interest for *near*-orthogonal spreading codes. Some examples finally selected for GNSS are the following:

(1) Gold codes, used in the GPS L1 C/A signal and further discussed in Chapter 2.

(2) Maximum-length sequences (M-sequences), used in GLONASS and GPS L5 and outlined in Chapters 3 and 8.

(3) Memory codes, which are hand-selected, predefined codes stored in large memory tables used in Galileo signals and introduced in Chapter 4.

(4) Weil codes, based on Legendre sequences introduced in a need for finding codes of length $N_c = 10{,}230$ chips, used in the modernized GPS L1C civil signal.

Autocorrelation and Cross-Correlation

Near orthogonality means that the cross-correlation between different spreading codes is not perfectly zero but small enough to be fairly neglected.[10] It is important to note that orthogonality is assessed through the *circular* cross-correlation[11] so that for two different spreading codes, namely $v^{(p)}[m]$ and $v^{(q)}[m]$ with $p \neq q$, we have

$$R_{v^{(p)}v^{(q)}}[k] = \frac{1}{N_c} \sum_{m=0}^{N_c-1} v^{(p)}[m] v^{(q)}[m-k]_{N_c} \approx 0, \qquad \text{for } p \neq q, \qquad (1.14)$$

and through the circular autocorrelation of the same spreading code, making sure that it is nearly zero outside of the central correlation lag. That is,

$$R_{v^{(p)}}[k] = \frac{1}{N_c} \sum_{m=0}^{N_c-1} v^{(p)}[m] v^{(p)}[m-k]_{N_c} \approx \delta[k], \qquad \text{for any } p, \qquad (1.15)$$

with $\delta[k]$ being the discrete-time Kronecker delta, namely $\delta[k] = 1$ for $k = 0$ and $\delta[k] = 0$ for $k \neq 0$. In the above expressions, the subscript N_c indicates that the time indexation of the sequence $v[m - k]$ must be cyclically shifted with the period N_c whenever $m - k < 0$. Finally, it is also worth noting that the correlation definitions in Eqs. (1.14)–(1.15) assume that $N_c \gg 1$ so that the spreading code sequence can fairly be approximated by a power-type signal.[12] This will facilitate the mathematical manipulations in the limit for $N_c \to \infty$, which is often invoked to achieve the perfect orthogonality that practical sequences with finite N_c cannot provide. This assumption will help in simplifying the results when computing the correlation and the PSD of the pulse-shaped spreading code signal.

An example of the circular auto- and cross-correlation of a primary spreading code sequence is shown in Figure 1.7. The results were obtained for two of the Gold codes used in GPS L1 C/A corresponding to PRNs 3 and 7. The lack of perfect orthogonality can be seen by the presence of a small noise-like contribution in both the auto- and cross-correlation functions. Note that despite the lack of perfect orthogonality, Gold codes serve well for the purpose of selecting the signal from a given satellite while

---

[10] This approximation needs to be revisited when many satellites are present, since it may happen that the aggregation of all their residual cross-correlations ends up being relevant. This is actually a key metric to be assessed in the design of new GNSS signals that may share the same frequency band with legacy ones.

[11] The correlation operation is often defined with one of its factors being complex conjugated, but this is ignored herein for the sake of simplicity under the assumption that spreading codes are real-valued.

[12] A signal is power-type if its average power is finite and greater than zero, even when observed over an infinite period of time [4, Section 8.1.4]. Sinusoids are examples of power-type signals.

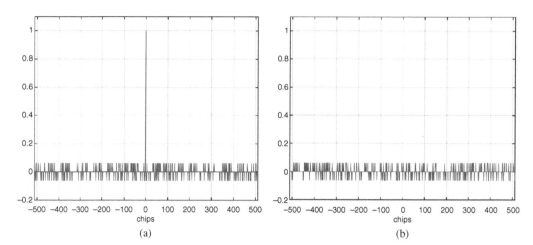

Figure 1.7 Correlation properties of the Gold spreading code with PRN $p = 3$ used in GPS L1 C/A. (a) Circular autocorrelation $R_{v(p)}[k]$. (b) Circular cross-correlation $R_{v(p)v(q)}[k]$ for PRN $p = 3$ and $q = 7$.

disregarding all the rest. This can be seen in the high peak obtained at the autocorrelation lag $m = 0$ when correlating with the same Gold code and in the small residual for any correlation lag m between two different codes.

Details about Gold codes will be provided in Section 2.2.2, but it is interesting to note that for a generic random code the main correlation peak is on average $\sqrt{N_c}$ times larger than the residual (i.e. secondary) peaks. In terms of power, this represents an average distance ratio of N_c. This means that for a random code with the same $N_c = 1{,}023$ length as the one depicted in Figure 1.7, the main peak would have 30 dB more power than the secondary peaks. The result is a useful indication for the value that may be achieved with other spreading codes, such as the Gold codes to be described later in Chapter 2. The interested reader may refer to [4, Section 9.2.1] for further details on the analysis of random code sequences for GNSS.

The autocorrelation properties discussed so far are related to the primary spreading code, which is nothing but a discrete-time sequence. However, this sequence is actually not transmitted as it is by the GNSS satellites. Instead, it is first pulse-shaped in order to obtain a continuous-time signal that can actually modulate the carrier wave. To do so, the chips of the spreading code are pulse amplitude modulated using a pulse shape with typically the same duration as the chip period. The result was shown in Eq. (1.9), and it is reproduced below for the sake of clarity:

$$c(t) = \sum_{m=0}^{N_c-1} v[m]p(t - mT_c). \tag{1.16}$$

The spreading code signal in Eq. (1.16) is actually the one transmitted by the GNSS satellites, so it is of interest to obtain its autocorrelation. After some manipulations and assuming the chip shaping pulse $p(t)$ to have a finite duration T_c, it can be found that[13]

$$R_c(\tau_t) = N_c \sum_{k=-N_c+1}^{N_c-1} R_v[k]R_p(\tau_t - kT_c), \tag{1.17}$$

where $R_p(\tau_t)$ stands for the autocorrelation of the time-limited chip pulse $p(t)$ defined as

$$R_p(\tau_t) = \int_{-\infty}^{\infty} p(t)p(t - \tau_t)dt. \tag{1.18}$$

An example of Eq. (1.17) is shown below in Figure 1.8 for the Gold code with PRN 3 used in GPS L1 C/A.

As can be seen in Figure 1.8, the main peak of $R_c(\tau_t)$ is determined by the autocorrelation of the chip pulse shape $p(t)$. Since we are considering the GPS L1 C/A signal, a rectangular chip pulse is considered having one chip duration and the autocorrelation becomes a triangular pulse spanning from $-1 \le \tau_t/T_c \le 1$ chips (see zoomed view in the bottom plot of Figure 1.8). Note that the secondary peaks in $R_c(\tau_t)$ observed in the upper plot are due to the lack of perfect orthogonality in the spreading code sequence. This effect could be mitigated by letting $N_c \to \infty$, as often done in mathematical derivations to obtain a more manageable expression for $R_c(\tau_t)$. In that case, we have that $\lim_{N_c \to \infty} R_v[k] = \delta[k]$ and then $R_c(\tau_t)$ in Eq. (1.17) simplifies to

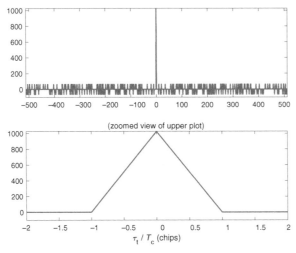

(zoomed view of upper plot)

τ_t / T_c (chips)

Figure 1.8 Autocorrelation of the spreading code signal, $R_c(\tau_t)$, for a GPS L1 C/A signal using the PRN 3 and a rectangular chip pulse with unit energy. Infinite bandwidth is assumed.

---

[13] The subindex $_t$ is used in τ_t to emphasize that it is a continuous-time delay measured in seconds, so it can be distinguished from its discrete-time version denoted simply by τ and measured in samples, which will appear when the signal is sampled.

$$\lim_{N_c \to \infty} \frac{1}{N_c} R_c(\tau_t) = R_p(\tau_t). \tag{1.19}$$

1.3.3   Power Spectral Density

The PSD provides information on how the power content of a signal is distributed in the frequency domain. This feature is important to understand how different signals behave, whether they have low- or high-frequency content (i.e. whether they exhibit slow or fast time variations, respectively), whether they incur a significant overlap or not (i.e. inter-band interference), whether they are strictly confined to the allocated frequency band or not (i.e. out-of-band interference), etc. But the PSD also plays a key role in determining how good a given signal is for ranging purposes. The reason is that the accuracy of time-delay measurements used for ranging depends on the mean square or *Gabor* bandwidth of the signal, which is actually measured through the second-order spread of its PSD [5, Section 3.11].

The PSD of a signal is related to its autocorrelation function through the Fourier transform, so for a signal $x(t)$, its PSD denoted by $S_x(f)$ is given by

$$S_x(f) = \mathcal{F}[R_x(\tau_t)], \tag{1.20}$$

with $\mathcal{F}[\cdot]$ being the Fourier transform operator and $R_x(\tau_t)$ being the autocorrelation of $x(t)$. The solution to Eq. (1.20) can easily be found for a GNSS signal component by understanding this signal as a PAM one, as already discussed in Section 1.3.1. To do so, we can take advantage of the following property, which states that for a generic PAM signal in the form

$$x(t) = \sum_{m=-\infty}^{\infty} \alpha[m]q(t - mT), \tag{1.21}$$

with some amplitude-modulating symbols $\alpha[m]$, shaping pulse $q(t)$ and symbol period T, the PSD is given by [6, Section 8.2]

$$S_x(f) = \frac{1}{T} \left[\sum_{k=-\infty}^{\infty} R_\alpha[k] e^{-j2\pi f k T} \right] S_q(f), \tag{1.22}$$

where $R_\alpha[k] = \mathrm{E}\,[\alpha[m]\alpha^*[m - k]]$ is the autocorrelation of the amplitude-modulating symbols and $S_q(f)$ is the energy spectral density (ESD) of the pulse, which is defined as

$$S_q(f) = \mathcal{F}[R_q(\tau_t)] = |Q(f)|^2 \tag{1.23}$$

with $Q(f) = \mathcal{F}[q(t)]$ being the Fourier transform of the pulse $q(t)$.

The spectral density in Eq. (1.23) is expressed in terms of *energy* because the pulse has a limited time duration and thus it becomes an energy-type signal.[14] In contrast, the whole PAM signal is ideally an infinite-length signal with random symbols being

---

[14] A signal $q(t)$ is energy-type if it has finite energy, that is, $\int_{-\infty}^{\infty} |q(t)|^2 dt < \infty$, even when observed over an infinite period of time [4, Section 8.1.4]. Rectangular pulses are examples of energy-type signals.

transmitted periodically every T seconds, and it is thus a random signal. Energy-type signals have ESD, while power-type and random signals have PSD. However, and for the sake of simplicity, we often refer to both spectral densities indistinguishably by simply the *spectrum*. Note as well that the definition of autocorrelation used for $R_\alpha[k]$ with the expectation operator $E[\cdot]$ is different from that used in Eq. (1.18) for the autocorrelation of the pulse. This is because $\alpha[m]$ is a (discrete-time) random signal, while $p(t)$ in Eq. (1.18) is an energy-type signal. For a discussion on the different definitions of autocorrelation for energy-type, power-type or random signals, the interested reader is referred to [6, Sections 2.3, 4.2].

The result in Eq. (1.22) is very insightful because it shows that the PSD of a GNSS signal depends on two terms. The first one is given by the autocorrelation of the amplitude-modulating symbols, which are composed of the data bits or pilots, the secondary code, the primary code, and eventually the alternating signs within the chip pulse of a subcarrier-modulated signal. The second one is given by the spectrum of the pulse. So both terms contribute in shaping the overall PSD, not only the spectrum of the pulse. Nevertheless, the latter is the only contribution for signals with uncorrelated symbols.

It is also worth emphasizing that the result in Eq. (1.22) facilitates the derivation of the PSD for GNSS signals. Two different approaches can be followed to do so depending on whether the starting point is the PAM expression in Eq. (1.7) or the one in Eq. (1.13). The latter is often simpler to substitute in Eq. (1.22) provided that the aggregated amplitude-modulated symbols $a[k]$ are zero-mean and uncorrelated. The uncorrelatedness assumption fits well when considering the spreading codes to be arbitrarily long. While this is not true in practice, one can assume that it is, approximately, in order to obtain a simplified and pedagogical result on how the PSD of a GNSS signal looks like. This approach is the easiest one for deriving the PSD of a GNSS signal, and it is often exposed first for an easy-to-understand explanation.

The second approach to take advantage of the result in Eq. (1.22) is by using the PAM expression in Eq. (1.7) as the starting point. In that case, the pulse is given by a tiered waveform, and therefore, the pulse ESD $S_q(f)$ in Eq. (1.22) is more tedious to obtain because one needs to recursively determine the ESD of each constituent waveform. The interested reader is referred to [7] for further details on this second approach.

Next, we will briefly discuss both of them in order to shed light on their advantages, their disadvantages and the difference in the final result that is obtained from each approach.

PSD for Arbitrarily Long Spreading Codes

This first approach assumes that the GNSS signal is expressed as in Eq. (1.13) so that the result in Eq. (1.22) can readily be applied by setting $\alpha[k] = a[k]$ and $q(t) = p(t)$. Considering the spreading codes to be arbitrarily long and with equiprobable random binary ± 1 values, the amplitude-modulating symbols become uncorrelated due to the orthogonality properties of spreading code sequences. This means that they are totally random with no relationship between one symbol and any other, so that their autocorrelation becomes $R_a[0] = 1$ and $R_a[k] = 0$ for any other $k \neq 0$. The resulting PSD simplifies to

$$S_x(f) = \frac{1}{T_c} S_p(f), \qquad (1.24)$$

and therefore, it is completely determined by the spectrum of the chip pulse $p(t)$.

For the case of GNSS signals relying on BPSK modulation, the chip pulse is a rectangular one with duration equal to the chip period T_c, as discussed in Eq. (1.10). The spectrum of this pulse, applying the squared modulus of the Fourier transform as indicated in (1.23), is given by

$$S_p(f)_{|BPSK} = T_c^2 \mathrm{sinc}^2(fT_c), \qquad (1.25)$$

where the $\mathrm{sinc}(\cdot)$ function is defined herein as[15] $\mathrm{sinc}(x) = \frac{\sin(\pi x)}{\pi x}$. The PSD of a BPSK-modulated GNSS signal is then given by

$$S_x(f)_{|BPSK} = T_c \mathrm{sinc}^2(fT_c). \qquad (1.26)$$

In case a BOC(1,1) modulation was considered instead, the chip pulse would be given by Eq. (1.12) and its spectrum would become

$$S_p(f)_{|BOC(1,1)} = T_c^2 \mathrm{sinc}^2(fT_c/2) \sin^2(\pi fT_c/2), \qquad (1.27)$$

which comes from the direct application of the Fourier transform to Eq. (1.12). Therefore, the PSD of a BOC(1,1)-modulated GNSS signal would become

$$S_x(f)_{|BOC(1,1)} = T_c \mathrm{sinc}^2(fT_c/2) \sin^2(\pi fT_c/2). \qquad (1.28)$$

It is interesting to note how the $\sin^2(\cdot)$ term in Eqs. (1.27)–(1.28), due to the alternating sign of the two short pulses forming the BOC(1,1) chip, is responsible for the spectral null that the BOC(1,1) PSD exhibits at the zero frequency. This effect can be clearly observed in the PSD plot shown in Figure 1.2. It is also worth pointing out that despite having a slightly different expression, the result in Eq. (1.28) is exactly the same as that often used by some authors and originally derived in [1], which is given by

$$S_x(f)_{|BOC(1,1)} = T_c \mathrm{sinc}^2(fT_c) \tan^2(\pi fT_c/2). \qquad (1.29)$$

PSD for Finite-Length Spreading Codes

Finite-length spreading codes are implemented in real-life GNSS signals, so strictly speaking, the simplifications discussed in the previous section cannot be applied. Finite-length spreading codes fit well into the tiered structure of GNSS signals introduced so far, since these codes repeat sooner or later along the GNSS signal, and thus a periodic correlation pattern does appear. In these circumstances, it is preferable to compute the PSD of the GNSS signal using Eq. (1.22) and substituting with the ele-

---

[15] This definition follows the convention mostly used in the literature (e.g. [6, 8]), even though the alternative definition $\mathrm{sinc}(x) = \frac{\sin(x)}{x}$ is preferred by some authors (e.g. [4]).

ments of $x(t)$ in Eq. (1.7). Then we can proceed in a tier-by-tier basis with the ESD of the finite-length tiered waveform $g(t)$, the spreading code waveform $c(t)$ and finally the chip pulse $p(t)$.

Regarding the first tier, we can start with the expression for $x(t)$ in Eq. (1.7), which is a PAM signal with binary random symbols being transmitted every T_d seconds with a waveform $g(t)$. According to Eq. (1.22), the PSD of such GNSS signal becomes

$$S_x(f) = \frac{1}{T_d} S_g(f) = \frac{1}{T_d} |G(f)|^2. \tag{1.30}$$

In order to find $G(f)$, the Fourier transform of the tiered waveform $g(t)$, it is interesting to express $g(t)$ as a function of the basic element being repeated, which is the spreading code $c(t)$. This can be done by expressing $g(t)$ as $g(t) = c(t) * u(t)$ with $u(t) = \sum_{k=0}^{N_r-1} u[k]\delta(t - kT_{code})$ the signal containing the secondary code values only, and $\delta(t)$ the Dirac delta function. Its Fourier transform can easily be obtained as

$$G(f) = \mathcal{F}[g(t)] = \sqrt{N_r} C(f) U(f), \tag{1.31}$$

where $C(f)$ is the Fourier transform of the spreading code signal $c(t)$ and $U(f)$ is the Fourier transform of the secondary code sequence,

$$U(f) = \frac{1}{\sqrt{N_r}} \sum_{k=0}^{N_r-1} u[k]e^{-j2\pi f k T_{code}}. \tag{1.32}$$

The same procedure can be followed for the spreading code signal $c(t)$, whereby we obtain $c(t) = p(t) * v(t)$ with $v(t) = \sum_{m=0}^{N_c-1} v[m]\delta(t - mT_c)$ the signal containing the primary code values only. This results in

$$C(f) = \mathcal{F}[c(t)] = \sqrt{N_c} P(f) V(f), \tag{1.33}$$

where $V(f)$ is the Fourier transform of the primary code sequence,

$$V(f) = \frac{1}{\sqrt{N_c}} \sum_{m=0}^{N_c-1} v[m]e^{-j2\pi f m T_c}. \tag{1.34}$$

Replacing these results into the preliminar PSD expression in Eq. (1.30), we finally get

$$S_x(f) = \frac{1}{T_c} |U(f)|^2 |V(f)|^2 |P(f)|^2, \tag{1.35}$$

with $U(f)$ being the Fourier transform of the secondary code sequence, $V(f)$ the Fourier transform of the spreading code sequence and $P(f)$ the Fourier transform of the chip pulse that can be obtained as previously explained.

The result in Eq. (1.35) is very insightful. It states that the PSD of the GNSS signal does not only depend on the spectrum of the chip shaping pulse, $|P(f)|^2$, but also on the spectrum of the primary and secondary code sequences. For instance, GPS L1 C/A signals could be understood as having a secondary code with $u[k] = 1$ for $k = 0, 1, \ldots, N_r - 1$ and $N_r = 20$ even though, formally speaking, no secondary code is mentioned in the ICD of this signal. But for practical purposes, we could think of the repetition of primary codes within the GPS L1 C/A symbol period as the aforementioned all-ones secondary code. This involves that $|U(f)|^2$ is then given by

$$|U(f)|^2 = \frac{1}{N_r} \left| \frac{\sin(\pi f N_r T_{code})}{\sin(\pi f T_{code})} \right|^2, \qquad (1.36)$$

which is known as the Fejér kernel. It is a comb-like periodic function in the frequency domain, with period $1/T_{code} = 1$ kHz for the case of GPS L1 C/A. It introduces spectral lines at 1 kHz due to the repetition of 1 ms length spreading codes every symbol period. This is an interesting observation that was ignored in the PSD of a BPSK modulated signal previously derived in Eq. (1.26), and it has some implications, for instance, on the impact of continuous-wave interference in GPS L1 C/A signals (see [4, Section 9.7.2]). Actually, frequencies where these periodic spectral lines are placed are more sensitive to the presence of a continuous wave (CW) signal. For a wider analysis on the spectral properties of finite-length spreading codes, the reader is referred to [7].

Impact of PAM Modulation on PSD

As previously mentioned, the interpretation of a GNSS signal in terms of a PAM modulation provides a compact expression that facilitates the mathematical derivation of the correlation and PSD. As for the latter, it is well-known that the PSD of a PAM signal has a shape that depends on two different contributions: the first one is provided by the spectrum of the shaping pulse, while the second one is provided by the correlation of the amplitude-modulating symbols [6, Section 8.2]. In our case, we followed two different approaches to apply this result. One of the approaches made direct use of Eq. (1.7), which in the GNSS case, depends on an intricate pulse $g(t)$ and the navigation data symbols as the amplitude-modulating symbols. Despite being a bit intricate at first glance, we showed how to proceed for the case of finite-length spreading codes, which led to the result in Eq. (1.35). The other approach made use of the expression in Eq. (1.13) that depends on the chip shaping pulse $p(t)$ and some aggregated amplitude-modulating symbols. We could easily apply this result to the case of arbitrarily long spreading codes, due to the fact that these aggregated symbols could be assumed to be all uncorrelated, thus leading to the PSD in Eq. (1.24).

This second approach helped us to emphasize that the PSD of a GNSS signal primarily depends on the spectrum of the chip shaping pulse, plus some additional contribution due to the correlation of the aggregated modulating symbols $a[m]$ in Eq. (1.13). Therefore, not only the chip shaping pulse but also the aggregated symbols have an impact on how the resulting power is distributed in the frequency domain [7].

This principle was actually driving the design of BOC signals by taking into account that the BOC chip pulse itself could be understood as a PAM signal as well, but with correlated symbols. Actually, alternating ones as shown in Eqs. (1.11)–(1.12). Such pattern leads to a zero-mean pulse, at least for even N_{scc}, and a zero-mean signal has always a null in the frequency domain at the zero frequency. This statement demonstrates why the BOC(1,1) spectrum shown in Figure 1.2 has a null at the zero frequency, which is compensated by shifting the power content at both sides of that frequency.

1.3.4   BOC Modulation

This section briefly introduces Binary Offset Carrier (BOC) modulations, already mentioned in this chapter, and which have become an essential part of modern GNSS signals. The key feature of BOC modulations is their ability to move signal power to specific parts of the allocated frequency band. This can be used to reduce interference with other signals and to improve the timing estimation accuracy by increasing the resulting mean square bandwidth. Redundancy in the upper and lower sidebands of BOC modulations also offers practical advantages for signal acquisition, code tracking, carrier tracking and data demodulation. However, some signal processing complexity and peak ambiguities are introduced as well, as discussed in detail in Chapter 10.

BOC signals are the result of modulating a GNSS signal component with a square wave subcarrier obtained from the sign of a sinusoid. The BOC modulation of the signal component in Eq. (1.7) would result in either of the two following implementations,

$$x_{\mathrm{sinBOC}}(t) = x(t)\mathrm{sign}\left(\sin(2\pi F_{\mathrm{sc}}t)\right) \tag{1.37}$$

$$x_{\mathrm{cosBOC}}(t) = x(t)\mathrm{sign}\left(\cos(2\pi F_{\mathrm{sc}}t)\right), \tag{1.38}$$

depending on whether the subcarrier is adopting a sine or a cosine waveform, which leads to the so-called sinBOC and cosBOC, respectively. A "BOC" signal, for short, often refers to a sinBOC signal. This was also the convention used in Eqs. (1.11) and (1.12), where a sinBOC was actually considered. Two parameters (m, n) control the generation of BOC signals, which are therefore referred in general as BOC(m, n) signals. The parameter

$$m = F_{\mathrm{sc}}/F_0, \tag{1.39}$$

is the ratio between the subcarrier frequency F_{sc} and the reference frequency $F_0 = 1.023$ MHz. In turn, the parameter

$$n = R_{\mathrm{c}}/F_0, \tag{1.40}$$

is the ratio between the chip rate R_{c} and the reference frequency. For instance, a BOC(6,1) means a 6.138 MHz subcarrier frequency and a 1.023 MHz chip rate. This signal together with a BOC(1,1) is transmitted in the E1-C component of the Galileo E1 signal. Their PSD are illustrated in Figure 1.9 and compared with the PSD of the signal transmitted in the GPS L1 C/A component.

As can be seen in Figure 1.9, even for a low-order BOC signal-like BOC(1,1), one can see that the power is split into two lobes slightly away from the L1 central frequency where the main lobe of the GPS L1 C/A PSD is placed. This confirms the ability of BOC signals to move the power content in the frequency domain. This feature can intuitively be understood by taking the ratio m/n, which provides the number of subcarrier cycles contained within a chip period. For a BOC(1,1), this ratio is equal to one, which means that the chip period contains one full subcarrier cycle. For a sinBOC(1,1), this would mean that the chip shaping pulse $p(t)$ is actually composed by

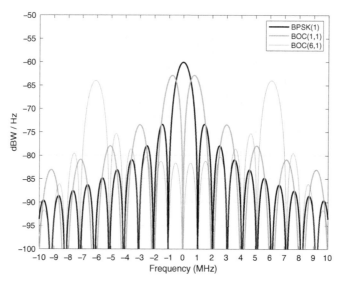

Figure 1.9 Power spectral density of the BPSK(1) modulation used in GPS L1 C/A and the BOC(1,1) and BOC(6,1) modulations used in Galileo E1. The center frequency $f = 0$ corresponds to 1575.42 MHz.

two shorter rectangular pulses, each one with a duration of half the chip period and sign-reversed. The result is a composite pulse $p(t)$ with zero-mean, which translates into the frequency domain to a pulse spectrum with a null at $f = 0$, at the expense of slightly increasing its power content at higher frequencies, as shown in the two sidelobes of the BOC(1,1) signal in Figure 1.9.

1.4   GNSS Signal Propagation, Reception and Conditioning

Before the signal is processed at the receiver, it has to be generated and then transmitted by the satellite antenna, travel a long distance until reaching the receiver antenna and be conditioned and digitally sampled in the receiver's Radio Frequency Front End (RFFE). This section briefly describes these steps.

1.4.1   Signal Propagation, Link Budget and Received Signal Strength

Link budget is the term used to calculate the power gains and losses from signal transmission to reception. A signal transmitted at a certain power P_T is then amplified by the satellite antenna, pointing to the Earth with a gain G_T. Usually, the power resulting from the term $P_T G_T$ is known as EIRP (Effective Isotropic Radiated Power, sometimes Equivalent Isotropic Radiated Power), where G_T is the maximum antenna gain, usually at *nadir*. Note that, in GNSS, the radiated power is concentrated on the Earth surface, and approximately, the same power should arrive to all users on Earth. This

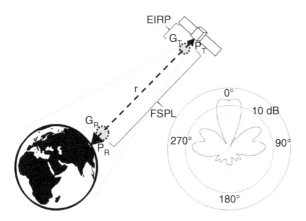

Figure 1.10 GNSS link budget. (Left) Signal path and link budget terms, including EIRP and FSPL. (Right) 2D antenna gain pattern of a Galileo satellite in the E1 (L1) frequency, 1575.42 MHz, from −15 to 20 dB, where the gain in the angle covering the Earth surface is between 10 and 15 dB [9].

leads to the antenna gain pattern shown in Figure 1.10 (right), with a tooth shape in its main lobe.

Signal polarization is another relevant property of the signals. It defines the direction of propagation of the electrical field. If we depict the electromagnetic wave as a spring between the transmitter and the receiver, if the spring departs to the right, the signal is Right-Hand Circularly Polarized (RHCP), and if it departs to the left, it is Left-Hand Circularly Polarized (LHCP). GNSS signals are RHCP. For an optimum transmission, the transmitting and receiving antennas should have the same polarization. Many receiver antennas are linearly polarized, instead of RHCP. This leads to a 3-dB loss. On the other hand, receiver antennas are often hemispherical, offering an approximate 3-dB gain for signals above the horizon, especially above a certain elevation.

After being amplified in the satellite antenna, the signal travels toward the Earth incurring free-space propagation losses, and then, it impinges on the receiver antenna, which also has some gain (G_R). This link budget calculation is expressed in the following formula:

$$P_R = \frac{P_T G_T G_R \lambda^2}{(4\pi r)^2}, \tag{1.41}$$

where λ is the wavelength and the term $(\lambda/4\pi r)^2$ is usually called free space path loss (FSPL). It encompasses two effects: the power spatial density, which decreases by a factor of $1/4\pi r^2$ as the distance r increases, and the effective area of the receiving antenna, which adds a factor of $\lambda^2/4\pi$. For example, a signal transmitted in the L1 band at 30 W, with a satellite antenna gain of 10, from a satellite at 20,200 km altitude (i.e. the approximate altitude of GPS at the zenith) and no receiver antenna gain will be received with the following power:

$$P_R = \frac{30 \cdot 10 \cdot 1 \cdot 0.19^2}{(4\pi \cdot 20.2 \cdot 10^6)^2} = 1.68 \cdot 10^{-16} \text{ W}. \tag{1.42}$$

Usually, power amounts are expressed in dB watts, as follows:

$$P_R(\text{dBW}) = 10\log_{10}(30) + 10\log_{10}(10) + 10\log_{10}(1) + 10\log_{10}((0.19/4\pi20.2\cdot10^6)^2)$$
$$= 14.77 + 10 + 0 - 182.5 = -157.74 \text{ dBW}, \tag{1.43}$$

which is within the range of typical GNSS signals.[16] Note that 30 W is a very low power, in the order of that of a small traditional bulb, and thus, the received power is extremely low, slightly higher than a billionth of a billionth of the transmitted power. We do not take into account signal propagation losses other than FSPL, nor receiver processing losses of around 0.5 dB. They are studied later. For a wider description of signal transmission, including GNSS antennas and propagation losses, see [9], [10] or [11].

In this section, we will also explain three relevant magnitudes of GNSS receivers for measuring the strength of the received GNSS signals: SNR, Carrier-to-Noise Density Ratio (C/N_0) and Bit Energy-to-Noise Density Ratio (E_b/N_0). We will explain how they relate to each other at the different stages of the receiver.

Signal-to-Noise Ratio (SNR)

SNR is the ratio between the signal power and the noise power, the latter mostly dominated by the thermal noise introduced by the same receiver. Sometimes it also referred to as S/N, although here we will use SNR. It is thus defined as:

$$\text{SNR} = \frac{P_s}{P_n}, \tag{1.44}$$

where P_s is the signal power and P_n is the noise power. We will show how to calculate the SNR before the correlation takes place at the signal processing stage. In our example, we assume the power of current GNSS signals, as received on Earth, is $P_s = -160$ dBW. In order to estimate the noise power P_n, we need the noise power *density*, or N_0. The noise power density expresses the noise power per frequency Hz and can be calculated as $N_0 = K_{\text{boltz}} \cdot T_e$, where K_{boltz} is the Boltzmann constant $(1.3806488 \cdot 10^{-23})$ and T_e is the effective temperature at the receiver. The effective temperature at the receiver is mainly driven by the ambient temperature T_0 (290° K). In reality, T_e is higher than T_0 due to the receiver antenna temperature and the different front-end stages, which include amplifiers and filters, especially the first stage. These stages increase the T_e with respect to T_0, and this leads to a noise power increase in the order of 1–2 dBs. However, for this example, we assume $T_0 = T_e$. A detailed explanation of how the receiver front-end affects the noise level is provided in [12, Chapter 6].

If $T_e = T_0$, $N_0 = 4 \cdot 10^{-21}$ W/Hz, which expressed in dB is −204 dBW/Hz. To get the noise power in the receiver, P_n, we just have to multiply by the receiver two-sided bandwidth:

---

[16] Note that, in this section, we use magnitudes in dB, as it is customary for power measurements expressed in a logarithmic scale. We assume that the reader is familiar with conversions to and from dB. In this section, they consist of using the $10\log_{10}$ conversion. We use dBW for expressing watt in a logarithmic scale and dBm for expressing milliwatt in a logarithmic scale. For example, −160dBW = −130 dBm.

$$P_n(\text{W}) = N_0(\text{W/Hz}) \cdot B(\text{Hz}). \tag{1.45}$$

If we have a two-sided bandwidth of $B = 2$ MHz to process GPS L1 C/A BPSK(1) signal, we multiply the noise power density by a bandwidth of $2 \cdot 10^6$ Hz, which involves adding 63 dB, thus obtaining a noise power P_n of -141 dBW. Therefore, the SNR becomes

$$\text{SNR[dB]} = -160 \text{ dBW} - (-141 \text{ dBW}) = -19 \text{ dB}. \tag{1.46}$$

In contrast to most communication systems, where a positive SNR is typically obtained, here we obtain a negative SNR when the signal reaches the receiver. This means that GNSS signals, when received and sampled, are buried below the noise level. This can be apparent by looking at the samples after the ADC, where there is no trace yet of a GNSS signal. This is a consequence of using an SS modulation, and the reason why we need to correlate the signal with its replica to raise it above the thermal noise level.

Carrier-to-Noise Density Ratio

C/N_0 is just the ratio between the signal power and the noise power density,

$$C/N_0 = \frac{P_s}{N_0}. \tag{1.47}$$

As we said before, $P_n = N_0 \cdot B$, so therefore,

$$C/N_0 = \text{SNR} \cdot B. \tag{1.48}$$

If we start with a noise density power of -204 dBW/Hz, as just explained, and a signal power of $C \equiv P_s = -160$ dBW at the receiver's antenna, this yields a C/N_0(dB-Hz) of $-160 - (-204) = 44$ dB-Hz.

In order to raise the signal above the noise level, we have to perform the correlation described in detail in the next section. Following our example, we will assume that a full 1-ms GPS L1 C/A code of 1,023 chips is correlated. We can depict the correlation process as the multiplication sample by sample of the signal with the replica. For example, assuming that 2,000 samples are correlated, obtained by sampling the code at approximately $N_{sc} = 2$ samples per chip, we would have a $10\log_{10}(2000) \approx 33$-dB gain. Note, however, that noise samples within the chip period may be correlated, and the gain does only depend on the number of truly uncorrelated samples [12]. So the actual correlation gain will be close to the 30 dB obtained by using $N_{sc} = 1$ sample per chip.[17] This 30-dB gain raises the signal above the noise level, leading to a positive post-correlation ratio (SNR_{corr}), and allowing processing of the signal. Note that the values in this example are theoretical and do not take into account the increase in the noise level due to the receiver front-end stages mentioned earlier, and the correlation

---

[17] Note also that this 30-dB gain is the result of two factors: the signal power gain, of $(1,023)^2$, of around 60 dB, and the noise gain, of 1,023, or about 30 dB, assuming that the noise power of our signal is normalized, or $\sigma_n = 1$. The resulting correlation gain is therefore $60 - 30 = 30$ dB.

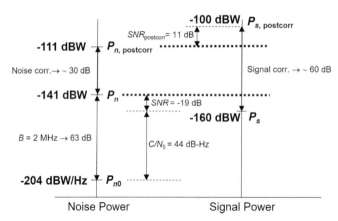

Figure 1.11 Example of noise and signal power at different stages of the receiver.

losses due to filtering, sampling and quantization, which subtract a few dBs from the correlation gain.

Carrier-to-noise density ratio (C/N_0) is used to measure the strength of a received GNSS signal with respect to the noise level, including that generated by the receiver itself in the front end. It is typically preferred to SNR in GNSS, contrary to what happens in communication systems, because SNR depends on the receiver bandwidth, while C/N_0 is bandwidth-independent. Since the bandwidth of a GNSS receiver is not fixed (e.g. it may range from whatever value from 2 MHz to 20 MHz or more), using the SNR as a performance metric can be confusing and difficult to compare among different receivers.

Bit Energy-to-Noise Density Ratio

A useful metric related to the C/N_0 is the bit energy to noise density, E_b/N_0. Since it comprises the energy per bit, it is obtained by multiplying the C/N_0 with the bit period, T_b, thus leading to

$$\frac{E_b}{N_0} = \frac{C}{N_0}T_b. \tag{1.49}$$

It gives the ratio between the energy in a bit and the noise density. The energy can be expressed in joules (or watts multiplied by seconds). The noise density is measured in watts per hertz, which can be expressed also in joules, so E_b/N_0 is dimensionless. An important feature of E_b/N_0 is that it allows us to calculate the Bit Error Rate (BER) of the received bits, before taking into account any coding scheme, through the following formula:

$$\text{BER} = \frac{1}{2}\text{erfc}\left(\sqrt{\frac{E_b}{N_0}}\right), \tag{1.50}$$

where erfc is the complementary error function.

Because of the relationship in Eq. (1.49), the BER depends on the product between the C/N_0 and the bit period. This is an interesting interplay because it means that no matter how weak the received signal power is, the data bits can still be recovered with an arbitrarily low error probability by increasing the bit period accordingly. This is a

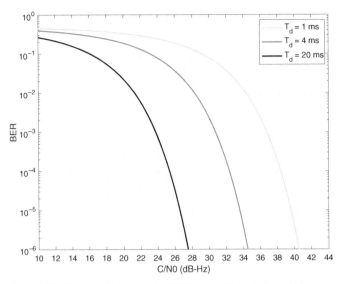

Figure 1.12 Uncoded bit error rate (BER) for different bit periods.

well-known principle in deep Space missions, where probes sending data to the Earth use very long bit periods in order to compensate for the very large propagation losses that cause the signals to be received on Earth with extremely weak power levels. As a result, the data rate is very small, just a few bits or a few tenths of bits per second.

The situation in GNSS is kind of similar and the bit period is actually quite long, for instance, 20 ms in GPS L1 C/A thus leading to a bit rate of just 50 bps. The resulting BER is shown in Figure 1.12. It corresponds to the *uncoded* BER because no encoding has been considered. In spite of this, one can see that for a low C/N_0 such as 28 dB-Hz, the BER is already below 10^{-6} or one bit in error per one million bits. If we consider the nominal C/N_0 of 44 dB-Hz already discussed in the previous section, corresponding to outdoors reception with open-sky conditions, then the BER is actually negligible if we extrapolate the results in Figure 1.12. It is so small that one could even think of deciding whether a bit is either -1 or $+1$ before the end of the bit period. It is for this reason that Figure 1.12 provides the resulting BER using a shorter period of time for deciding the bit sign, namely 4 and 1 ms. It can be seen that even when observing the received signal for just 1 ms, which would correspond to one code period in GPS L1 C/A, the BER in outdoors open-sky conditions is still very good, less than 10^{-6}.

1.4.2   Received Signal Model

As discussed in the previous section, the GNSS received signal is severely attenuated in its journey from the GNSS satellites down to the user receiver. This makes the received signal an extremely weak one, so specific countermeasures are needed to compensate for that at the receiver end. But apart from severe attenuation, the signal is received with a time delay due to the distance travelled during its journey, with a frequency shift due to the relative movement between the satellite and the user, and with a phase shift caused by both the time delay and the frequency shift onto the carrier. Additional disturbances such as ionospheric scintillation, multipath replicas or potential interference signals might be encountered as well, but they will be omitted here for the sake of simplicity. The main goal at this point is to focus on the model representing the signal received from a given satellite of interest.

We will first consider the Doppler effect experienced by the signal due to the satellite movement. This effect has an impact on the whole spectrum of the signal, with every frequency f being shifted to $(1 + \frac{s_r}{c})f$, with s_r the radial speed along the line of sight (LOS) between the satellite and the receiver, or relative speed of the satellite as seen by the receiver, and where c is the speed of light. This means that for a signal placed at the central frequency F_c, such as the one in Eq. (1.2), the signal ends up being placed at the new central frequency $(1 + \frac{s_r}{c})F_c$. Thus, it experiences a Doppler *frequency* shift given by[18]

$$\nu_t = \frac{s_r}{c}F_c. \qquad (1.51)$$

In the sequel, we will follow the convention whereby the Doppler shift is considered to be positive when the satellite is approaching the user receiver, while it is considered to be negative when the satellite is moving away.

The impact of the Doppler effect onto the central or carrier frequency is the most visible one, but actually the same effect is experienced by every single frequency of the signal spectrum. For a signal $s(t)$ with frequency representation $S(f)$, the presence of Doppler makes the latter be converted into $S(f/(1 + \frac{s_r}{c}))$. This means that the spectral content originally at $f = F_c$ for $S(f)$ is now found at the Doppler-shifted frequency $f = (1 + \frac{s_r}{c})F_c$ for $S(f/(1 + \frac{s_r}{c}))$. This confirms that the spectral content was actually moved.

The effect is translated into the time domain by having the signal $s(t)$ now given by $s((1+s_r/c)t)$, or equivalently $s((1+\nu_t/F_c)t)$ in terms of the Doppler frequency shift.[19] The result involves that the time-domain signal is compressed (i.e. when $\nu_t > 0$) or expanded (i.e. when $\nu_t < 0$) as a function of the Doppler shift. This effect will need to be accounted for at the receiver because the local replicas of the code and carrier waveforms to be generated need to match the compression or expansion of the received signal accordingly.

Let us now incorporate the Doppler effect onto the received signal. To do so, we will consider the bandpass or RF signal in Eq. (1.2) transmitted by a GNSS satellite. Only one of the two signal components of $s(t)$ is processed by the receiver at a time. For instance, GPS satellites transmit in the L1 signal $s(t)$ both the C/A and the P(Y) components, one in quadraphase with the other. But in our receiver, we are likely interested in the C/A component only. It is for this reason that we consider henceforth

---

[18] The subindex t is used in ν_t to emphasize that it is a frequency shift in the continuous time measured in Hz, so it can be distinguished from its discrete-time version denoted simply by ν and unitless, which will appear once the signal is sampled.

[19] The result is obtained in virtue of the property of the Fourier transform whereby $s(at) \overset{\mathcal{F}}{\leftrightarrow} \frac{1}{|a|}S(\frac{f}{a})$ for some constant a.

only one signal component in the GNSS received signal, the one to be processed by the receiver. This leads to the following received signal model:[20]

$$r'(t) = \sqrt{2C}x((1 + v_t/F_c)t - \tau_t)\cos(2\pi(F_c + v_t)t + \theta) + w'(t),\qquad(1.52)$$

where C is the received power P_R in Eq. (1.41) including the receiver implementation losses, τ_t is the time delay due to the signal propagation from the satellite to the user receiver and $w'(t)$ is some bandpass noise that encompasses the contribution of the rest of satellites in the same band as well as the background and receiver noise. We can see that the signal is affected by three main parameters: the Doppler frequency shift v_t, the propagation time delay τ_t and the carrier phase θ, the latter encompassing the effect of the time delay onto the carrier.[21]

It is also worth noting that the signal model in Eq. (1.52) is applicable to any of the two signal components in Eq. (1.2) just by adding a 90° phase offset to θ. So we are considering the carrier using the cosine function, but we could have used the sine function irrespectively, since we do not know the value of θ either. Note as well that the Doppler effect is present in both the signal component $x(t)$ and the carrier. At the carrier level, we already mentioned that Doppler introduces a frequency shift v_t, and at the signal component level, it makes the signal waveform compress or expand depending on the sign of the Doppler shift. This effect is referred to as code Doppler, and it is very subtle because the radial speed of the satellite is orders of magnitude smaller than the speed of light, so that $v_t/F_c \approx 0$. Effects of code Doppler are only perceived when the received signal is observed for a long period of time. This is not the case in most GNSS applications, where just a few milliseconds of received signal are often enough to determine the user's position. In this case, the received signal model can be further simplified to

$$r'(t) = \sqrt{2C}x(t - \tau_t)\cos(2\pi(F_c + v_t)t + \theta) + w'(t).\qquad(1.53)$$

Code Doppler will be revisited later when addressing the use of long observation periods for high-sensitivity applications, that is, when the received signal power is extremely weak. For the time being, the expression in Eq. (1.53) serves well for most applications, so we will stick to it unless otherwise stated. The interested reader can find further details on the impact of code Doppler in [13, Section 16.3.5].

1.4.3   Received Signal Conditioning

The signal impinging onto the user receiver antenna is modeled according to Eq. (1.53), which is a bandpass signal that needs to be properly conditioned so that it can be processed by a GNSS receiver. Signal conditioning takes place at the RF front end where filtering, amplification, down conversion and digitization are carried out. The output of the RF front end is typically a stream of digital samples that can readily

---

[20] The upperscript ′ is used in $r'(t)$ to indicate that this is the bandpass received signal, so it can be distinguished from its down converted version denoted simply by $r(t)$.

[21] The signal model assumes that all parameters are fixed for the sake of simplicity, but in practice, they all vary due to the satellite movement.

be processed by a digital GNSS receiver, either in a hardware or software implementation. A brief summary of the tasks carried out by the RF front end is described next.

Down Conversion

As already described in Section 1.3.1, the GNSS signal components are multiplied by a carrier wave to enable radio transmission over the L band allocated to radionavigation services. Once the signal arrives at the receiver, the carrier must be removed so that the GNSS-modulated signals are recovered back. Removal of the carrier frequency is done through a process known as *down conversion*. The received carrier is converted from a high frequency down to a low frequency, which is known as *intermediate frequency* (IF), or even down to the zero frequency, which means that no carrier frequency is present anymore and the resulting signal is purely a baseband signal. Down conversion can be done gradually, in consecutive steps, or directly in a single step. Gradual down conversion is the conventional approach in most RF front ends because at each step, the resulting signal is filtered and amplified. Step-by-step filtering and amplification helps in a better rejection of the undesired image replicas that appear in the down conversion process, a better isolation between weak (input) and strong (output) signals and a relaxation of the requirements for the hardware components. The resulting architecture is known as a *superheterodyne* receiver, while the one for the single-step down conversion is known as a *direct conversion* receiver.

The fundamental concept behind down conversion is the same already used for signal modulation. It relies on the fact that multiplication of a signal by either a cosine or a sine term causes the spectrum of the signal to be shifted (i.e. *modulated*) in the frequency domain as shown in Eq. (1.54)[22]. We will adopt the cosine case even though a similar principle applies for the sine as well. Let us assume a real-valued signal component $x(t)$ with Fourier transform $X(f)$ that has a generic amplitude A and it is multiplied by a cosine term. The resulting frequency-domain representation becomes:

$$Ax(t)\cos(2\pi F_c t) \xrightarrow{\mathcal{F}} \frac{A}{2}X(f + F_c) + \frac{A}{2}X(f - F_c). \tag{1.54}$$

This operation is also referred to as *up conversion*, and it results in a bandpass signal whose spectral content is placed in the neighborhood of the carrier frequency F_c, as seen in the upper part of Figure 1.13.

At the receiver side, we have two options to compensate for the carrier frequency. The first one is to shift the spectrum of the bandpass signal in Eq. (1.54) down to a smaller carrier frequency, F_{IF}, which makes the resulting signal much easier to manipulate than with the high frequency of the initial carrier. This can be done by multiplying the signal in the left-hand side of Eq. (1.54) by $B\cos(2\pi(F_c - F_{IF})t)$ for a generic amplitude B, thus leading to

$$Ax(t)\cos(2\pi F_c t)B\cos(2\pi(F_c - F_{IF})t) \xrightarrow{\mathcal{F}} \frac{AB}{4}X(f + 2F_c - F_{IF})$$

$$+ \frac{AB}{4}X(f - 2F_c + F_{IF}) + \frac{AB}{4}X(f + F_{IF}) + \frac{AB}{4}X(f - F_{IF}). \tag{1.55}$$

---

[22] Down-conversion can also be understood as an example of the general trigonometric formula $\cos(a)\cos(b) = \frac{1}{2}[\cos(a - b) + \cos(a + b)]$.

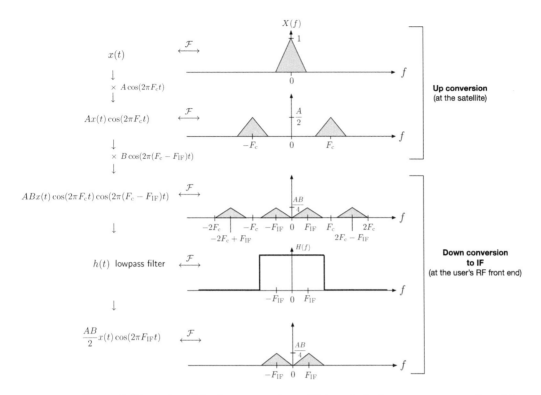

Figure 1.13 Illustration of the down conversion to IF through the frequency representation of the signal at each step of the process.

The result is illustrated in the central part of Figure 1.13 where the frequency representation in the right-hand side of Eq. (1.55) is shown. As can be seen, two replicas of the signal spectrum are placed at $\pm F_{IF}$, which is typically a much lower frequency (i.e. in the order of a few MHz) compared to the initial carrier in the L band. A lowpass filter is then applied to remove the aliased replicas at high frequency and to keep the low-frequency replicas only, as shown in the bottom part of Figure 1.13.

Down conversion to IF is an intermediate step to recover the signal of interest $x(t)$ because the result is not $x(t)$ itself but a bandpass version of it, at a smaller carrier frequency than the initial one. Many RF front ends provide at their output an already digitized down-converted signal at IF. One of the advantages of doing so is that the IF signal continues to be real-valued and therefore can be sampled with a single analog-to-digital converter (ADC). This was the case of the RF front end accompanying the preceding book, where the output signal was digitized as real-valued samples at an IF of either 1.364 or 4.092 MHz [14].

The second option for down conversion is to fully down convert to baseband so that the spectrum of the signal of interest is placed at the zero frequency. This would be the reverse operation to the up conversion carried out in Eq. (1.54), and it involves multiplying the received signal again by a cosine term with the same frequency, $B\cos(2\pi F_c t)$, where a generic amplitude B is considered. The result becomes

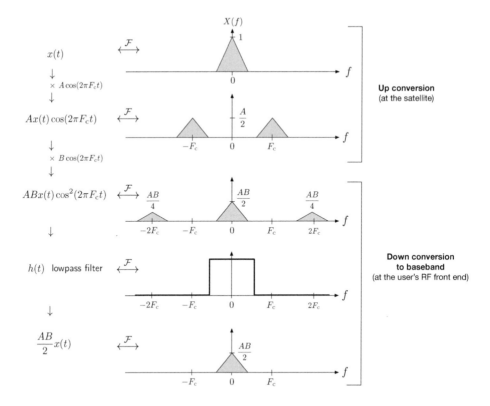

Figure 1.14 Illustration of the down conversion to baseband through the frequency representation of the signal at each step of the process.

$$ABx(t)\cos^2(2\pi F_c t) \overset{\mathcal{F}}{\longleftrightarrow} \frac{AB}{4}X(f+2F_c) + \frac{AB}{2}X(f) + \frac{AB}{4}X(f-2F_c), \quad (1.56)$$

where the original spectrum $X(f)$ is recovered and two high-frequency replicas are obtained as shown in the central part of Figure 1.14. These replicas are then removed by applying a lowpass filter, and the result is shown in the bottom part of Figure 1.14. The result in Eq. (1.56), once the lowpass filtering is applied to remove the high-frequency aliases, provides the signal of interest.

The previous example serves well for illustration purposes and it assumes that the original signal was up converted using a cosine term, and then down converted using the same cosine term aligned in frequency and phase. But this situation hardly ever occurs in practice. In particular, the received signal in Eq. (1.53) is affected by a Doppler frequency shift and a phase offset that are both unknown to the receiver. So it is impossible for the front end to perfectly know beforehand how the cosine term to be implemented should be. If down conversion is implemented considering the nominal frequency, that is, using the term $\cos(2\pi F_c t)$, the output would then be given by the expression at the output of the upper branch in Figure 1.15. As can be seen, the down-converted signal is still affected by a cosine term given by $\cos(2\pi \nu_t + \theta)$. To illustrate the situation, let us assume that no Doppler shift is present such that $\nu_t = 0$. Then, if

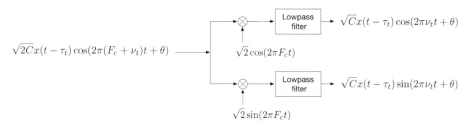

Figure 1.15 Implementation of down conversion to baseband where both cos and sin branches are considered. An amplitude $\sqrt{2}$ has been used in the cos and sin multipliers so that the power of the complex baseband signal coincides with that of the bandpass or RF received power C.

θ happens to be zero or π, the signal of interest will appear in the upper branch, but if θ happens to be equal to $\pi/2$ or $-\pi/2$, the signal at the output of this upper branch would be zero. It is for this reason that both the cosine and sine branches need to be implemented in practice when performing down conversion to baseband as shown in Figure 1.15, regardless of the fact that the original signal in Eq. (1.54) had only a cosine term. Once the outputs of both branches are available, they are merged together and a complex baseband signal is obtained, as already discussed in Eq. (1.3). Since the baseband down-converted signal is complex, in contrast to the IF down-converted one, two ADC are needed to sample the in-phase (I) and quadrature (Q) components of the resulting signal.

Taking the aforementioned considerations into account, and following the down conversion approach in Figure 1.15, the received signal in Eq. (1.53) would be down converted by the RF front end and result in the following complex-valued baseband signal,

$$r(t) = \sqrt{C}x(t - \tau_t)\cos(2\pi\nu_t t + \theta) + j\sqrt{C}x(t - \tau_t)\sin(2\pi\nu_t t + \theta). \tag{1.57}$$

Adding the contribution of the noise, which has been omitted so far for the sake of simplicity, the complex-valued baseband signal can be expressed for convenience as

$$r(t) = \sqrt{C}x(t - \tau_t)e^{j(2\pi\nu_t t + \theta)} + w(t), \tag{1.58}$$

where $w(t)$ contains the complex-valued baseband contribution of the noise, which is obtained following a similar reasoning to that of the signal obtained at the output of Figure 1.15. The formulation in Eq. (1.58) follows the same approach as in [4, Eq. (11.20)], where the received and down-converted signal is formulated as a complex-valued signal, in contrast to some other contributions in the GNSS literature where the same signal is formulated as two independent and real-valued signals (i.e. the in-phase and quadrature components) that are processed separately.

The presentation provided herein is just a glimpse of the fundamentals behind the down conversion process going on at the RF front end. The interested reader will find a complete and comprehensive discussion on the topic in [4, Section 11.1], [14, Chapter 4] or [13, Section 14.3].

Sampling and Quantization

Once down conversion takes place, the next step at the RF front end is to convert the resulting signal into a *digital* signal so that it can be processed by either a digital hardware or a software receiver implementation. Digital hardware implementations have been the *de facto* standard for a long time already due to their reliability, simplicity and flexibility compared to analog implementations. Rapid advances in very large-scale integration (VLSI) circuits during the last decades have also contributed to dramatically reducing the manufacturing cost and the device size. But when it comes to a software receiver implementation, as we are targeting in this book, digital signal processing is implicitly required since all operations are carried out in a digital signal processor (DSP) or a multipurpose personal computer, and thus, implemented in the digital domain. It is for this reason that the input signal to a software receiver is always a file with digital signal samples that are obtained at the output of an RF front end.

A *digital* signal is actually the result of two different operations. The first one involves sampling a continuous-time signal so that it is converted into a discrete-time one. The latter is a signal having samples of the original signal at discrete-time instants only, and thus, being discrete in the time domain. The second operation involves quantizing the value of each sample so that it can be represented with fixed precision using a finite number of bits. Sampling and quantization are both carried out by an analog-to-digital (ADC) converter at the receiver front end, and they are controlled by the sampling rate and the number of bits.

As for the first operation, sampling is typically done on the down-converted signal. Two main considerations must be borne in mind. The first one is that the selected sampling frequency should always fulfill the Nyquist criterion whereby $F_s \geq 2B_{bb}$, with B_{bb} the baseband (i.e. one-sided) signal bandwidth or $F_s \geq B$, with B the bandpass (i.e. two-sided) signal bandwidth. The second consideration is that one should carefully choose F_s such that it is not an integer multiple of the chip rate R_c in order to achieve sub-sample accuracy in the time-delay estimation of the received signal [15, 16].

In the recent years, some RF front ends have been implementing both down conversion and sampling into the same step. This is the so-called *IF or bandpass sampling*, a process whereby the bandpass-received signal is directly sampled at a wisely chosen sampling frequency [4, Section 11.1.3], [13, Section 14.3.3], [17]. This is done in order to benefit from the spectral aliases that naturally appear as a result of the sampling process, which actually constitute down-converted versions of the original signal obtained for free. That is, without having to explicitly implement a down conversion stage with an RF mixer. The reason why such spectral aliases appear is due to the well-known property in digital signal processing whereby sampling a signal in a given domain (i.e. time or frequency) results in aliased replicas in the other domain (i.e. frequency or time, respectively). This means that replicas of the spectrum of a GNSS signal at a central frequency F_c will appear at frequencies $F_c \pm mF_s$, for any integer m, once the signal is sampled at a sampling frequency F_s.

The key point of bandpass sampling is then to wisely choose F_s such that: (i) it avoids aliasing between consecutive replicas of the signal spectrum and (ii) it places one of these replicas as close as possible to the zero frequency. This would provide a discrete-time signal without aliasing while at the same time already down-converted, ready to be processed by the subsequent stages of the receiver. The penalty to be paid is that some degradation is incurred by the superposition of noise aliases in the frequency domain, which lead to a slight increase in the noise floor. The interested reader can find a detailed description of bandpass sampling for GNSS in [4, Section 11.1.3].

Regarding the quantization process, which is the second task carried out by the ADC, the key parameter is the number of bits to represent the value of each sample. Due to the extremely weak received power of GNSS signals, actually only one bit would suffice. The losses would be just 2 dB with respect to the case without quantization [18, Section III.C.4]. But in practice it is often customary to work with two bits per sample (e.g. sign and magnitude), which incurs just 0.5-dB losses, and at the same time, it is easy to handle by storing four signal samples into one byte. Beyond four bits per samples, the quantization losses are negligible. The only reason to move further beyond would be the case when interference signals are present. In that case, since terrestrial interferences have power levels much higher than the received GNSS signals, a larger dynamic range is needed at the ADC to avoid overload distortion. In these situations, it is also interesting to disable the automatic gain control (AGC) of the RF front end to avoid that the interfered GNSS signals might be further attenuated by the AGC when accommodating the dynamic range of the interference plus GNSS-received signal.

1.4.4 Discrete-Time Received Signal Model

In the sequel, and for the sake of simplicity, we will ignore quantization in the mathematical formulation unless otherwise stated. We will also assume that the signal is fully down-converted and placed at baseband and that a proper sampling frequency has been chosen. This results in the discrete-time counterpart of the complex-valued baseband received signal already introduced in Eq. (1.58), leading to:

$$r[n] = \sqrt{C}x[n - \tau]e^{j(2\pi\nu n + \theta)} + w[n], \tag{1.59}$$

where $r[n] = r(nT_s)$ and $x[n] = x(nT_s)$ are the discrete-time versions of the received signal and the signal component of interest, respectively, whereas $w[n] = w(nT_s)$ is the discrete-time complex-valued noise contribution. The parameters $\{\tau, \nu\}$ are now the discrete-time counterparts of the continuous-time delay and the aggregated frequency error including the Doppler effect and the frequency offset of the local oscillator. That is,

$$\tau = \tau_t/T_s, \qquad \nu = \nu_t/F_s, \tag{1.60}$$

with T_s being the sampling time and $F_s = 1/T_s$ the sampling frequency.

In the discrete-time domain, the data symbol period in Eq. (1.7), the code period in Eq. (1.8) and the chip period in Eq. (1.9) become the number of samples per data symbol N_{sd}, the number of samples per code N_{scode} and the number of samples per chip N_{sc}, respectively, which are defined as:

$$N_{sd} = T_d/T_s, \tag{1.61}$$

$$N_{scode} = T_{code}/T_s, \tag{1.62}$$

$$N_{sc} = T_c/T_s. \tag{1.63}$$

Finally, a piece of received signal of T seconds will be processed at the receiver, first for signal acquisition, and then for each measurement generated at the signal tracking stage. This leads to a snapshot of N_T received signal samples in the discrete-time domain with

$$N_T = T/T_s. \tag{1.64}$$

In contrast to the previous parameters in Eqs. (1.61)–(1.63), the snapshot length N_T will always be an integer number in practice because we cannot get a fraction of sample from the ADC output of the RF front end. In practice, we will be using $\lfloor N_T \rfloor$ samples, where $\lfloor x \rfloor$ stands for the *floor* of x. That is, the greatest integer smaller than or equal to x.

1.5   Receiver Architecture

Once the received signal has been conditioned and converted into the discrete-time domain by the RF front end, the output stream of digitized samples is ready to be processed by the GNSS receiver. The first task to be carried out by the receiver is to align its local code and carrier replicas to the actual code and carrier of the received signal. Such alignment is done following a two-stage approach. In the first stage, the correlation between the received signal and the local spreading codes is implemented for each satellite using coarse tentative delays and Doppler frequencies. This stage is considered to be successful when the satellite number, identified by its PRN, and a coarse estimate of its delay and Doppler frequency, are identified. In the second stage, the receiver implements a fine alignment to precisely track any possible time variation of the aforementioned parameters due to the satellite's movement and the user's dynamics. These two stages correspond to the so-called acquisition and tracking, whose blocks are shown next in Figure 1.16 as part of the architecture of a GNSS receiver.

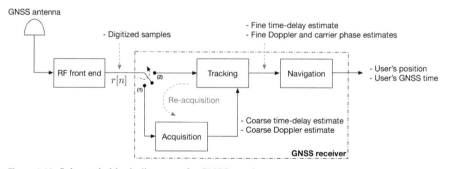

Figure 1.16 Schematic block diagram of a GNSS receiver.

Acquisition is performed first on a satellite-by-satellite basis, meaning that typically only one satellite is searched at a time. When the satellite of interest is successfully acquired, the receiver switches to tracking mode where it remains until this satellite is lost. In some situations, the satellite may still be present in the sky, but the reception is blocked due to an obstacle in the LOS. In that case, the acquisition stage is restarted for this satellite and its signal is reacquired, if still visible. After the tracking stage, the signal received from the satellite is decoded and the navigation data bits are extracted. Hardware receivers implement the acquisition and tracking of each satellite using parallelized hardware elements or *channels*, such that each channel acquires and tracks one satellite and many channels are simultaneously run in parallel. The acquisition and tracking stages will be discussed in Section 1.6 and Section 1.7, respectively.

Finally, when the observables of time delay and Doppler frequency are available and the navigation data have been decoded, the receiver is able to determine the satellite positions and then to estimate the user's position based on solving the so-called navigation equations. This task is carried out at the navigation module shown in Figure 1.16 where the user's position as well as the GNSS time are obtained. The procedure to do so will be described in Sections 1.8 and 1.10.

1.6   Acquisition

In order to compute the range measurements and to get the satellite data, the receiver first needs to be synchronized in frequency and time with the signal coming from the visible satellites. This involves determining, at least coarsely, the frequency shift v and the propagation delay τ in Eq. (1.59), a task that is referred to as *acquisition*. The procedure is similar to that of tuning the radio of our car to a radio station. But in contrast to radio signals, GNSS signals are buried below the noise floor and exhibit large frequency shifts due to the satellite's motion, so the acquisition problem is much more difficult to solve.

Discrete-time formulation will be considered henceforth motivated by the fact that all GNSS signal processing is nowadays implemented in the digital domain. That is, using either digital hardware, such as an ASIC (Application-Specific Integrated Circuit) or an FPGA (Field-Programmable Gate Array), or digital software implementations (e.g. software receiver running on a host computer), the latter being the case of the present book. The discrete-time received signal $r[n]$ in Eq. (1.59) is processed by the acquisition stage following a three-step approach as illustrated in Figure 1.17:

1. First, the received signal is correlated with a local replica (i.e. each sample is multiplied and the result is summed to the previous multiplications) using tentative values of time delay and frequency shift. These tentative values are denoted as $\{\tau', v'\}$, respectively, in order to distinguish them from the true values $\{\tau, v\}$ being sought. When the received signal power is too low, the correlation time must be increased accordingly in order to be able to detect the signal. This is due to the fact that signal detection depends on the received energy, and energy is the product of power by time. So for a fixed received power, the collected energy

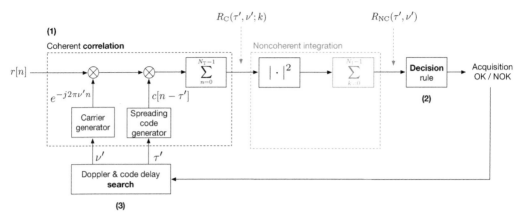

Figure 1.17 Architecture of the acquisition stage where the three steps discussed above are indicated: correlation with the received signal (either just coherently or in combination with noncoherent accumulations); decision rule to decide whether acquisition is successful for the pair of tentative $\{\tau', \nu'\}$ values; and finally, the search for a new pair of tentative values when those being used did not lead to a successful acquisition.

is increased by lengthening the observation time as well. Extending the correlation over a long time period suffers from problems that will be discussed later, so in practice the correlation time cannot be increased without bound. It is for this reason that the total correlation time must be split into a set of N_{I} shorter coherent correlations $R_{\mathrm{C}}(\tau', \nu'; k)$ for $k = 0, 1, \ldots, N_{\mathrm{I}} - 1$ that are noncoherently accumulated, that is, squared and summed, as indicated in Figure 1.17, resulting in the noncoherent correlation $R_{\mathrm{NC}}(\tau', \nu')$. This feature will be described later and allows the receiver to increase the sensitivity. Receivers implementing this feature are known as *high-sensitivity* GNSS receivers. For receivers operating in good visibility conditions and thus receiving a high enough power, a short correlation time is enough and noncoherent accumulations are not needed. This is the reason why this block is shaded in Figure 1.17, to highlight that it is optional.

2. When the tentative $\{\tau', \nu'\}$ are close enough to the true values, the correlation of the received signal with the local replica results in a large value compared to the case when the local replica is poorly aligned. A result similar to the central peak in Figure 1.7(a) should be obtained. So the second step of the acquisition stage involves detecting whether the correlation output is large enough to consider that the acquisition is successful. The latter means that the satellite being searched is present, and the local replica is well enough aligned to it. To do so, a decision rule must be applied on the correlation output value, which essentially boils down to comparing this value with a predefined threshold.

3. In case acquisition is not successful, which means that the output correlation value is not large enough, then the receiver must come back to step (1) and try again by correlating the received signal with a new pair of tentative $\{\tau', \nu'\}$ values. The search for the right values for $\{\tau', \nu'\}$ can be done sequentially for both time and frequency, or in parallel for time but sequentially for frequency. This search is the

most consuming part of the acquisition stage because many tentative values for $\{\tau', \nu'\}$ need to be tested. These values are taken by discretizing the search space into bins of $\Delta\tau'$ width on the time axis and $\Delta\nu'$ width on the frequency axis. Significant savings can be obtained in case assistance information is available, so the receiver can narrow down the search space, reduce the number of time–frequency bins to test, and consequently speed up the acquisition time [12].

Once the three main steps of the acquisition stage have been introduced, we will discuss in more detail each of them, while providing some insights into their practical implementation.

1.6.1 Correlation of the Received Signal with the Local Replica

As already introduced, the first step of the acquisition stage is to correlate the received signal with a local replica of the satellite signal being searched. The replica is generated by the receiver at both code and carrier levels. For the code level, let us omit at this time the presence of data bits and the secondary code, so that we can concentrate essentially on the correlation with the primary spreading code only. This would lead the receiver to implement the following correlation,

$$R_C(\tau', \nu') = \frac{1}{N_T} \sum_{n=0}^{N_T-1} r[n] c^*[n - \tau']_{N_{\text{scode}}} e^{-j2\pi\nu'n} \tag{1.65}$$

where we can see the replica of the discrete-time spreading code $c[n] = c(nT_s)$ at a tentative discrete-time delay τ', and the replica of the discrete-time complex carrier at a tentative discrete-time frequency shift ν'. Note that since the received signal $r[n]$ and the local carrier replica in Eq. (1.65) are both complex-valued, so is the result of the correlation $R_C(\tau', \nu')$. Furthermore, and because of the finite length of the spreading code, the indexation of its replica $c^*[n-\tau']$ in Eq. (1.65) must ensure that $0 \le |n-\tau'| < N_{\text{scode}} - 1$. This is achieved by indexing the code replica in a cyclic or modular manner with period N_{scode}. That is why the subindex N_{scode} is indicated in the code replica in Eq. (1.65), as done for the circular correlation in Eqs. (1.14)–(1.15). For the same reason, the time delay τ that we are looking for can only be resolved within a spreading code period. This means that our estimates for τ are just the fractional part of the true (absolute) time delay within one code period.

In the sequel, we will define the error between the tentative and the true values of the time delay and frequency shift as

$$\begin{array}{llll} \tau_{t,\epsilon} = \tau_t - \tau'_t & \text{(seconds)} & \rightarrow & \tau_\epsilon = \tau - \tau' \quad \text{(samples)} \\ \nu_{t,\epsilon} = \nu_t - \nu'_t & \text{(Hz)} & \rightarrow & \nu_\epsilon = \nu - \nu' \quad \text{(unitless).} \end{array} \tag{1.66}$$

It is important to mention that the correlation in Eq. (1.65) can either be implemented in the time domain as in most hardware implementations (e.g. following a matched filter approach as in Figure 1.17) or in the frequency domain as in most software implementations. The latter takes advantage of the property of the Fourier transform whereby the correlation in the time domain can be implemented as a multiplication in the frequency domain. The Fast Fourier Transform (FFT) is used in practice to speed up the computations, as discussed in Section 1.6.3. Examples of time-domain implementations can be found in [19, 20, 21], whereas examples of frequency-domain implementations such as the classical FFT-based acquisition, the half-bit acquisition method or the double block zero padding method can be found in [22] and [23], respectively.

The squared mean value of Eq. (1.65) leads to the so-called *ambiguity* function, which provides information on the signal power that is obtained at the correlator output as a function of the tentative values τ' and v'. Therefore, it provides information on how *sensitive* the correlator output is to changes on these tentative values. Ideally, we would like the correlator output to be very sensitive to τ' and v' because this would allow us to precisely determine the best (i.e. the most accurate) tentative value to be used. It is for this reason that the ambiguity function is widely used in signal design, as it is the case, for instance in radar signal processing, for determining the best time–frequency properties of new signal waveforms. For our application at hand, where the signal waveform is already given to us, the ambiguity function is used as an indicator function to determine how time–frequency misalignments of the local replica do affect the received signal power. In this way, we can use the ambiguity function as a cost function to be maximized as a function of τ' and v'. The tentative values of τ' and v', where the ambiguity function becomes maximum, are those values for which the local replica is properly aligned with the received signal and thus the maximum received signal power is obtained. In other words, the values of τ' and v' for which the ambiguity function is maximized become the estimates of the actual time delay and frequency shift of the received signal.

We will address the incurred losses later on, but for the time being, let us concentrate on the ambiguity function. The latter is nothing but the squared magnitude of the complex-valued correlation in Eq. (1.65) as a function of the difference between the true and the tentative time delay and frequency shift. That is,

$$\Psi(\tau_\epsilon, v_\epsilon) = \left| R_C(\tau', v') \right|^2, \tag{1.67}$$

where τ_ϵ and v_ϵ are the time delay and frequency shift errors defined in Eq. (1.66). Substituting $R_C(\tau', v')$ and ignoring the noise contribution, we obtain

$$\Psi(\tau_\epsilon, v_\epsilon) = \frac{C}{N_T^2} \left| \sum_{n=0}^{N_T - 1} c[n - \tau] c^*[n - \tau']_{N_{\text{scode}}} e^{j 2\pi v_\epsilon n} \right|^2 \tag{1.68}$$

$$= \frac{C}{N_T^2} \left| \frac{R_p(\tau_\epsilon)}{R_p(0)} \frac{\sin(\pi v_\epsilon N_T)}{\sin(\pi v_\epsilon)} \right|^2. \tag{1.69}$$

A long enough spreading code has been assumed when deriving Eq. (1.69) so that the autocorrelation of the spreading code signal could be approximated by the autocorrelation of the chip shaping pulse, only. Note that for $v_\epsilon = 0$, the ambiguity function simplifies to

$$\Psi(\tau_\epsilon, 0) = C \left| R_p(\tau_\epsilon) / R_p(0) \right|^2, \tag{1.70}$$

which means that in the absence of frequency errors, the accuracy for determining the time delay of the received signal is determined by the shape of the chip pulse auto-correlation, R_p. This is a very important result because it means that the sharper the chip pulse autocorrelation, the easier it will be to detect small variations on $\tau_\epsilon \approx 0$ and thus, the more accurate the time-delay estimation will be. A sharper autocorrelation involves a wider spectrum, so this observation links time-delay accuracy with the bandwidth of the signal.[23]

On the other hand, even when the local code replica is time aligned to the received signal, the presence of residual frequency errors degrades the correlation in Eq. (1.65). This can be observed by setting $\tau_\epsilon = 0$ in Eq. (1.69), which leads to[24]

$$\Psi(0, \nu_\epsilon) = \frac{C}{N_T^2} \left| \frac{\sin(\pi \nu_\epsilon N_T)}{\sin(\pi \nu_\epsilon)} \right|^2 \overset{|\nu_\epsilon| \ll 1/2}{\simeq} C \operatorname{sinc}^2(\nu_{t,\epsilon} T). \tag{1.71}$$

The result in Eq. (1.71) indicates that whenever the product $\nu_\epsilon N_T$, or its continuous-time counterpart $\nu_{t,\epsilon} T$, becomes an integer number, then $\Psi(0, \nu_\epsilon) = 0$ and the correlation output in Eq. (1.65) collapses. This observation demonstrates that the coherent sum being done in Eq. (1.65) is strongly affected by the presence of residual frequency errors.

An example of ambiguity function is illustrated in Figure 1.18 to better show how this function looks like in practice.

The time domain is presented in continuous time through the code delay error $\tau_{t,\epsilon}$ in chips (i.e. $\tau_{t,\epsilon}/T_c$), while the frequency domain is presented in continuous time as well, through the frequency error $\nu_{t,\epsilon}$ in Hz. This is done for the sake of clarity because continuous-time magnitudes are easier to understand than discrete-time ones, which many times do not even have units (e.g. as the discrete-time frequency). Notwith-standing, one can move from one domain to the other by using the relationships in Eq. (1.60). The example in Figure 1.18 considers a GPS L1 C/A signal using a coher-ent integration time of $T = 1$ ms. One can see that the envelope on the time-domain axis is given by the autocorrelation of the chip pulse, as indicated in Eq. (1.70). GPS L1 C/A uses rectangular chip pulses thus giving rise to a triangular autocorrelation shape. On the frequency-domain axis, one can see that the envelope corresponds to the Fejér kernel in Eq. (1.71) having a sinc-like shape. Nulls appear when the time–frequency product $\nu_\epsilon N_T$, or equivalently $\nu_{t,\epsilon} T$, is an integer number. In this example where $T = 1$ ms is considered, it corresponds to integer multiples of $1/T = 1$ kHz. This suggests that the residual frequency error $\nu_{t,\epsilon}$ should be smaller than 1 kHz to prevent the correlation from collapsing. As a rule of thumb, typically one half this value is con-sidered, so the receiver should make sure that $\nu_{t,\epsilon} \leq 1/2T = 500$ Hz. As mentioned

---

[23] Actually, the mean square bandwidth rather than the conventional bandwidth plays a role here, as discussed in detail in [5, Example 3.13].

[24] The result coincides with the Fejér kernel, which is the squared modulus of the Dirichlet kernel given by $\sin(\pi N x)/\sin(\pi x)$. The representation is very similar to that of the sinc function already defined as $\operatorname{sinc}(x) \doteq \sin(\pi x)/\pi x$, except for the fact that the Fejér and Dirichlet kernels are periodic functions of x, while the $\operatorname{sinc}(x)$ function is not.

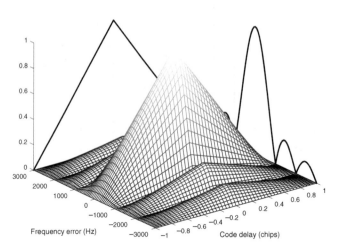

Figure 1.18 Ambiguity function normalized to unit amplitude for a GPS L1 C/A signal using $T = 1$ ms. The plot shows the square root of the ambiguity function to facilitate the visual identification of the triangular shape of the GPS L1 C/A correlation function. Infinite bandwidth is assumed.

earlier, the interplay between coherent integration time and residual frequency errors will be further discussed next when addressing the limitations of coherent integration.

It is worth mentioning that the result in Figure 1.18 assumes that the received signal has an infinite bandwidth and therefore the correlation with the local replica leads to a perfect triangle. In practice, GNSS receivers have a band-limited front end that cuts the received signal PSD down to a certain bandwidth. For example, for GPS L1 C/A signals, this bandwidth can be as low as 2 MHz. This provides significant computational savings at the expense of slightly degrading the sharpness of the correlation peak, and thus, slightly degrading the time-delay estimation accuracy. Some examples on the resulting correlation function using a band-limited GPS L1 C/A signal are shown in Figure 1.19. Two different bandwidths have been considered, $B = 2.046$ MHz (i.e. $B = 2R_c$) and $B = 4.092$ MHz (i.e. $B = 4R_c$). The bandwidth is left in general as $B = 2R_c$ and $B = 4R_c$ because the results for BPSK signals are the same regardless of the specific value of R_c, either it is $R_c = 1.023$ as in GPS L1 C/A or $R_c = 10.23$ MHz as in GPS L5. The correlation with infinite bandwidth is also shown in Figure 1.19 as a benchmark.

SNR Gain of Coherent Integration

The presence of a residual frequency shift has already been mentioned as a source of degradation for the coherent correlation in Eq. (1.65). Such a degradation can easily be assessed by observing the losses incurred in the ambiguity function in terms of ν_ϵ. For instance, when both the code and carrier replicas are perfectly aligned to the input signal, we have $\nu_\epsilon = 0$ and $\tau_\epsilon = 0$, and the ambiguity function becomes $\Psi(0,0) = C$. The result does not depend on the coherent integration length because the coherent correlation $R_C(\tau', \nu')$ was already normalized by the integration length

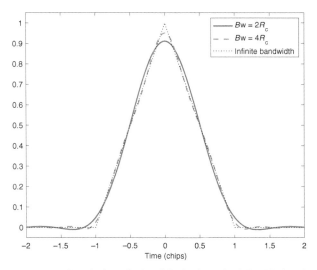

Figure 1.19 Correlation of a band-limited received GNSS signal using rectangular chip pulses at chip rate R_c with its ideal local replica. An example would be the GPS L1 C/A signal, which has $R_c = 1.023$ MHz, or the GPS L5 signal, which has $R_c = 10.23$ MHz.

N_T in its definition in Eq. (1.65). If we take the noise contribution into account, ignored so far, it is not difficult to see that the output noise power decreases as a function of N_T/N_{sc}, which is the total number of uncorrelated samples at the coherent correlation output.[25].

Based on these two results, we can compute the resulting SNR at the coherent correlation output. On the one hand, we have just mentioned that the signal power remains constant as a function of the coherent integration length N_T, according to the way that the coherent correlation output has been defined in Eq. (1.65), and as shown in $\Phi(0,0)$. On the other hand, we have that the noise power decreases linearly with N_T/N_{sc}. So merging both results together, we have that the SNR at the coherent correlation output increases linearly with N_T/N_{sc}, which is the number of uncorrelated samples being coherently integrated. For this result, we are assuming the optimistic case where the local replica is perfectly aligned. However, we can extend it for any τ_ϵ and ν_ϵ resulting in the following SNR gain at the coherent correlation output,

$$\Delta\text{SNR}_{\text{coh}}(\tau_\epsilon, \nu_\epsilon, N_T) = 20\log_{10}\left|\frac{\sin(\pi\nu_\epsilon N_T)}{\sin(\pi\nu_\epsilon)}\right| + 10\log_{10}\frac{N_T}{N_{sc}}, \quad \text{(dB)} \quad (1.72)$$

which is always contained within the range

$$0 \leq \Delta\text{SNR}_{\text{coh}}(\text{dB}) \leq 10\log_{10}\frac{N_T}{N_{sc}}. \quad (1.73)$$

---

[25] This is because for a set of N uncorrelated random variables $x_0, x_1, \ldots, x_{N-1}$ with the same variance σ_x^2, the sample mean $\bar{x} = \frac{1}{N}\sum_{n=0}^{N-1} x_n$ has a variance $\sigma_{\bar{x}}^2 = \sigma_x^2/N$ [24, p. 188].

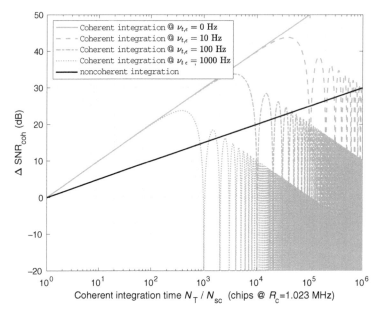

Figure 1.20 SNR gain at the coherent correlation output as a function of the integration time in units of chips, at $R_c = 1.023$ MHz.

An example of the SNR gain in Eq. (1.72) is shown in Figure 1.20 for different values of the frequency shift error. The SNR gain is shown as a function of the coherent integration length N_T normalized to the chip period N_{sc}, so that the result becomes the coherent integration length in units of chips. A chip rate of $R_c = 1.023$ MHz has been assumed in this example. The results in Figure 1.20 first show the ideal case when no frequency shift errors are present, so that $v_{t,\epsilon} = 0$. In this case, we can see how the SNR gain increases linearly without bound. Notwithstanding, when $v_{t,\epsilon} \neq 0$, the linear increase of the SNR gain is limited up to a point where it starts to decrease and it actually collapses. The collapse occurs whenever $v_{t,\epsilon}T$ or equivalently $v_\epsilon N_T$ are integer numbers, as previously discussed when introducing the ambiguity function. So for the case of $v_{t,\epsilon} = 1,000$ Hz, it means that the SNR first collapses for $T = 1/1,000 = 1$ ms, as confirmed in Figure 1.20. This integration length corresponds to 1,023 chips at the chip rate $R_c = 1.023$ MHz considered in Figure 1.20.

Finally, it is interesting to note that Figure 1.20 does also consider the case of noncoherent integration, where the accumulation in Eq. (1.65) would be carried out using the squared absolute value of each sample. This operation has no practical sense because we do need to integrate coherently at least for one code period in order to despread the received signal. But the curve in Figure 1.20 for noncoherent integration serves well for illustrating the advantage of coherent versus noncoherent integration. When integrating for 1,000 chips, one can see that there is a 30-dB gain with coherent integration, while just a 15-dB gain with the noncoherent one. So in practice there is a trade-off between coherent and noncoherent integration. The longer the coherent integration time, the higher the SNR gain but the smaller the residual frequency errors that can be tolerated. Noncoherent integration provides a means to compensate for the latter problem, but it comes at the expense of a worse capability to reject noise, as unveiled by the smaller SNR gain.

Constraints on the Coherent Integration Time

The minimum coherent integration time is usually set to one code period, so we have $T = T_{\text{code}}$ and thus $N_T = N_{\text{scode}}$ in Eq. (1.65). By doing so one is able to despread the received signal and to benefit from the orthogonality properties already discussed in Section 1.3.2. The maximum coherent integration time, though, could ideally be unbounded. Actually, the larger this value, the better the sensitivity of the receiver because more energy is accumulated. But one cannot let the coherent integration time increase without bound because of some limiting factors such as: (i) the presence of data modulating symbols; (ii) residual frequency errors still present in Eq. (1.65); and (iii) clock instabilities introduced by the front-end local oscillator.

The presence of data-modulating symbols restricts the coherent integration time down to the symbol period, so that $T \leq T_d$. Otherwise, integrating beyond this value may face the presence of symbol transitions that lead to the sum of destructive pieces of samples. So instead of having a gain by integrating over a longer time, we may end up having a severe signal loss whenever sign-reversed symbols are contained within the integration time. The only way to increase the coherent integration time beyond the symbol period is by knowing in advance what these symbols are, so one can compensate for them while doing the coherent accumulation in Eq. (1.65). This approach is actually the one that motivates the introduction of the so-called pilot or data-less components in modernized GNSS signals. These signals carry no data-modulated symbols, and thus, they are completely known to the receiver by the time they are received. It is for this reason that once the starting point of both the primary and secondary code is found, the rest of the signal is known and thus long coherent integration times can be implemented without the risk of suffering from unexpected symbol transitions. This involves, however, determining the starting point of the secondary code present in the pilot signal. The extension of Eq. (1.65) to cover this case will be briefly discussed next, but the interested reader can find more details in [4, Section 16.4].

The second limiting factor for the coherent integration time is the presence of residual frequency errors, which appear when wiping off the input carrier at frequency v with the local replica at the tentative frequency v'. Because of this frequency mismatch, the resulting signal still has some residual frequency error $v_\epsilon = v - v'$ that causes the ambiguity function in Eq. (1.69) to collapse whenever $v_\epsilon N_T$, or $v_{t,\epsilon} T$ in continuous time notation, is an integer number. This effect was already discussed in the previous section, and it could easily be seen in Figure 1.18 and also as a degradation of the SNR gain in Figure 1.20. The rule of thumb to deal with residual frequency errors is to limit the coherent integration time up to $T \leq \frac{1}{2|v_{t,\epsilon}|}$. This condition can be achieved by two different means: either by shortening T for a fixed frequency error $v_{t,\epsilon}$ or by reducing $v_{t,\epsilon}$ for a fixed coherent integration time T. The latter approach is the one usually considered in practice because T is a design parameter that is set to meet certain performance requirements, such as the receiver sensitivity. Once T is set, the acquisition stage must be designed so that the frequency search is fine enough to ensure that the remaining $v_{t,\epsilon}$ is compliant with the aforementioned condition.

Finally, the third limiting factor to deal with is the clock stability. This is often measured through the Allan variance $\sigma^2(T_A)$, a dimensionless measure that provides the variance of the random frequency fluctuations of the clock over a time period T_A [25]. When the Allan variance is referred to a given carrier frequency F_c in Hz, one can obtain the one-sigma frequency deviation of the clock as $\sigma_A(T)F_c$ in Hz. The coherent integration of this fluctuating frequency error results in an effect similar to that of the residual frequency error just mentioned above, causing the coherent integration in Eq. (1.65) to collapse when T is too long. Applying a similar principle, one should make sure that $\sigma_A(T)F_c \ll \frac{1}{2T}$ to minimize the impact of clock instabilities.

The three conditions to be fulfilled for setting the correlation time T are therefore:

$$T \le T_d, \quad T \le \frac{1}{2|v_{t,\epsilon}|}, \quad T \ll \frac{1}{2\sigma_A(T)F_c}. \tag{1.74}$$

The first condition depends on the GNSS signal, and there is nothing the receiver can do unless it has access to the navigation data symbols via an alternative means (e.g. using an assistance GNSS server), or a pilot component is processed. In contrast, the second and third conditions depend on the receiver implementation and can thus be improved at the expense of increasing complexity and cost. Complexity is due to the need to make $v_{t,\epsilon}$ small, thus increasing the number of tentative values for v' covering the same frequency range. Cost is due to the need to replace the clock with a better one, in case the stability does not allow for a long enough correlation. To have an idea of the order of magnitude, conventional temperature-controlled crystal oscillators allow T to be up to one hundred milliseconds, chip-scale atomic clocks (CSAC) up to a few hundred milliseconds and oven-controlled crystal oscillators (OCXO) up to one second [26]. Further details on the clock at the RF front end are provided at the end of Section 11.2.2.

Noncoherent Extension of the Coherent Integration Time

It now becomes clear that the coherent integration time cannot be increased without bound due to the presence of symbol transitions, residual frequency errors and clock instabilities that hinder the integration in Eq. (1.65). Among all of them, symbol transitions are often the most limiting factor because it is difficult for a standalone receiver to know in advance these symbols, so that they could be compensated. Assuming that symbol alignment has already been achieved at the receiver, the symbol period leaves just a few milliseconds of margin to safely perform coherent integration without the risk of experiencing a sign transition. For instance, the symbol period for GPS L1 C/A is just 20 ms long, which provides a sensitivity of ~35 dBHz of C/N_0. This value may not be enough for receivers operating in urban canyons, soft indoor environments or even in Space, so extending the integration time beyond the symbol period becomes a real need. In essence, the longer the correlation time, the larger the energy that is collected, and thus, the weaker the signals that can be detected.

The solution to this problem is to extend the correlation time by noncoherently accumulating N_I coherent correlations [12, 27]. The extended noncoherent correlation can thus be computed as follows:

$$R_{\mathrm{NC}}(\tau', v') = \sum_{k=0}^{N_1-1} \left| R_{\mathrm{C}}(\tau', v'; k) \right|^2, \qquad (1.75)$$

where the k-th coherent correlation $R_{\mathrm{C}}(\tau', v'; k)$ is derived from Eq. (1.65) as

$$R_{\mathrm{C}}(\tau', v'; k) = \sum_{n=0}^{N_T-1} r[n + k N_T] c^*[n - \tau']_{N_{\mathrm{scode}}} e^{-j2\pi v' n}. \qquad (1.76)$$

Note that the total correlation time in Eq. (1.75) is now:

$$T_{\mathrm{tot}} = N_{\mathrm{I}} T \quad \text{(seconds)}, \quad N_{\mathrm{tot}} = N_{\mathrm{I}} N_T \quad \text{(samples)}. \qquad (1.77)$$

The problem with noncoherent integration is that it is not as effective as the coherent one when it comes to mitigating the noise. This can be seen in Figure 1.20 where the solid black line shows the SNR gain when performing the full integration noncoherently. This will never be the case in practice because we need to coherently correlate at least one code period, so that we can despread the code in the received signal. But the example illustrates well how less efficient noncoherent integration is. While coherent integration has an SNR gain growing linearly with the integration time, noncoherent integration has an SNR gain that barely grows with the square root of the integration time. It is for this reason that one should always try to use the longest possible coherent integration time before noncoherent integration comes into action. At least, when sensitivity is being targeted. For accuracy purposes, this is not the case because the only important parameter for precise time-delay estimation is the total integration time, no matter how it is split into coherent or noncoherent integrations. But for detecting the satellites with the best possible sensitivity, one should always set the longest possible coherent integration time and then extend this time if needed, noncoherently.

It now becomes clear that the receiver design must face a trade-off between coherent and noncoherent integration at the correlation implemented by the acquisition stage. On the one hand, the longer the coherent integration, the better the noise rejection (i.e. the higher the receiver sensitivity), but the more sensitive the correlation is to residual frequency shifts. On the other hand, using short coherent integrations relaxes the requirement on the frequency search, because larger residual frequency shifts are tolerated. But this comes at the expense of a poor noise rejection and thus a poor sensitivity, so the receiver will need to operate in better sky visibility conditions than when long coherent integrations are used.

In practice one needs to carefully assess the receiver requirements and the implementation constraints in order to find the proper combination of coherent integration time and number of noncoherent integrations. This is because for the same total integration time T_{tot}, the performance can be quite different. To demonstrate this statement, Figure 1.21 shows an example for the sensitivity or minimum C/N_0 that would be required for detecting a GPS L1 C/A signal with a relatively high confidence (probability of detection $P_\mathrm{d} = 0.9$ and probability of false alarm $P_{\mathrm{FA}} = 10^{-4}$). Details on

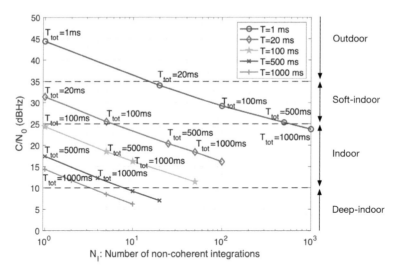

Figure 1.21 Minimum required C/N_0 (i.e. sensitivity) to detect a GPS L1 C/A signal with probability of detection $P_d = 0.9$ and probability of false alarm $P_{FA} = 10^{-4}$ for different combinations of coherent and noncoherent integrations. It is interesting to observe how the same total integration time T_{tot} provides different sensitivities when different combinations of T and N_I are used.

how to obtain these metrics will be discussed later in step 3 of the acquisition stage and with more detail in Section 9.2.4.

The results in Figure 1.21 show, for instance, that an integration time of 20 ms has ~3 dB better sensitivity when it is performed as a single coherent integration rather than as a noncoherent integration of 20 coherent integrations of 1 ms each. The difference is even larger when we move to longer integration times. For example, a single coherent integration of 1,000 ms length has ~10 dB better sensitivity than when the same integration is carried out by noncoherently accumulating 1,000 coherent integrations of 1 ms each. These results are illustrative of how different the sensitivity can be for the same total correlation time T_{tot}, when different combinations of T and N_I are set.

Finally, it is worth mentioning that based on the results shown so far, one may think that noncoherent integration is the answer to let the integration time increase without bound. It solves the presence of symbol transitions by removing the sign changes and it solves the clock issues by letting the coherent integration be performed on a short enough interval, short enough to prevent clocks instabilities to significantly degrade the coherent correlation output. But despite these advantages, the total integration cannot be increased without bound either. The last impairment to be coped with is actually the first one already introduced in Eq. (1.52), namely the code Doppler. It causes the received signal to be compressed or expanded depending on the experienced Doppler shift, thus causing the received spreading code not to match with the local replica being used at the receiver. Actually, the Doppler shift is still unknown at this stage of the receiver, so we cannot compensate for it beforehand. Assuming that the frequency search is done in steps of $\Delta v'$ width, the maximum residual frequency shift is then given by half this bin width, $\Delta v'/2$. So similarly to code Doppler contribution in Eq. (1.52), the residual frequency shift will introduce an additional time delay of $\delta \tau_t = \frac{\Delta v'_t/2}{F_c} T_{tot}$ seconds when the signal is observed over an interval of T_{tot} seconds long. If such additional time delay wants to be kept smaller than a given maximum value, $\delta \tau_{max,t}$, the total correlation time must then be limited to

$$T_{tot} \leq 2\delta\tau_{max,t} \frac{F_c}{\Delta v'_t} \tag{1.78}$$

which becomes the upper bound on the maximum total correlation time that can be used by the receiver.

Correlation in the Presence of Data Bits or Secondary Code

The coherent correlation in Eq. (1.65), which is also the basis for the noncoherent correlation in Eq. (1.75), was derived under the assumption that no data symbols or secondary code. If any of them are present, then the starting point of the symbol transition or the starting point of the secondary code must be searched as well. For instance, the symbol period in the GPS L1 C/A signal is 20 ms long, and it is composed of the repetition of $N_r = 20$ spreading codes with 1-ms duration each. Determining the starting point of the symbol period allows the receiver to implement a safe coherent integration up to $T = 20$ ms, without the risk of experiencing sign transitions in-between. Symbol or bit synchronization is often implemented by collecting the correlation peaks every 1 ms and then checking their signs to confirm whether they belong or not to the bit pattern of 20 consecutive equal signs. Variations of this approach can be found in [28, 29], and an implementation using a sliding window is in Section 9.2.8.

Pilot signals do not carry data-modulating symbols but incur a similar problem because they are implementing a modulated sequence on top of the spreading code, which is actually the secondary code. Even though the secondary code sequence is known, the starting point of this sequence in the received signal is unknown to the receiver. So eventually, the problem is the same as determining the data-modulated symbol transition, except for the fact that we are now searching for a known pattern of alternating signs. It is for this reason that many of the previous methods for symbol synchronization can be adapted for secondary code synchronization as well, just by changing the repetition pattern. The interested reader will find a detailed discussion on the topic in [11, Section 14.3.6] and [13, Section 16.5].

Alternatively, one could decide not to search for the symbol transition and proceed with the coherent integration in Eq. (1.65) using the received signal samples straightaway. In that case, there is a certain probability of experiencing a symbol transition somewhere within the coherent integration period. The symbol transition will split this period into two blocks that will add destructively in Eq. (1.65). The result can either collapse when the two blocks have equal size or barely suffer any degradation when one block is much larger than the other. In the end, the presence of symbol transitions can be modeled as a loss in signal power. This problem was analyzed in [30] where it was shown that the average losses for a total integration time $T_{tot} = T N_I$ amount at

$$L_{\text{bit,mean}} = -10 \log_{10} \left(1 - \frac{T}{3T_{\text{d}}}\right), \qquad \text{(dB)} \tag{1.79}$$

with T_{d} the symbol period, while the worst-case losses can reach up to

$$L_{\text{bit,max}} = -10 \log_{10} \left(1 - \frac{T}{T_{\text{d}}}\right). \qquad \text{(dB)} \tag{1.80}$$

These results are valid for any total integration time $T_{\text{tot}} \geq T_{\text{d}}$ provided that the coherent integration time is $T < T_{\text{d}}$. As can be seen, the results do not depend on the number of noncoherent integrations but just on the coherent integration time. The worst-case losses asymptotically increase as $T \to T_{\text{d}}$, but for a shorter coherent integration time of just 10 or 5 ms, the maximum losses are limited down to 3 or 1.25 dB, respectively. This means that in the absence of symbol synchronization, it is convenient to use a shorter coherent integration time T in order to keep the losses due to such transitions under control. This in turn will incur a sensitivity loss, but it can be compensated up to a great extent by increasing N_{I} and thus having a longer total integration time.

Otherwise, one always has the option of trying to detect the sign of the spreading codes present in the coherent integration period T. This would be like understanding spreading codes as if they were data-modulated symbols, and trying to detect and compensate their sign before adding them all together. The results in Figure 1.12 indicate that this is feasible even for spreading codes of 1 ms provided that the C/N_0 is above 35 dB-Hz, leading to a probability of error smaller than 10^{-2}.

1.6.2   Signal Detection

So far it has been considered that the noncoherent correlation $R_{\text{NC}}(\tau', v')$ constitutes the basis for deciding whether the local replica is correctly aligned to the received signal or not. As a by-product, it serves for deciding whether the satellite being searched is present or not, as well. These two possible hypotheses are often denoted as $\mathcal{H}_1$, that is, satellite present and well-aligned replica, and $\mathcal{H}_0$, that is, satellite absent or badly aligned replica. Different variations on $R_{\text{NC}}(\tau', v')$ can alternatively be used for this purpose. For instance, by using the absolute value or a fractional exponent instead of the square shown in Eq. (1.75) [31]. In order to formulate the signal detection problem, let us in general denote by Z_i the resulting metric, using either Eq. (1.75) or any other variant. Let us also assume that the search space is composed of $N_{\tau'}$ bins of width $\Delta \tau'$ in the time axis, and $N_{v'}$ bins of width $\Delta v'$ in the frequency axis. So we have a total of $N_{\tau'} N_{v'}$ bins to test, and thus Z_i spans for $i = 1, \ldots, N_{\tau'} N_{v'}$. Because of the bi-dimensional time–frequency search, it is often customary to store all the resulting Z_i values in a matrix with dimensions $(N_{\tau'} \times N_{v'})$. Each Z_i value is therefore often referred to as a test *cell*. Finally, the decision test statistic is formed through some function $\Phi(\cdot)$ of all the Z_i results,

$$Z = \Phi(Z_1, Z_2, \ldots, Z_{N_{\tau'} N_{v'}}). \tag{1.81}$$

Since the Z_i values are ultimately based on the correlation between the received signal and a tentative local replica, it is not difficult to see that a large value will be

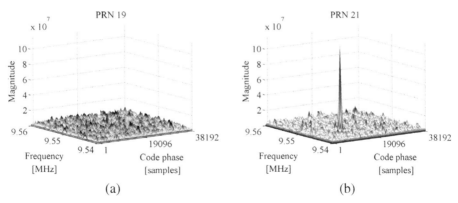

Figure 1.22 Sample acquisition output (a) PRN 19 is not visible, so no peak is present. (b) PRN 21 is visible, so a significant peak is present. The peak is at code delay 13404 samples and frequency 9.5475 MHz.

obtained whenever the local replica is well-enough aligned to the received signal. So finding the best tentative local replica, and thus the best pair of (τ', ν') values, boils down to finding which of the Z_i values is large enough. It is for this reason that the function $\Phi(\cdot)$ can be the maximum of its arguments, the ratio between the largest value and the second largest value separated apart by a certain interval, or any other criterion that may help in finding the correct cell. In the end, the procedure consists of comparing the decision test statistic Z with a given value or threshold γ. That is,

$$
Z \underset{\mathcal{H}_0}{\overset{\mathcal{H}_1}{\lessgtr}} \gamma,
\tag{1.82}
$$

being the outcome of this comparison, the answer to the decision question on whether either $\mathcal{H}_1$ or $\mathcal{H}_0$ is the correct hypothesis. An example of the $(N_{\tau'} \times N_{\nu'})$ matrix containing all Z_i values under analysis is shown in Figure 1.22. On the left-hand side, we can see the case when the satellite signal of interest is not present, so all cells contain noise only. On the right-hand side, we can see the case when the signal is present, observing a cell having a much larger value than the rest of cells in the search space.

As mentioned previously, finding the correct cell provides twofold information. First, that the tentative local replica is well-enough aligned to the received signal. Second, as an obvious consequence of the former, that the satellite being searched is actually present in the received signal. In that case, the acquisition is said to be *successful* and the tracking stage is then initialized with the corresponding pair of (τ', ν') as coarse estimates of the code delay and frequency shift. So we can declare that a given satellite is present, while at the same time we can provide rough estimates on the time delay and frequency shift values. The decision may be correct or wrong, but in the way the threshold γ is designed, the probability of correct decision and the probability of false alarm are both under control.

The test statistic can be based on either a data or a pilot signal component, or a combination of both data and pilot components. The combining methods can be either noncoherent or coherent. Few research papers have addressed the issue on how to combine the data and pilot components in forming the decision statistic [32, 33, 34]. The general approach is the noncoherent one where the correlation outputs from the data and pilot components are squared and then accumulated in order to form the decision variable. Alternatively, coherent combining methods are proposed in [32] for both E1 and E5 Galileo signals. The main idea of coherent combining methods is to add and subtract the pilot and data components to form two decision variables Z_p and Z_d, and then form a decision variable based on the maximum between the two outputs. According to the results reported in [32], the coherent combining scheme slightly outperforms the noncoherent one, but only if the signal level is strong enough. At moderate and low signal levels, the noncoherent and coherent combinations show similar performance.

The problem of detecting a signal in noise has been extensively presented in [35], where the reader is referred to for further information on the underlying principles. The application to GNSS involves some additional considerations because multiple test cells, on the order of thousands, need to be evaluated and this requires some search strategy planning, as discussed in the next step on the search process. At high level, signal detection in GNSS can be performed either in parallel by evaluating all the testing cells at the same time, that is, as indicated in Eqs. (1.81)–(1.82), or sequentially by evaluating one test cell at a time. In the latter case, one moves from one test cell to the next until detection is declared successful. When this occurs, the search stops and acquisition is declared successful.

The search strategy has an impact on the decision rule in Eq. (1.81) because the statistics of a large set of values are different from the statistics of just a single value. Parallel detection involves evaluating all the testing cells simultaneously. It is often implemented in software receivers because it fits well with the use of the FFT. It will be analyzed in detail in Sections 1.6.3 and 9.2.4 for snapshot receivers. The alternative is sequential detection, which involves evaluating a single cell at a time. The evaluation of the decision rule in Eq. (1.81) can be done in a single stage or in multiple stages with various coherent and noncoherent integration times. The single-stage detection is known as single-dwell detection, see [36, 37]. Multiple-stage detection is also known as multidwell detection [36, 38, 39], where we gain more confidence on the presence or absence of the satellite under analysis as we move on to subsequent stages. In this way, we do not need to wait until the whole signal is processed, but rather take partial decisions dwell by dwell. By doing so we can stop whenever we are confident enough on the decision to be made, or continue to the next dwell in case we are not confident enough yet and would like to double check. The concept of multidwell detection is illustrated in Figure 1.23. Each dwell is characterized by the so-called *dwell time*, which is simply the integration time used in the correlation of that dwell. Then a decision statistic $Z_{i,k}$ is formed at the k-th dwell for the i-th time–frequency bin, and the result is compared with a threshold γ_k. If the decision statistic is larger than the threshold, we move to the next $(k + 1)$-th dwell, with a larger dwell time in order to cross-check the decision just taken. At each dwell or decision stage, the same steps are undertaken. The final delay and frequency estimates are obtained after the last

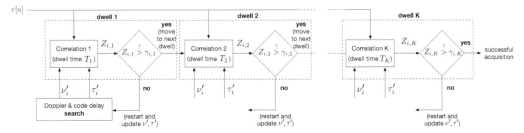

Figure 1.23 Multidwell structure for the acquisition of GNSS signals. Detail showing the dwells for the i-th testing cell using the pair of tentative values (v_i', τ_i').

dwell. Ideally, the larger the number of dwells, the better the trade-off between a high detection probability and a small false alarm probability we can get, though this might not always be the case [40]. Moreover, a higher number of dwells means an increased complexity of the acquisition stage.

Coming back to the global decision rule in Eq. (1.81), the probability of detection is obtained as

$$P_D(\gamma) = \int_{\gamma}^{\infty} f(Z|\mathcal{H}_1)dZ, \tag{1.83}$$

whereas the probability of false alarm, that is, the probability of declaring acquisition successful when actually no signal but only noise was present, is obtained as

$$P_{FA}(\gamma) = \int_{\gamma}^{\infty} f(Z|\mathcal{H}_0)dZ, \tag{1.84}$$

with $f(Z|\mathcal{H}_m)$ the probability density function (PDF) of Z under hypotheses $\mathcal{H}_m$, where $m = \{0,1\}$. Note that both the probability of detection and the probability of false alarm have been expressed as a function of the detection threshold γ, which is the parameter that allows us to adjust these probabilities to match the receiver requirements. In case of sequential detection using multiple dwells, the decision rule at each individual dwell is characterized by the corresponding probability of detection $P_{d,k}(\gamma_k) = \int_{\gamma_k}^{\infty} f(Z_k|\mathcal{H}_1)dZ_k$ and probability of false alarm $P_{fa,k}(\gamma_k) = \int_{\gamma_k}^{\infty} f(Z_k|\mathcal{H}_0)dZ_k$, for $k = 1, \ldots, K$. Based on these individual probabilities, the global detection and false alarm probabilities are computed as $P_D(\gamma_k) = \prod_{k=1}^{K} P_{d,k}(\gamma_k)$ and $P_{FA}(\gamma_k) = \prod_{k=1}^{K} P_{fa,k}(\gamma_k)$, respectively. Multidwell structures have not been so extensively studied in GNSS, even though some references on the topic can be found in [41, 42, 43].

For the detection problem in Eq. (1.81), the common practice is to be given a requirement in terms of P_D for a given working condition (i.e. at some minimum C/N_0) and in terms of P_{FA}. Based on these inputs, we will need to tune the detection stage so that these probabilities are ultimately achieved. The tuning is done by adjusting the detection threshold γ in Eqs. (1.83)–(1.84) accordingly.

The relationship between the probability of detection, probability of false alarm and the detection threshold is schematically illustrated in Figure 1.24. For instance, the probability of false alarm is the probability of declaring acquisition successful when

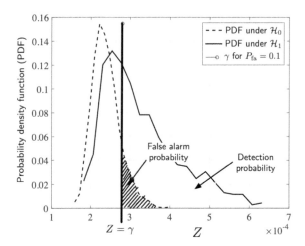

Figure 1.24 Threshold choice at the acquisition stage. $\mathcal{H}_0$ is the null hypothesis (no signal), and $\mathcal{H}_1$ is the alternative hypothesis (signal present).

it is actually not. This corresponds to the integral in Eq. (1.84), which is graphically illustrated in Figure 1.24 by the area below the right tail of the PDF under $\mathcal{H}_0$, beyond the point $Z = \gamma$. Since the PDF under $\mathcal{H}_0$ does only depend on the noise statistics, and not on the signal one, the probability of false alarm is directly related to the detection threshold γ. So once P_{FA} is provided as a requirement to be met, we can obtain γ from either Eq. (1.84), in case an analytical expression is available, or from the histogram of the empirical PDF for $\mathcal{H}_0$ by moving γ until the shaded area in Figure 1.24 coincides with P_{FA}. Once the detection threshold has been set, then we need to make sure that for the minimum C/N_0 that has been specified, the area below the right tail of the PDF under $\mathcal{H}_1$ is greater than or equal to the specified P_D. Finally, it can easily be seen in Figure 1.24 that the larger the detection threshold γ, the smaller the false alarm probability, which is good, but the smaller as well the detection probability, so a trade-off needs to be found.

A performance criterion for the acquisition stage is the mean acquisition time (MAT), which refers to the average time needed to be spent in the acquisition stage until the delay and frequency estimates are obtained [44]. Naturally, different combinations between search and detection mechanisms will provide different MAT. Alternative performance criteria can also be found in [45]. Typically, the trade-off is between the complexity of the acquisition unit (number of parallel correlators) and the acquisition times. Generic methods to compute MAT for DS-SS signals can be found in [37, 38]. For Galileo, the MAT values have been analyzed in [39, 43, 46, 47]. Some techniques to improve MAT both at the search and at the detection stages have been discussed in [48].

More details on the process of detecting a GNSS signal will be discussed in Chapter 9 when addressing snapshot receiver implementations.

1.6.3 Time–Frequency Search

This step is in charge of evaluating the correlation $R_{\mathrm{NC}}(\tau', \nu')$ for all tentative values of time delay and frequency shift within the search space. As mentioned previously, the search space is composed of $N_{\tau'}$ bins in the time domain with resolution or bin width $\Delta\tau'$, and $N_{\nu'}$ bins in the frequency domain with resolution or $\Delta\nu'$. The total search space is thus composed of $N_{\tau'}N_{\nu'}$ bins.

Time Search

Regarding the time-delay search, the received signal comes with samples equispaced every T_s seconds, as given by the sampling rate. The local code replica is thus sampled at the same rate in order to perform the correlation. Despite the one sample time resolution, it is important to keep in mind that the time resolution of the correlation output, in terms of τ', does not depend on the sampling rate but on the resolution of the actual τ' used for generating the local replica. This resolution, $\Delta\tau'$, can actually be as fine as desired. However, it has the disadvantage that, by doing so, the number of tentative time-delay bins $N_{\tau'}$ dramatically increases. In practice it is often preferred to use a conservative time-delay resolution, typically half the one-sided width of the chip autocorrelation peak. By doing so one can trade off with the number of tentative delay bins that are required to explore. It is interesting to note that once acquisition is successful and the correlation peak of the received signal is found, sub-sample accuracy can always be obtained by interpolating the samples in the neighborhood of the maximum peak [49] or by recomputing the correlation peak once an estimate of the frequency shift is available as well [50]. In the end, one should keep in mind that the accuracy of time-delay estimation does only depend on the received signal bandwidth and the received C/N_0, but not on the sampling rate of the receiver, provided that it fulfills the Nyquist criterion. The impact of the time–frequency resolution in terms of C/N_0 losses at the correlation output is thoroughly discussed in [13, Section 16.2.2].

So in practice, the time-delay resolution $\Delta\tau'$ is required to be smaller than the one-sided width of the chip pulse autocorrelation peak. This, in turn, determines the minimum sampling rate needed. Indicative values for $\Delta\tau$ are provided in Table 1.2 for some representative GNSS signals.

The impact of the time-delay mismatch between the local replica and the received signal is shown in Figure 1.25(a). The results are shown in terms of the incurred loss in signal power. A rectangular chip pulse with duration equal to the chip period, such

Table 1.2 Maximum time delay resolution $\Delta\tau'$ at the acquisition search step, for some representative GNSS signals.

Modulation	One-sided width of the chip pulse autocorrelation main peak [chips]	Time-delay resolution $\Delta\tau'$ of the acquisition search [chips]
BPSK (any order)	1	0.5
BOC(1,1)	0.34	0.175
sinBOC(10,5)	0.14	0.07
cosBOC(10,5)	0.11	0.055

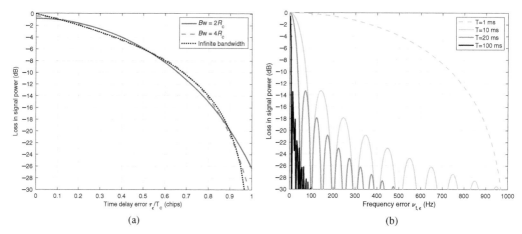

Figure 1.25 Losses in signal power due to either (a) a time mismatch or (b) a frequency mismatch in the correlation between the received signal and the local replica. A rectangular chip pulse with duration equation to the chip period $T_c = 1/R_c$ is considered.

as in GPS L1 C/A, is considered. One can verify that when the local replica is offset by the half chip period with respect to the received signal, the correlation output corresponds to the mid-point of one side of the triangular peak. So being the mid-point, its amplitude should be one half of that of the peak, thus incurring in a 6-dB loss in signal power. This is actually the value observed in Figure 1.25 for a time-delay error of 0.5 chip. Now let us come back to the results in Table 1.2. The time-delay resolution $\Delta\tau'$ for the time-delay search was mentioned to be (at least) half of the one-sided chip pulse autocorrelation peak width. Therefore, for the example at hand, this means using at least $\Delta\tau'/N_{sc} = \Delta\tau_t'/T_c = 0.5$. That is, two samples per chip. On the other hand, searching in steps of $\Delta\tau'$ incurs a maximum time delay error of $\tau_{\epsilon,max} = \Delta\tau'/2$. This translates into a maximum time delay error of $\tau_{\epsilon,max}/N_{sc} = \tau_{t,\epsilon,max}/T_c = 0.25$ chip for this example, which is found to incur a ~2-dB loss according to Figure 1.25. The conclusion is that in the worst case, the losses due to the bin size of the time delay search amount up to ~2 dB. Finally, once the time resolution is known, the time-delay search spanning within one code period of N_{scode} samples results in a total of $N_{\tau'}$ time bins where the correlation peak may appear,

$$N_{\tau'} = \left\lfloor \frac{N_{scode}}{\Delta\tau'} \right\rfloor. \tag{1.85}$$

Frequency Search

Regarding the frequency search, it is interesting to review first the three main elements contributing to the frequency shift in the received signal. The first one is the satellite motion, which amounts to a maximum of ±5 kHz frequency shift, as per GPS satellites at L1.[26] The second one is the frequency offset of the receiver oscillator, which

---

[26] The frequency shift of other GNSS satellites in ordinary MEO orbits is similar, but this is not the case for satellites in IGSO orbits, where Doppler shift is approximately half of that of MEO orbits; GEO orbits,

adds ± 1.5 kHz per each ppm (part per million) of frequency stability when observed at L1. The third one is the user's velocity, which adds ± 1.5 Hz per each 1 km/h of speed [12, Chapter 3].

When no a-priori knowledge is available and an oscillator with 1 ppm frequency stability is used, the acquisition stage should account for a maximum frequency search of $v_{t,\max} = 6.5$ kHz. If the frequency offset of the receiver oscillator was known, for instance, by some prior calibration[27] means, then the maximum frequency search could be reduced down to $v_{t,\max} = 5$ kHz. If on top of the oscillator calibration we also had a-priori information on the satellite Doppler, for instance, thanks to the use of assistance information, then there would be no need to perform the coarse frequency search. In that case, the only remaining frequency shift would be due to the user's dynamics, and this can normally be accommodated within the span of a single frequency bin, which reaches up to $\pm \Delta v_t'/2$.

While the maximum span of the frequency search is now clear, it remains to determine the actual bin size $\Delta v'$. We will use the continuous-time notation $\Delta v_t'$ instead, because it is a more intuitive metric than its discrete-time counterpart $\Delta v'$, which is actually unitless. This will facilitate the discussion hereafter.

Some details on the tolerable frequency error were already unveiled when introducing the coherent correlation in Eq. (1.65), which was found to collapse whenever $v_{t,\epsilon}T = 1$ or multiples of it. The effect was clearly visible in the ambiguity function in Eq. (1.69) and shown therein in Figure 1.18. As a rule of thumb, the maximum frequency error was set at $v_{t,\epsilon,\max} = 1/2T$ with T the coherent integration time. This would correspond to a frequency search using a bin width of $\Delta v_t' = 1/T$, assuming that the errors are uniformly distributed within this width. The impact of having such frequency error can now be seen in Figure 1.25(b). For a coherent integraton time of $T = 1$ ms, the frequency bin width would be 1 kHz and the maximum frequency error would then become 500 Hz. Such error incurs a $\sim$4-dB loss in signal power as observed in Figure 1.25, which may be too high to afford in certain working conditions. A more conservative approach is to set the frequency bin width to

$$\Delta v_t' = 1/2T, \tag{1.86}$$

which then leads to a maximum frequency error,

$$v_{t,\epsilon,\max} = 1/4T. \tag{1.87}$$

For the same example above, this now would lead to a frequency search bin width of $\Delta v_t' = 500$ Hz, a maximum frequency error of 250 Hz, and just a $\sim$1-dB loss in signal power. Finally, once the frequency resolution is known, the frequency search comprised in the range within $\pm v_{t,\max}$ is then divided into $N_{v'}$ frequency bins to test, where

---

which is much lower; or the eccentric Galileo eccentric satellites (E18 and E14), which reach a higher Doppler shift than standard MEO satellites.

[27] Note that such calibration may need to be repeated from time to time to account for temperature variations and ageing of the oscillator.

$$N_{\nu'} = \left\lfloor \frac{2\nu_{t,\max}}{\Delta\nu_t'} \right\rfloor + 1 \qquad\qquad (1.88)$$

or equivalently using the frequency bin width in Eq. (1.86),

$$N_{\nu'} = \lfloor 4T\nu_{t,\max} \rfloor + 1. \qquad\qquad (1.89)$$

The results in Figure 1.25(b) provide many insights on the interplay between coherent integration time and maximum tolerable frequency error at the correlation stage. One can see how increasing the coherent integration time, as required in high-sensitivity applications, places stringent demands on the frequency search, since the residual frequency error must be decreased accordingly. For instance, let us assume that we are fine with the ~4-dB loss of signal power when using $T = 1$ ms and a frequency bin width of $\Delta\nu_t' = 1$ kHz. Let us now assume that we would like to preserve the same loss but when implementing a long coherent integration time of $T = 100$ ms, as it could be needed for instance in some high-sensitivity applications. The 4-dB loss for $T = 100$ ms is found in Figure 1.25 to be obtained for a frequency error of just 5 Hz, which implies a frequency search with a bin width of $\Delta\nu_t' = 10$ Hz. This is an extremely demanding requirement. To make it clear, let us assume that the receiver at hand is operated in standalone mode, so assistance information is not available. According to the discussion above, the frequency search should span at least for $\nu_t' \in [-\nu_{t,\max}, +\nu_{t,\max}] = [-6.5, +6.5]$ kHz. Covering this range in steps of $\Delta\nu_t' = 10$ Hz leads to $N_{\nu'} = 1,301$ frequency bins to evaluate. On the time-delay search, assuming a spreading code with 1,023 chip and thus a time delay bin width of $\Delta\tau_t'/T_c = 0.5$ chip, we would have $N_{\tau'} = 2,046$ time bins to evaluate. Putting everything together, we would need a time–frequency search space composed of a total of $N_{\tau'}N_{\nu'} \sim 2.6 \cdot 10^6$ bins, which is extremely demanding. This example shows how high-sensitivity receivers requiring long coherent integration times, and thus, a very fine frequency search must operate with assistance information in order to make the acquisition process feasible in practice. The same principle applies to any receiver in the sense that assistance information does always help in reducing the frequency search span, and thus, in reducing the search space. But for demanding applications where long coherent integration times are needed, assistance information becomes a must.

Search Strategies

So far we have set the grounds for the time–frequency search, determining the search space, the search bin width and the incurred losses. The next step is to determine how to actually go through the search space and evaluate all the time–frequency cells. One or several cells can be analyzed at each trial, resulting in three different approaches for implementing the time–frequency search:

– *Serial search.* In this case, the decision metric of each cell, which is given by the correlation output in Eq. (1.75), is computed and tested individually right away [51, 52, 14]. Decisions are taken on a cell-by-cell manner, and thus, the search process follows sequentially with the rest of the cells in the search space as illustrated in Figure 1.26(a).

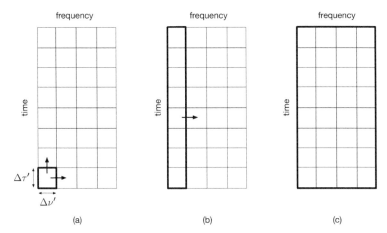

Figure 1.26 Illustration of different search strategies for going through the whole time–frequency acquisition matrix: (a) serial search, (b) hybrid search and (c) full parallel search.

- *Hybrid search.* All time-domain cells are used together to form a decision metric. Typically, this is done when using the FFT to compute the code correlation for a given tentative frequency. The result of the FFT simultaneously provides all the correlation samples, which are used to detect whether the signal is present or not, as illustrated in Figure 1.26(b). Then the search proceeds sequentially by implementing the FFT for a new tentative frequency, and the detection process is repeated again until the whole frequency search space is scanned [53, 54, 39, 40].
- *Full parallel search.* This is an extension of the hybrid search where the FFT implementing the correlation is computed for all the tentative frequencies, and the results are stored in a time–frequency matrix. Then, all cells of this matrix are simultaneously used to detect whether the signal is present or not. This is done for instance by taking the largest value of the whole time–frequency acquisition matrix, and then comparing this maximum value with a given detection threshold [55, 56, 57].

Example of Acquisition Technique: Full Parallel Search

We provide next an algorithmic description of the full parallel search, which is actually the technique implemented in the multi-GNSS software of this book, and therefore, it serves well for the purpose of illustrating an example of acquisition technique. From a high-level perspective, the implementation of the parallel search comprises the following steps:

1. Generate the local spreading code replica $c[n]$ for the satellite to be acquired.
2. Apply the FFT to $c[n]$ (see Section 9.2.9 for more details) and take the complex conjugate of the result. The complex conjugate of the FFT is needed because we seek to implement the correlation operation in the frequency domain. This is possible in virtue of the convolution theorem of the Fourier transform, whereby

the product of Fourier transforms corresponds to the convolution in the time domain [8, Section 2.9.6]. The convolution of two discrete-time signals $x_1[n]$ and $x_2[n]$, defined as $x_1[n] * x_2[n] = \sum_{k=-\infty}^{\infty} x_1[k]x_2[n-k]$, is very similar to the correlation we are looking for, but not identical. The correlation of these two signals is defined as $R_{x_1,x_2}[n] = \sum_{k=-\infty}^{\infty} x_1[k]x_2^*[k-n]$ where the complex conjugate $(\cdot)^*$ has been considered for the sake of completeness, in case the signals were complex-valued. So we can see that convolution and correlation are related as follows, $R_{x_1,x_2}[n] = x_1[n] * x_2^*[-n]$. That is, we need one of the signals to be complex conjugated and time reversed so that by doing the convolution we actually obtain the correlation. Both complex conjugate and time reversal in the time domain imply the complex conjugate only of the Fourier transform [8, Section 2.8]. This is the reason why we are taking the complex conjugate of the code replica FFT in the present step, so we can obtain the correlation of the received signal with the local replica by means of implementing their convolution in the frequency domain.

3. Two options are possible here:
   (a) Multiply the received signal $r[n]$ with a local carrier wave $e^{-j2\pi v'n}$ at the tentative frequency value v' to tentatively remove the frequency shift v present in $r[n]$. Take the FFT of the result.
   (b) Alternatively, the code replica $c[n]$ could be the one multiplied by the local carrier, in this case, by $e^{j2\pi v'n}$. This could be done in the frequency domain by circularly rotating the FFT samples of the code replica, which are already available from step 2. In that case, the frequency shift bin width $\Delta v'$ is determined by the frequency resolution of the FFT. Care must be taken when choosing the size of the FFT in order to match the FFT frequency resolution with the required $\Delta v'$.

4. Obtain $R_C(\tau', v'; k)$ in Eq. (1.76). If step 3(a) was implemented, $R_C(\tau', v'; k)$ is obtained by multiplying the result of step 2 with that of step 3(a) and applying the IFFT. If step 3(b) was implemented, $R_C(\tau', v'; k)$ is obtained by multiplying the FFT of $r[n]$ with the result of step 3(b) and applying the IFFT.

5. Repeat steps 2-4 for $k = 0,\ldots N_I - 1$ and accumulate $|R_C(\tau', v'; k)|^2$ in order to form $R_{NC}(\tau', v')$ in Eq. (1.75).

6. The satellite is detected whenever $\max_{\tau',v'} R_{NC}(\tau', v') > \gamma$ for a given threshold γ. The coarse estimates for the time delay and frequency error, denoted by $(\hat{\tau}, \hat{v})$, are obtained as the pair of time–frequency values (τ', v') where the maximum above is achieved. That is,

$$(\hat{\tau}, \hat{v}) = \arg\max_{\tau',v'} R_{NC}(\tau', v') > \gamma. \tag{1.90}$$

As can be seen, the implementation makes extensive use of the FFT operation in order to efficiently implement the correlation operation, and it is therefore the preferred approach in software receiver implementations. In the subsequent chapters, we will refer to this technique when addressing the implementation details for each specific GNSS signal. Further details on the FFT implementation are discussed in Section 9.2.9.

1.7   Tracking

The acquisition stage is completed once the satellite of interest is detected and coarse estimates of its code delay and frequency shift are obtained. These values are then fed as an input to the tracking stage, whose role is to provide fine estimates of both the code and carrier phase and to keep track of any potential variation over time. If the receiver has a snapshot architecture, no tracking stage is then implemented. Instead, the coarse estimates coming from the acquisition stage are provided as the output observables straightaway. Snapshot architectures are specifically addressed in Chapter 9. In the rest of this chapter, we will focus on conventional GNSS receivers, which implement both acquisition and tracking stages following the architecture shown in Figure 1.16. Thanks to the tracking stage, the receiver can continuously retrieve the bits from the navigation message and measure the satellite ranges that are needed to compute the user's position.

The tracking stage is composed of two separate modules, each of them in charge of code tracking and carrier tracking, respectively. Both modules are implemented using a closed-loop architecture. This involves comparing the input signal with a local replica, determining their relative error, and then using this error to correct the local replica that will be compared again with the incoming signal in the next time instant. Both code and carrier tracking will briefly be described next with the aim of providing a high-level overview of their operation principle.

1.7.1   Carrier Tracking

Carrier tracking is required at the receiver in order to synchronize the local carrier replica with the incoming signal, and thus to completely wipe off the carrier. Once the carrier is properly removed, all subsequent operations such as despreading the PRN code of the satellite under analysis, measuring the time-delay and retrieving the navigation data bits can all be safely carried out as long as the SNR allows us to do so. As can be inferred, carrier tracking is intimately related to code tracking because actually one is depending on the other. Carrier tracking allows the carrier to be wiped off and thus to proceed safely with the coherent correlation for code despreading. But at the same time, carrier tracking needs the code to be despread first, because otherwise the signal of interest is buried well below the noise floor and becomes imperceptible. It is for this reason that both carrier and code tracking operate simultaneously, and indeed, carrier tracking is often used to assist code tracking under the so-called *carrier aiding*. This strategy takes advantage of the capability of carrier tracking to precisely keep track of the user dynamics. Assisting the code tracking stage with the carrier dynamics allows a simpler implementation of the code tracking, since user dynamics are already provided and the emphasis can be placed on just filtering out the thermal noise.

Carrier tracking is implemented following the rationale of the block diagram in Figure 1.27. When carrier tracking is first initiated, the incoming signal is correlated with the local code and carrier replicas using the coarse code delay and frequency shift estimates $(\hat{\tau}, \hat{\nu})$ provided by the acquisition stage. Later, once tracking is ongoing, code delay and carrier phase values are recursively computed within the tracking loop. These values are used to align the local code and carrier replicas to those in the received signal, and the goodness of such alignment is assessed through the correlation of both terms as previously done at the acquisition stage with Eq. (1.65). This leads to the so-called *prompt* correlation, whose expression is given by

$$y_P[k] = \sum_{n=0}^{N_T-1} r[n + kN_T]c^*[n - \hat{\tau}]_{N_{\text{scode}}} e^{-j\hat{\theta}[n;k]}, \tag{1.91}$$

for the k-th prompt epoch, with the locally generated carrier phase given by $\hat{\theta}[n; k] = 2\pi\hat{\nu}[k]n + \hat{\theta}[k]$. Substituting the received signal $r[n]$ with the signal model in Eq. (1.59), the expected value of the prompt correlator results in

$$E\left[y_P[k]\right] = \sqrt{C} \sum_{n=0}^{N_T-1} c[n + kN_T - \tau]_{N_{\text{scode}}} c^*[n - \hat{\tau}]_{N_{\text{scode}}} e^{j\theta_\epsilon[n;k]}, \tag{1.92}$$

where data symbols or the values of the secondary code in pilot signals have been omitted for simplicity. The carrier phase error $\theta_\epsilon[n; k]$ results from the difference between the carrier phase of the received signal $\theta[n; k]$ and the carrier phase of the local replica $\hat{\theta}[n; k]$. Note that k is indexing the block of N_T received samples being processed at the coherent correlation, whereas n is the sample by sample indexing. Note also that the carrier phase of the received signal can be further decomposed as $\theta[n; k] = 2\pi\nu(n + kN_T) + \theta = 2\pi\nu n + \theta[k]$ for some carrier phase offset $\theta[k]$. So the error between both carrier phases that appears in Eq. (1.92), namely $\theta_\epsilon[n; k]$, is thus given by

$$\theta_\epsilon[n; k] = \theta[n; k] - \hat{\theta}[n; k] = 2\pi\nu_\epsilon[k]n + \theta_\epsilon[k], \tag{1.93}$$

where $\nu_\epsilon[k] = \nu - \hat{\nu}[k]$ stands for the frequency error and $\theta_\epsilon[k] = \theta[k] - \hat{\theta}[k]$ stands for the phase offset error at the k-th prompt correlator epoch. When the summation in Eq. (1.91) takes place, the presence of a residual frequency error $\nu_\epsilon[k]$ leads to the Fejér kernel already mentioned when introducing the ambiguity function in Eq. (1.69), causing the output amplitude of the prompt correlator to degrade. The expected value in Eq. (1.92) thus results in

$$E\left[y_P[k]\right] = \sqrt{C}\frac{R_p(\tau_\epsilon)}{R_p(0)}\frac{\sin(\pi\nu_\epsilon[k]N_T)}{\sin(\pi\nu_\epsilon[k])}e^{j\theta_\epsilon[k]}, \tag{1.94}$$

which is a complex value. In the sequel, we will refer to the real (i.e. in-phase) and imaginary (i.e. quadrature) components of the prompt correlator output by $I_P[k]$ and $Q_P[k]$, respectively, so that

$$I_P[k] = \text{Re}[y_P[k]], \quad Q_P[k] = \text{Im}[y_P[k]]. \tag{1.95}$$

Once some preliminaries have been introduced, we are now in position to describe how carrier tracking is implemented in a GNSS receiver following the architecture in Figure 1.27. As can be seen, the aforementioned prompt correlator output samples are fed to a carrier discriminator. This block is in charge of providing an error signal $\epsilon[k]$ that, on average, is proportional to the (frequency or phase) error between the

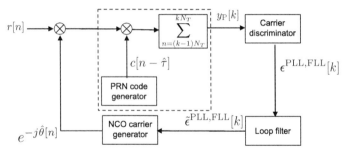

Figure 1.27 Generic block diagram for carrier tracking in a GNSS receiver.

received signal and the local carrier replica. A tracking loop making use of a *phase* discriminator is referred to as a Phase-Locked Loop (PLL), while a tracking loop making use of a *frequency* discriminator is referred to as a Frequency-Locked Loop (FLL). Regardless of the type of discriminator, the output error signal needs to be filtered by a loop filter to obtain a smoothed error signal. The latter is finally used to control the generation of the carrier replica for the next epoch, which is carried out by a numerically controlled oscillator (NCO). The resulting closed-loop architecture is able to align the local carrier to the received one by subsequently iterating until the resulting error signal is driven to zero.

Phase-Locked Loop Discriminators

It is important to bear in mind that carrier phase tracking needs to be insensitive to the 180° phase shifts that, from time to time, are introduced by the symbol transitions in data-modulated GNSS signals. A way to circumvent this limitation is by using a *Costas* discriminator, which is a type of discriminator insensitive to sign transitions. The baseline Costas phase discriminator relies on the atan of the prompt correlation samples so that the signal error is generated as

$$\epsilon_{\text{atan}}^{\text{PLL,Costas}}[k] = \text{atan}\left(\frac{Q_P[k]}{I_P[k]}\right) = \theta_\epsilon[k] \in [-90°, 90°]. \tag{1.96}$$

The result in Eq. (1.96) actually provides the phase error to be corrected regardless of the presence of data-modulating symbols. The main drawback is that it relies on a non-linear function (atan), and thus, its hardware implementation may pose some troubles. Alternative Costas discriminators have been proposed to circumvent this problem at the expense of shrinking the linear region of their input-output response. Some examples are shown below along with the region in which they operate linearly. That is, the region where their output coincides with the phase error $\theta_\epsilon[k]$ that is present at their input and that we are looking for.

$$\epsilon_{I \times Q}^{\text{PLL,Costas}}[k] = I_P[k]Q_P[k] = \sin(2\theta_\epsilon[k]) \approx \theta_\epsilon[k] \in [-22.5°, 22.5°] \tag{1.97}$$

$$\epsilon_{\text{DD}}^{\text{PLL,Costas}}[k] = \text{sign}(I_P[k])Q_P[k] = \sin(\theta_\epsilon[k]) \approx \theta_\epsilon[k] \in [-45°, 45°].$$

Note that $\epsilon_{\text{DD}}^{\text{PLL}}[k]$ is designed for BPSK-modulated symbols. It implements a Decision Directed (DD) strategy for deciding the data-modulated symbol through $\text{sign}(I_P[k])$

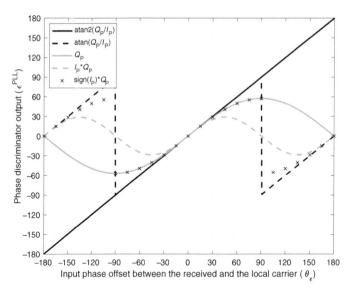

Figure 1.28 Output $\epsilon[k]$ provided by different PLL discriminators as a function of the input phase difference $\theta_\epsilon[k]$ between the received carrier and the local replica. Phase in units of degrees.

and then uses this decision for removing the contribution of that symbol on $Q_P[k]$. The expressions above assume that an AGC is applied to the prompt correlator in Eq. (1.91) so that it is normalized to unit power, $P_{y_P} = E\left[|y_P[k]|^2\right] = 1$. Otherwise, the discriminators in Eq. (1.97) turn out to depend on P_{y_P} and $\sqrt{P_{y_P}}$, respectively, so they should be normalized accordingly using an estimate of the prompt correlator power. Note also that for small arguments it is reasonable to assume $\sin(x) \approx x$, so the low-complexity Costas discriminators in Eq. (1.97) actually provide a valid correction for the NCO at the expense of a reduced input linear operation region. This can be clearly seen in Figure 1.28 where the input-output response of some Costas phase discriminators is shown, normalized so that their slopes are equal to one in the linear region. Such input-output representation is often referred to as the *S-curve*,[28] typically having a linear region for small inputs, then a zero slope for larger inputs, and sometimes, a negative slope for much larger inputs that may cause the curve to exhibit additional zero-crossings.

It is worth noting that in the absence of data-modulating symbols, such as when tracking pilot signals, the limited input range of Costas discriminators leads to an unnecessary degraded performance in the presence of severe noise. In these circumstances, it is preferred to avoid Costas discriminators and go for a PLL discriminator based on the full-range, or four-quadrant version atan2, which is widely used in software receiver implementations. In that case,

$$\epsilon_{\text{atan2}}^{\text{PLL}}[k] = \arg\left(y_P[k]\right) = \text{atan2}\left(\frac{Q_P[k]}{I_P[k]}\right) = \theta_\epsilon[k] \in [-180°, 180°], \qquad (1.98)$$

---

[28] The S-curve is also used for code tracking discriminators, as later shown in Figure 1.32.

where arg($\cdot$) stands for the complex argument (i.e. angle) of a complex number. A simpler implementation is possible recalling that $\arg(x) \approx \text{Im}[x]$ for a small and complex number x. Then we have

$$\epsilon_Q^{\text{PLL}}[k] = \text{Im}\left[y_P[k]\right] = Q_P[k] \approx \theta_\epsilon[k] \in [-45°, 45°]. \qquad (1.99)$$

Similarly to the previous low-complexity Costas discriminators in Eq. (1.97), the PLL discriminator in Eq. (1.99) also needs to be normalized to avoid fluctuations caused by variations on the received signal strength. In this case, normalized by $\sqrt{P_{y_P}}$.

Phase-Locked Loop Performance and Filter Bandwidth

The carrier phase tracking performance can be assessed through the variance of the estimated phase at the NCO output. Assuming that the tracking loop operates in the linear region, such variance is given by (see for instance [4, Section 12.3.6])

$$\sigma_{\hat{\theta}}^2 = \frac{B_L}{\frac{C}{N_0}}\left(1 + \frac{1}{2\frac{C}{N_0}T}\right) \quad (\text{rad}^2), \qquad (1.100)$$

with B_L the bandwidth of the loop filter in charge of smoothing the discriminator output and T the coherent correlation time corresponding to the N_T samples used in Eq. (1.91) to obtain the prompt correlation output. Note that the term $\frac{C}{N_0}T$ in Eq. (1.100) becomes the SNR at the output of the prompt correlator. The overall term in brackets in Eq. (1.100) is known as the squaring losses and accounts for the performance degradation that is experienced at medium- to low-SNR due to nonlinear effects caused by the increased noise. Carrier phase tracking is a well-known topic that has extensively been addressed in the existing literature. An exhaustive coverage of this topic is thus out of the scope of the present book. Just the fundamental concepts have been addressed here to support the understanding and operation of the companion software to this book. The interested reader can find further details on this topic in [58, 59, 60] or [10] for a comprehensive overview. For details on the loop filter design and performance analysis, the interested reader is referred to [4, Section 12.3].

Frequency-Locked Loops

Regarding FLL implementations, they provide an alternative way of implementing carrier tracking by using frequency measurements instead of phase ones. By doing so, FLLs provide much more robust performance than PLLs in the presence of high user dynamics, owing to the fact that frequency variations are always much smaller than phase variations. This allows the tracking loop to operate in a more relaxed manner and avoid the abrupt jumps that are typically observed in phase measurements. The price to be paid is that there is no access to the actual phase of the received signal, but only to its frequency. This means that the phase of the locally generated carrier replica is always behind the phase of the received signal by a constant phase offset. This is not a major problem because such offset can be resolved at the demodulation stage of the receiver. However, the common practice in many GNSS receiver implementations is to first start the tracking stage using an FLL and once in steady-state, switch to a PLL to perform the finer tracking and resolve the pending phase offset ambiguity.

Most FLL discriminators are based on the observation that the product between two consecutive prompt correlator outputs provides a complex value whose argument is proportional to the frequency error. This motivates the following baseline FLL discriminator,

$$\epsilon_{atan2}^{FLL}[k] = \frac{1}{2\pi N_T} \arg\left(y_P[k]y_P^*[k-1]\right),$$ (1.101)

which is often expressed in terms of in-phase and quadrature formulation as follows:

$$\epsilon_{atan2}^{FLL}[k] = \frac{1}{2\pi N_T} atan2\left(\frac{I_P[k-1]Q_P[k] - I_P[k]Q_P[k-1]}{I_P[k]I_P[k-1] + Q_P[k]Q_P[k-1]}\right).$$ (1.102)

The four-quadrant arctangent discriminator above (atan2) can also be implemented in a decision-directed manner by using the two-quadrant arctangent discriminator instead (atan), as done in the Costas PLL shown in Eq. (1.96). By doing so, we get rid of the 180° phase shifts due to symbol transitions in the prompt correlator samples.

In the sequel, we will consider Eq. (1.101), though, because it allows a more intuitive interpretation. Actually, Eq. (1.101) is providing the phase difference taking place in the complex product of two consecutive prompt correlator outputs, namely $y_P[k]y_P^*[k-1]$. Note that being a phase difference, it has the sense of frequency, and thus, it is consistent with being a frequency discriminator. It is not difficult to see that the expression in Eq. (1.101) is equivalent to:

$$\epsilon_{atan2}^{FLL}[k] \equiv \frac{\theta_\epsilon[k] - \theta_\epsilon[k-1]}{2\pi N_T} = v_\epsilon[k],$$ (1.103)

which is the frequency error between the received signal and the local carrier replica, $v_\epsilon[k]$. Actually, what we were looking for.

Similarly to the PLL, low-complexity FLL discriminators have been proposed to circumvent the nonlinear function for retrieving the angle of a complex number. Some examples are the following:

$$\epsilon_{I \times Q}^{FLL}[k] = Im\left[y_P[k]y_P^*[k-1]\right] = I_P[k-1]Q_P[k] - I_P[k]Q_P[k-1]$$

$$\epsilon_{DD}^{FLL}[k] = sign\left(\epsilon_{I \times Q}^{FLL}[k]\right)\epsilon_{I \times Q}^{FLL}[k].$$ (1.104)

Finally, it is interesting to note that the phase difference taking place in $y_P[k]y_P^*[k-1]$ can be extended to the product of two prompt correlations separated apart by more than one epoch. The general case would be products in the form $y_P[k]y_P^*[k-m]$ for $m \geq 1$. The combination of several of these terms helps in mitigating the noise increase that is experienced when two noisy samples are multiplied, as it is the case in the FLL discriminators presented so far. This approach has its roots on autocorrelation-based frequency estimators such as the Kay, Fitz and Luise & Reggiannini methods widely adopted in digital communication receivers. The interested reader is referred to [61, Section 3.2] for a detailed while comprehensive discussion on the topic.

1.7.2   Code Tracking

Code tracking shares the same principle as carrier tracking, but now the magnitude being tracked is not the carrier phase or carrier frequency but the code offset of the local code replica with respect to the incoming signal. Contrary to carrier tracking, which uses only the prompt correlator, code tracking requires more than one correlator to ascertain whether the local code replica is truly aligned with the received one or not. In practice, two more correlators, each of them referred to as the *early* and the *late* correlator, often suffice for this purpose apart from the prompt one already used for carrier tracking. Each of these two additional correlators is placed on each side of the correlation time, at a distance $\delta/2$ from the central correlation peak. The separation δ is therefore referred to as *early-late separation*.

The generic block diagram of a code tracking architecture is shown below in Figure 1.29, where the three aforementioned correlators are denoted by $y_E[k]$, $y_P[k]$ and $y_L[k]$ for the early, prompt and late, respectively. The output of these correlators is a complex value indicating how well each specific code replica is aligned with the received signal. It is for this reason that code tracking architectures are often referred to as Delay-Locked Loops (DLL) . Following the same convention as for carrier tracking, their in-phase and quadrature components are referred herein as $I_{E,P,L}[k]$ and $Q_{E,P,L}[k]$ depending on the specific correlator under analysis. Note that the input signal for code tracking is the received signal $r[n]$ with the carrier already wiped off using the carrier replica generated by the carrier tracking module. This indicates that both carrier and code tracking must operate simultaneously, since the output of each of them has an impact on the other.

Delay-Locked Loop Correlator Spacing
Regarding the correlator spacing, typically $\delta = 1$ chip is used in most GNSS receiver implementations, especially for BPSK signals. A narrower separation provides better accuracy in determining the code delay of the input signal and is more robust to the presence of multipath. A narrower separation is needed as well to track BOC signals in order to monitor the central correlation peak, whose width is typically a fraction of chip. However, using a too narrow spacing is troublesome when rapid variations of the code delay are experienced, either because of the user's dynamics or because of rapidly

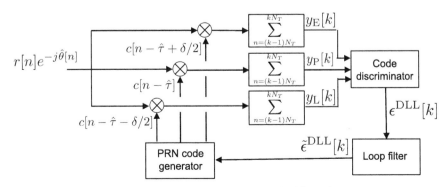

Figure 1.29 Generic block diagram for code tracking in a GNSS receiver.

fluctuating multipath. In these circumstances, it is beneficial to use a wider early-late separation. Modern GNSS receivers allow the early-late spacing to be adjusted while the receiver is tracking the signal. This has the advantage of providing an additional degree of freedom for the receiver to adapt the tracking stage to the working conditions, in order to cope with situations where the C/N_0 suddenly drops, and thus to reduce the risk of loss of lock.

To better illustrate the underlying concept of code tracking let us consider the following example. In Figure 1.30, right, the late code has the highest correlation, so the estimated code delay $\hat{\tau}$ must be increased. In Figure 1.30, left, however, the highest peak is located at the prompt replica, and the early and late replicas have equal correlation. It is in this situation where one can claim that the code delay is being properly tracked. The estimated code delay $\hat{\tau}$ is now aligned with the true delay τ of the received signal, and the resulting delay error τ_ϵ goes to zero. From Figure 1.30, one can also infer the correlation process as a multiplication of the sampled replica and signal, and accumulation. The black samples in the replicas contribute to the correlation gain, as they multiply a +1 by a +1, or a −1 by a −1. The grey samples do not, as they multiply a +1 by a −1, which happens when the replica and the code are not well aligned. Note that, in reality, the signal is buried in noise, so the correlation of several samples is needed to raise the signal above the noise level. Figure 1.30 also shows why it is convenient to select a sampling frequency F_s that is not a multiple of the spreading code frequency, to avoid aliasing, as otherwise several contiguous replicas would have the same gain, leading to an accuracy reduction.

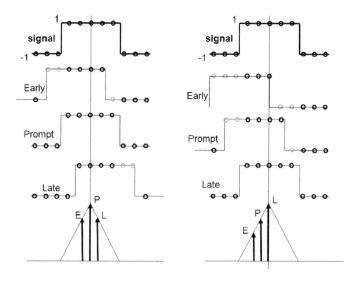

Figure 1.30 Example where the early, prompt and late replicas are correlated with the incoming signal. Left: The prompt code has the highest correlation and $I_E[k] \approx I_L[k]$, so the loop is perfectly tuned. Right: The late replica has the highest correlation, so the code delay must be increased, that is, the code must be delayed. Note that T_s is purposely not a multiple of T_c.

Another example is shown in Figure 1.31 with the evolution of the early, late and prompt correlator outputs as a function of time. Figure 1.31(a) corresponds to the case where the PLL is not locked. As can be seen, the early, late and prompt outputs of the DLL are not locked either. They are actually rotating in the complex plane, and their energy is moving from the in-phase (upper subplot) to the quadrature component (bottom subplot) and vice versa. In contrast, Figure 1.31(b) corresponds to the case where the PLL is locked so that the carrier is properly wiped off from the received signal. In that case, all the energy is now on the in-phase component of the DLL, as shown in the upper plot of Figure 1.31(b). The DLL is also found to operate as expected, with the early and late correlators having a similar magnitude, indicating that the code replica is well aligned with the received signal. The square value of the early, late and prompt correlators is shown in Figure 1.31. It is for this reason that the upper plot in Figure 1.31(b) has a magnitude of 0.25 for the early and late correlators, which is 0.5^2, being 0.5, half, the height of a unit-amplitude correlation peak when a rectangular chip pulse is considered.

Delay-Locked Loop Discriminators

Similarly to carrier tracking, the way in which the error signal is obtained from the correlation between the received signal and the local code replica plays an important role in the code tracking performance. For a DLL, such error signal is obtained by means of a code or DLL discriminator whose output $\epsilon^{DLL}[k]$ is fed back to the code replica generator, in order to compensate for the misaligment of the local replica with respect to the received signal. There are essentially two different types of DLL discriminators depending on their sensitivity to carrier phase mismatches. The first type is composed of the so-called *coherent* discriminators, which operate straight-away on the correlation output samples assuming that the carrier phase is perfectly tracked. This is the case of the early-minus-late (EML) discriminator given by

$$\epsilon^{DLL}_{EML}[k] = \frac{1}{2}\left(I_E[k] - I_L[k]\right),\tag{1.105}$$

which is the simplest form of DLL discriminator. It has the problem, though, that it depends on the signal amplitude and thus, variations in the received signal power cause the EML output to vary as well. In order to avoid this problem, the EML discriminator is often normalized by the magnitude of the early plus the late correlator outputs. This leads to the so-called normalized EML (NEML) discriminator as follows:

$$\epsilon^{DLL}_{NEML}[k] = \frac{1}{2}\frac{I_E[k] - I_L[k]}{|I_E[k]| + |I_L[k]|}.\tag{1.106}$$

Another problem of the EML discriminator is that it is sensitive to the presence of data-modulating symbols. To circumvent this limitation, the EML is often replaced by the *dot-product* (DP) discriminator, which takes the same expression in Eq. (1.105) but multiplies by the in-phase component of the prompt correlator. The latter contains the sign of the data-modulated symbols and thus multiplying by it, the symbol is effectively removed from the discriminator output. The expression of the DP discriminator is therefore,

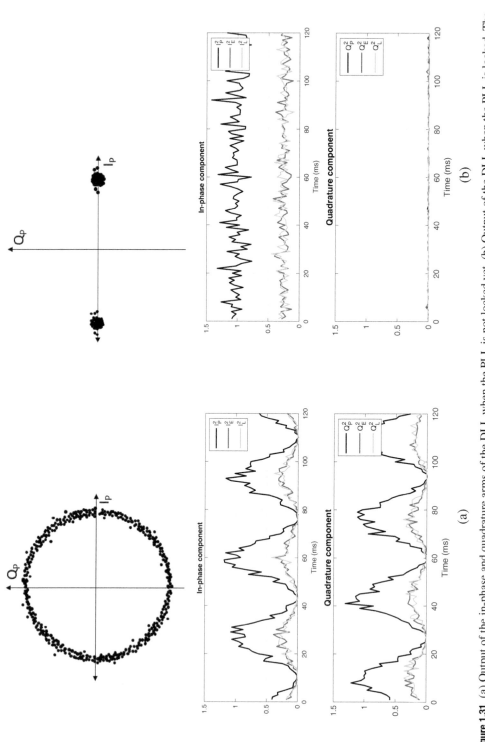

Figure 1.31 (a) Output of the in-phase and quadrature arms of the DLL when the PLL is not locked yet. (b) Output of the DLL when the PLL is locked. The upper part shows the prompt correlator outputs in the complex plane. A BPSK(1) signal is considered with an early-late separation of $\delta = 1$ chip.

$$\epsilon_{\text{DP}}^{\text{DLL}}[k] = \frac{1}{2} \left(I_{\text{E}}[k] - I_{\text{L}}[k] \right) I_{\text{P}}[k]. \tag{1.107}$$

The result is again affected by variations of the received signal power, so a normalized version of the DP discriminator is often used by dividing Eq. (1.107) with the square of the prompt correlator, which acts as an estimate of the instantaneous power. This, however, is only valid when the code delay error is small. Otherwise, the square of the prompt correlator is no longer a reliable estimate of the received power because it is affected by the code delay error, and thus, the output of the discriminator would become biased. According to the aforementioned discussion, the resulting normalized DP (NDP) discriminator is given by

$$\epsilon_{\text{NDP}}^{\text{DLL}}[k] = \frac{1}{2} \frac{I_{\text{E}}[k] - I_{\text{L}}[k]}{I_{\text{P}}[k]}. \tag{1.108}$$

The second type of DLL discriminators are the *noncoherent* ones, which typically take the square or the magnitude of the correlation outputs in order to become insensitive to residual phase errors and to the presence of data symbols. As can be inferred, such robustness comes at the price of increasing the noise due to the non-linear operation on the noisy input samples. The impact is often acceptable and the fact of being insensitive to the two aforementioned effects typically pays off. One of the most popular noncoherent DLL discriminators is the normalized early minus late power (NEMLP) discriminator,

$$\epsilon_{\text{NEMLP}}^{\text{DLL}}[k] = \frac{1}{4} \frac{(I_{\text{E}}^2[k] + Q_{\text{E}}^2[k]) - (I_{\text{L}}^2[k] + Q_{\text{L}}^2[k])}{(I_{\text{E}}^2[k] + Q_{\text{E}}^2[k]) + (I_{\text{L}}^2[k] + Q_{\text{L}}^2[k])}, \tag{1.109}$$

which relies on the squared modulus of the early and late correlators. An alternative using just the modulus instead of the squared value is the normalized noncoherent EML (NNEML), which can be seen as the noncoherent implementation of the EML in Eq. (1.105), and it is the discriminator actually used in the software receiver accompanying the present book. It is given by[29]

$$\epsilon_{\text{NNEML}}^{\text{DLL}}[k] = \frac{1}{2} \frac{\sqrt{I_{\text{E}}^2[k] + Q_{\text{E}}^2[k]} - \sqrt{I_{\text{L}}^2[k] + Q_{\text{L}}^2[k]}}{\sqrt{I_{\text{E}}^2[k] + Q_{\text{E}}^2[k]} + \sqrt{I_{\text{L}}^2[k] + Q_{\text{L}}^2[k]}}. \tag{1.110}$$

The output of the six discriminators discussed above is shown in Figure 1.32 for the case of an infinite bandwidth rectangular chip pulse. As can be seen, they all exhibit a linear behavior with a unitary slope when the input code delay error is small, that is, $\tau_\epsilon \to 0$. For larger values of the input code delay error, the widest linear region is provided by the EML-type discriminators, namely the EML, NEML and NNEML discriminators, spanning within the interval $\tau_\epsilon/T_c \in [-0.5, +0.5]$ chips. At least, in

---

[29] The $1/2$ scaling factor in Eq. (1.110) is merely used here for the sake of convenience to guarantee a unitary slope within the linear region of the discriminator, as shown in Figure 1.32. This scaling factor, as those in the rest of discriminators discussed herein, is actually not needed provided that its effect is accounted for in the design of the DLL tracking loops, as it was done in our software receiver.

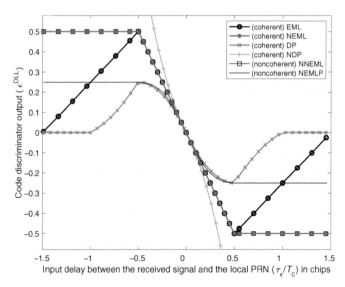

Figure 1.32 Output provided by different code discriminators as a function of the code delay difference τ_ϵ at their input. Infinite bandwidth is assumed.

the absence of noise, as considered herein. In contrast, the NDP and NEMLP discriminators do only provide a linear response over a much narrower interval of barely $\tau_\epsilon/T_c \in [-0.15, +0.15]$ chips. This involves that the two latter discriminators are more sensitive to the presence of input noise, since variations on the early and late correlators can easily lead the discriminator output to lie outside of the linear region and thus to incur in bias.

Delay-Locked Loop Performance and Filter Bandwidth
Regarding the performance of code tracking, the variance of the closed-loop estimated code delay for a coherent DLL discriminator such as the EML can be approximated for a BPSK signal by

$$\sigma_{\hat{\tau},\text{EML}}^2 \approx \frac{B_L \delta}{2\frac{C}{N_0}} \quad (\text{chips}^2), \tag{1.111}$$

with B_L the loop filter bandwidth. Similarly, the variance of the closed-loop estimated code delay for a noncoherent DLL discriminator such as the NEMLP can be approximated by

$$\sigma_{\hat{\tau},\text{NEMLP}}^2 \approx \frac{B_L \delta}{2\frac{C}{N_0}} \left(1 + \frac{2}{\frac{C}{N_0}T}\right) \quad (\text{chips}^2), \tag{1.112}$$

as shown in [4, Eq. (12.34)]. Finally, it is interesting to note that, as a rule of thumb, code tracking is assumed to be in lock when $\sigma_{\hat{\tau}} < \delta/12$ [13, Section 19.4]. This is a simple guideline that can be used to ascertain whether the receiver is in track at code level or the tracking has been lost and then re-acquisition may be needed. The interested reader can find the derivation of Eqs. (1.111) and (1.112) in [62, Chapters 7–8], [4, Chapter 12] or [13, Chapter 19].

1.7.3 C/N_0 Estimation

As already introduced in Section 1.4.1, the C/N_0 is a key metric to measure the strength of the received GNSS signal, and it is usually measured in the tracking block. However, the receiver cannot measure this ratio directly and it has to estimate it. A well-known technique for (C/N_0) estimation will be described here. It is called Narrow-band Wide-band Power Ratio (NWPR),[30] and is based on the ratio of the signal wideband power to its narrow-band power as mentioned in [62, Chapter 8]:

$$\left(\frac{C}{N_0}\right)[k] = 10\log_{10}\left(\frac{1}{T} \cdot \frac{\hat{\mu}_{\text{NP}}[k] - 1}{M - \hat{\mu}_{\text{NP}}[k]}\right), \tag{1.113}$$

where $\hat{\mu}_{\text{NP}}$ is the mean normalized power, as expressed in the following equation:

$$\hat{\mu}_{\text{NP}}[k] = \frac{1}{K}\sum_{k=1}^{K} \text{NP}[k], \tag{1.114}$$

where $\text{NP}[k]$ is the normalized power between narrow-band power and wide-band power, given by

$$\text{NP}[k] = \frac{\text{NBP}[k]}{\text{WBP}[k]}, \tag{1.115}$$

with $\text{NBP}[k]$ and $\text{WBP}[k]$ defined as follows:

$$\text{NBP}[k] = \left(\sum_{n=1}^{M} I_{\text{P}}[n + kM]\right)^2 + \left(\sum_{n=1}^{M} Q_{\text{P}}[n + kM]\right)^2, \tag{1.116}$$

$$\text{WBP}[k] = \sum_{n=1}^{M} I_{\text{P}}^2[n + kM] + Q_{\text{P}}^2[n + kM]. \tag{1.117}$$

In the equations above, M stands for the number of correlation output samples that are accumulated to get one sample of the ratio in Eq. (1.115). Typically $M = 20$ in GPS L1 C/A signals in order to implement a one symbol length coherent integration, assuming that symbol synchronization has already been achieved. Then, K is usually set to $K = 50$ in order to have one smoothed C/N_0 estimate at a rate of one value per second.

The NWPR estimation is useful for receivers tracking signals that can be coherently integrated for relatively long periods. However, this is not the case for tracking signals with frequent symbol transitions or snapshot receivers. For these two cases, this book proposes alternative C/N_0 estimators. Section 7.2.1 proposes a C/N_0 estimator where there are frequent symbol transitions, based on the signal-to-noise power ratio (SNPR), and Section 9.2.6 presents an estimator for snapshot positioning based on noncoherent integrations.

---

[30] This technique is implemented in the FGI-GSRx software companion of this book.

1.8   Navigation Message

The previous sections have explained how the signals can be acquired and tracked, allowing us to generate the range measurements and demodulate the symbols. This section focuses on the navigation message. The navigation message provides the information to calculate the satellite position, among other data. This section presents how the receiver can synchronise to the signal and start interpreting the symbols, how these symbols encode the navigation bits to enhance reception even in case of signal impairments and what are the main navigation message parameters.

1.8.1   Synchronization to the Signal

We have described how to track the carrier phase and the code delay through the PLL/FLL and DLL, respectively. However, none of these outputs allow, strictly speaking, calculation of the satellite-receiver range. The carrier phase measurement provides an estimation of the signal phase at a given cycle, but the number of cycles, or wavelengths, between the satellite and the receiver is unknown. Since GNSS wavelengths are in the order of 18 to 25 cm, this carrier phase integer ambiguity will remain unknown unless the receiver has very accurate error information and uses special techniques such as real-time kinematics (RTK) or PPP, which are out of the scope of this book (see, for example, Part B of [63] for an updated guide on these techniques). Similarly, the DLL output provides an estimate of the *code delay*, or the fractional measurement of the satellite-to-receiver signal delay within one code length. That is, a value between 0 and 1 ms (0 and 300 km) for GPS L1 C/A, and between 0 and 4 ms (0 and 1200 km), for Galileo E1-C or E1-B. In order to compute the full range measurement, the receiver needs to know the integer number of full codes to be added to the fractional code delay measurement, that is, the remaining part of the total distance. In order to do that, classic receivers need to synchronize to the navigation message. By this synchronization, the receiver can determine the *transmit time*, or when the code associated with the measured code delay was transmitted. This is achieved because the message time-tags symbol transitions in the data stream, and this allows determining the number of codes, or time, between our code delay measurement, and the symbol transition associated with the time reported in the signal, as explained in the next section.

The synchronization process starts with the receiver processing symbols until a fixed sequence is received. For GPS L1 C/A, this sequence is composed of the bits (10001011) and it is called *preamble*. It is within a field called Telemetry Word that is followed by another field called Hand-Over Word, that provides the GPS Time (GPST). These fields appear once every GPS L1 C/A *subframe*, or every 6 seconds. Often, after recognizing the preamble, receivers check the parity of the message portion in which it is embedded, to verify that the sequence is not due to a combination of bits from other sections of the message. From that point onward, the receiver can interpret the bits of the navigation message. The next step is, by interpreting these bits, to synchronize with the beginning of a 30-second *frame*. Once

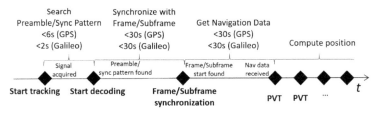

Figure 1.33 Receiver synchronization with the navigation message for GPS L1 C/A and Galileo I/NAV.

this is achieved, the receiver can wait for another 30 seconds to decode the satellite position data, or *ephemeris*, described later in this chapter. Receivers can save some time by combining data from two contiguous frames so the total decoding time can be around 30 seconds. The GPS L1 C/A navigation message lasts 12.5 minutes due to the *almanacs*, also explained later, but the receiver does not need this information to have a position. We call the time to compute the first position *Time To First Fix* (TTFF). The process for Galileo E1-B I/NAV is very similar. The message provides a 10-bit *synchronization pattern* (0101100000), every 2 seconds, similar to the GPS preamble. From that sequence onward, the receiver can process the navigation data. Galileo I/NAV also transmits the satellite position data and time every 30 seconds, or Galileo I/NAV *subframe*, so the TTFF process is similar. Figure 1.33 illustrates the receiver synchronization process for GPS L1 C/A and Galileo I/NAV.

1.8.2 Channel Encoding and Decoding

The navigation data are often received with errors due to the impairments in its path from the satellite to the receiver. Therefore, *channel encoding* schemes are added at the transmitter to encode the data bits, allowing the receiver to detect corrupted bits, and even recover them in some cases. The main error detection and correction principles used in GNSS are outlined in this section.

Error Detection

The navigation message has some extra bits allowing the receiver to check if it has been corrupted. This is commonly defined as a *checksum*. For example, the GPS L1 C/A signal provides six bits of parity, or check bits, for every 24 bits of information, through a (32,26) *Hamming code* [64]. Most messages, as for example the Galileo I/NAV, F/NAV and C/NAV, provide a cyclic redundancy check (CRC). The principle of CRC generation is to predefine a *generator polynomial* G of degree equal to the size of the CRC, create a polynomial including the bit stream to check, divide it by G and check that the reminder R obtained from the division is the same as the transmitted CRC. For example, in the case of Galileo I/NAV, M is the stream of 196 bits, that is expressed as a polynomial of degree 196. For I/NAV, F/NAV and C/NAV, $G(X)$ is a predefined polynomial of degree 24 which depends on the prime, primitive or irreducible polynomial $P(X)$:

$$G(X) = (1 + X)P(X)$$
$$P(X) = X^{23} + X^{17} + X^{13} + X^{12} + X^{11} + X^9 + X^8 + X^7 + X^5 + X^3 + 1. \qquad (1.118)$$

Then, the polynomial $m(X)X^{24}$, that is, the original message $m(X)$ with 24 zeros appended, is divided by $G(X)$, and the 24-bit reminder is the CRC.

Error Correction

Modern coding techniques allow not only error detection but also error correction. There are multiple error detection algorithms which are described in the various ICDs. For example, Galileo, GPS L5, NAVIC L5 and satellite-based augmentation system (SBAS) signals include forward error correction (FEC) convolutional encoding, by which n bits are encoded into k symbols, where $r = n/k$ is the code rate. The convolutional property implies that each symbol depends on a number of bits, which is the *constraint length* of the encoder. The encoding process for most of these signals is depicted in Figure 1.34,[31] where $r = 1/2$, the constraint length $c = 7$ and the encoding sequence is performed through two polynomials ($G1$ = 171 octal, or 1111001 binary, and $G2$ = 133 octal, or 1011011 binary; they are equivalent to the arrows pointing up and down, respectively, in Figure 1.34, where D expresses one bit shift, or delay). The encoded sequence alternates symbols from $G1$ and $G2$. Other error detection techniques are low-density parity check, for GPS L1C signals, and Reed Solomon, used for Galileo [65] and QZSS [66]. See Figure 1.34 as an example of FEC encoding.

As regards decoding in the receiver, the most common decoder used for convolutional codes is the *Viterbi decoder*. The main property of the Viterbi decoder is that it estimates the bits recursively until the sequence with the maximum likelihood is found. The Viterbi decoder can be implemented through *hard-decision* or *soft-decision* decoding, where the former takes as an input only binary symbol values and the latter takes as an input intermediate values. For more details on Viterbi decoders, see [67] or [68].

Interleaving

Due to the perturbations in the signal transmission channel, bit errors often come in bursts, affecting several continuous bits. In order to make the encoded sequence more

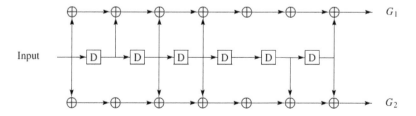

Figure 1.34 Convolutional encoder.

---

[31] For Galileo, the G_2 symbol is inverted at the end.

robust against bursts, the order of the symbols is changed before transmission through *interleaving*, which means that the encoded sequence is written in columns and then read in rows before transmission. Interleaving is normally implemented by using a two-dimensional buffer array. The data enter the array column by column and leave the array row by row. A burst of errors in the communication channel is spread out into single-symbol errors, which are easier to correct than consecutive errors.

We illustrate in the following example how an input data stream is written into an array column by column. Then, it is read out row by row, to achieve the bit interleaving operation:

- Data input: $x_1\ x_2\ x_3\ x_4\ x_5\ x_6\ x_7\ x_8\ x_9\ x_{10}\ x_{11}\ x_{12}\ x_{13}\ x_{14}\ x_{15}\ x_{16}$
- Interleaving array:

$$\begin{array}{cccc} x_1 & x_5 & x_9 & x_{13} \\ x_2 & x_6 & x_{10} & x_{14} \\ x_3 & x_7 & x_{11} & x_{15} \\ x_4 & x_8 & x_{12} & x_{16} \end{array}$$

- Interleaved data output for transmission:

$$x_1\ x_5\ x_9\ x_{13}\ x_2\ x_6\ x_{10}\ x_{14}\ x_3\ x_7\ x_{11}\ x_{15}\ x_4\ x_8\ x_{12}\ x_{16}$$

In this simple example, three consecutive errors in the interleaved transmitted data, let us say x_1, x_5, x_9, are translated into three isolated, single errors when interleaving is reversed at the receiver end.

More details on navigation message encoding are provided in the next chapters for each GNSS. For a more general description of channel encoding techniques, see [69].

1.8.3 Message Parameters

Once synchronized, our receiver can start interpreting the data transmitted by the satellite. The satellite data are generally structured in a *navigation message* that is repeated continuously for all users worldwide. The most relevant information that the receiver needs are the satellite position and time. This information is conveyed through what GNSS call satellite *ephemeris*, which include the satellite orbital parameters.

Satellite Orbits
Most GNSS satellites are describing almost circular orbits around the Earth. Their trajectories, as with any other bodies in space, follow Newton's Law of Universal Gravitation[32]:

$$F_{1,2} = \frac{Gm_1 m_2}{r^2}, \tag{1.119}$$

where $F_{1,2}$ is the gravitational force between the two bodies, m_1 and m_2 are their masses, r is the distance between them, and G is the universal gravitational constant ($G = 6.674 \cdot 10^{-11}\ \text{m}^3\ \text{kg}^{-1}\ \text{s}^{-2}$). Dividing the force by the satellite mass, we obtain the satellite acceleration, and rearranging terms and expressing Newton's law in vectors we obtain the following:

---

[32] Particularized here for a two-body problem with masses m_1 and m_2.

$$\ddot{\mathbf{r}} + \frac{\mu \mathbf{r}}{r^3} = 0, \tag{1.120}$$

where μ is G times the Earth mass $(5.972 \cdot 10^{24}$ kg$)$, or $3.98 \cdot 10^{14}$ m^3/s^2, also known as the *Earth's gravitational constant*; and $\mathbf{r}$ is the vector from the Earth center of mass to the satellite center of mass.[33] The expression in Eq. (1.120) is usually called the *fundamental orbital differential equation*. It already encompasses Kepler's three laws which, prior to Newton, described the planetary orbits as ellipses sweeping a constant area around one of its focus with a period proportional to $a^{3/2}$ (their semimajor axis at the power of 3/2). A body that follows a trajectory as per Newton's law will also fulfil Kepler's laws. However, for convenience, orbits are often described according to their so-called *Keplerian elements*, which are the following:

– the orbit semimajor axis (a),
– the orbit eccentricity (e),
– the orbit inclination with respect to the equatorial plane (i),
– the right ascension of the ascending node (RAAN, Ω), or angle between the point where the satellite orbit crosses to the northern hemisphere and the vernal equinox,[34] a reference point in the celestial sphere; it is also called *Longitude of the Ascending Node* (LAN),
– the argument of the perigee (ω), or the angle between the equatorial plane and the semimajor axis (perigee direction),
– the true anomaly (ν), which defines where the satellite is in the orbit; in particular, ν defines the angle between the perigee and the satellite, as seen from the Earth's center of mass.[35]

The first two parameters, a and e, describe the ellipse that the orbit follows. The second three (i, Ω, ω), describe how the orbit is positioned with respect to the Earth. The last parameter (ν) describes the location of the satellite in the orbit at a given *time*, which is sometimes considered as the seventh parameter.

The navigation message provides mostly these parameters, although with some alterations and additions. The complete table of parameters is provided in Table 1.3. This table provides first the time of applicability of the corrections, then the six (modified) Keplerian parameters, and finally, some linear and periodical corrections to the orbits.

[33] Note that, while the orbit parameters give the position at the center of mass of the satellite, the signal departs from the antenna phase center, which is some decimeters apart. The navigation message transmits the position of the antenna phase center, but high accuracy applications generally calculate the center of mass and correct this offset, and to correct it, they also need the satellite attitude and to compensate for antenna phase center offset variations.

[34] The vernal equinox can be defined by the intersection between the ecliptic, or rotation plane of the Earth, and the equatorial plane at the equinoxes. It used to point at the Ares constellation, therefore the symbol representing it in Figure 1.35. RAAN always refers to the vernal equinox, but the term *longitude* in LAN may relate to the usual longitude used in Earth-centered Earth-fixed systems, which relates to the Greenwich meridian, or another one, depending on how it is defined.

[35] Note that ν here does not refer to the signal Doppler shift used in the previous signal processing sections, but we use it here again in order to follow the standard Keplerian nomenclature.

Table 1.3 Ephemeris parameters.

t_{0e}	ephemeris reference time
$\sqrt{a}$	square root of the semimajor axis
e	eccentricity
i_0	inclination angle at reference time
Ω_0	right ascension of ascending node at reference time
ω	argument of perigee
M_0	mean anomaly at reference time
Δn	mean motion difference from computed value
$\dot{i}$	rate of change of inclination angle
$\dot{\Omega}$	rate of change of right ascension
C_{uc}, C_{us}	amplitude of cosine and sine corrections to argument of latitude
C_{rc}, C_{rs}	amplitude of cosine and sine corrections to orbit radius
C_{ic}, C_{is}	amplitude of cosine and sine corrections to inclination

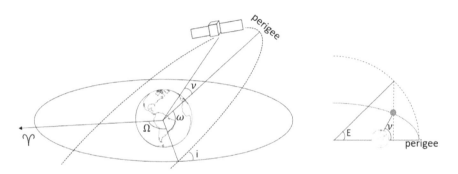

Figure 1.35 Satellite orbital parameters. Left: Keplerian parameters. Right: True anomaly versus eccentric anomaly.

Figure 1.35 shows the Keplerian parameters. The left plot depicts a satellite in its orbit, with respect to the Earth, while the right plot depicts the true anomaly v and the eccentric anomaly E. For more details on satellite and GNSS orbits, see [70] or [4]. You can also consult the SIS ICDs of GPS [2], Galileo [71] or BeiDou [72], implemented in our MATLAB receivers.[36]

Satellite Position Computation

Here, we explain how to determine the satellite position at a given time t from the ephemeris in the navigation message (t_{0e} in Table 1.3). This is the process that is applied to GPS and Galileo ephemerides according to their signal specifications [2] [71].

Our first task is to calculate the true anomaly at the reference time t, or v_t. Then we calculate the x, y satellite position in the orbital plane. Finally, we rotate the satellite position into an Earth-centered Earth-fixed (ECEF) coordinate system.

---

[36] GLONASS navigation message provides the satellite positions in a different format [73].

The GNSS Keplerian parameters do not transmit the true anomaly v but the *mean anomaly* at reference time, or M_0. The advantage of M_0 is that it allows to linearly extrapolate to the mean anomaly, M, at our time of interest. The disadvantage is that it requires to convert to the *eccentric anomaly*, E, depicted in Figure 1.35, and then to v. For this conversion, we need to use the semimajor axis a and the eccentricity $e = \sqrt{1 - a^2/b^2}$, where b is the semiminor axis. We also need to calculate the mean motion n, which is constant, but it is affected by Δn, provided in Table 1.3. We start by calculating the mean anomaly M_t at our target time t (1.121).[37] Then, we estimate the eccentric anomaly E_t by iteration in Eq. (1.122). With E_t we calculate the true anomaly v_t:

$$M_t = M_0 + \sqrt{\frac{\mu}{a^3}}(t - t_{0e}) \tag{1.121}$$

$$M_t = E_t - e\sin(E_t) \tag{1.122}$$

$$v_t = 2\operatorname{atan}\left(\sqrt{\frac{1+e}{1-e}}\tan\left(\frac{E_t}{2}\right)\right). \tag{1.123}$$

Once the satellite is located in the orbital plane by v_t, its position can be expressed in rectangular coordinates. For that, we first add the argument of perigee ω to get the argument of latitude Φ_t:

$$\Phi_t = v_t + \omega. \tag{1.124}$$

The receiver needs now to apply harmonic corrections from the message. These corrections are due to the gravitational force of third bodies not taken into account in our model, like the Sun and the Moon, the solar radiation pressure, and the nonuniformity of the Earth's gravitational field. We call the result the *corrected argument of latitude, u_t*:

$$\delta u_t = C_{us}\sin(2\Phi_t) + C_{uc}\cos(2\Phi_t) \tag{1.125}$$

$$u_t = \Phi_t + \delta u_t. \tag{1.126}$$

We now calculate the orbital radius r_t, to which we also add harmonic corrections from the broadcast message:

$$\delta r_t = C_{rs}\sin(2\Phi_t) + C_{rc}\cos(2\Phi_t) \tag{1.127}$$

$$r_t = a(1 - e\cos E_t) + \delta r_t. \tag{1.128}$$

With r_t and u_t we can express the satellite position in rectangular coordinates in the orbital plane, or *perifocal* coordinates:

$$x_t' = r_t\cos u_t \tag{1.129}$$

$$y_t' = r_t\sin u_t. \tag{1.130}$$

Now, the satellite position must be rotated into an Earth-centered inertial (ECI) reference frame, that is, a coordinate system centered at the Earth that does not take

---

[37] Note the appearance of Kepler's third law on the right side, second term.

Table 1.4 Clock correction parameters.

t_{0c}	clock correction reference time
a_{f0}	clock bias correction
a_{f1}	clock drift correction
a_{f2}	clock drift rate correction

Earth rotation into account, and then into an ECEF reference frame, solidary to the Earth rotation. For that, we need first to obtain the corrected inclination i and the corrected argument of latitude Ω_t, which already takes into account the Earth rotation angular speed, $\dot{\Omega}_e = 7.2921151467 \cdot 10^{-5}$ rad/s, to convert from ECI to ECEF:

$$\delta i = C_{is} \sin(2\Phi_t) + C_{ic} \cos(2\Phi_t) \tag{1.131}$$

$$i = i_0 + \dot{i}(t - t_{0e}) + \delta i \tag{1.132}$$

$$\Omega_t = \Omega_0 + (\dot{\Omega} - \dot{\Omega}_e)(t - t_{0e}) - \dot{\Omega}_e t_{0e}. \tag{1.133}$$

Finally, we can rotate our satellite position from perifocal coordinates and obtain our final ECEF position x_t, y_t, z_t at the time of interest t:

$$x_t = x'_t \cos(\Omega_t) - y'_t \cos(i) \cos(\Omega_t) \tag{1.134}$$

$$y_t = x'_t \sin(\Omega_t) + y'_t \cos(i) \cos(\Omega_t) \tag{1.135}$$

$$z_t = y'_t \sin(i). \tag{1.136}$$

Satellite Clock Corrections

As important as the satellite position is the satellite time. Even if the satellites have extremely accurate atomic clocks, they have deviations with respect to the GNSS time. These clock deviations usually remain low, but they get into the range measurement (a one-microsecond clock error would add an error of 300 m) and therefore need to be provided in the navigation message, so the receiver can correct them.[38] They are usually expressed as a polynomial with fixed, linear and quadratic parameters, according to Table 1.4. Notice that the time of applicability of the clock corrections, t_{0c}, usually coincides with that of the orbital parameters, t_{0e}.

The clock corrections are applied as follows:

$$dt(t) = a_{f0} + a_{f1}(t - t_{0c}) + a_{f2}(t - t_{0c})^2 + \Delta t_{rel}, \tag{1.137}$$

where $dt(t)$ is the clock correction to be applied at time t; $a_{f0,f1,f2}$ are the correction coefficients and Δt_{rel} is the *relativistic correction term*, an additional time error due to relativity.

Relativistic Effect

The theory of relativity affects GNSS in various ways: first, the satellite clocks must be adjusted to account for it. Due to their speed, and the special relativity, the clocks run slightly faster in orbit than on Earth, where they are built, but due to their lower

---

[38] For a performance characterization of clocks on-board GNSS satellites, see [74] or [11, Chapter 7].

gravitational force, they run at a slightly slower pace in orbit than on Earth. The net effect is a slight slowdown of a few tens of microseconds per day, depending on the GNSS orbit (e.g. 38 μs per day for GPS orbits [12]). This is already accounted for in the satellite clock manufacturing process, so the receiver does not have to care about it. However, this assumes a perfectly circular orbit. When the orbits have some eccentricity, the relativistic correction term needs to be taken into account, depending on where the satellite is located in its orbit:

$$\Delta t_{rel} = Fe \sqrt{a} \sin(E_t), \tag{1.138}$$

where e is the eccentricity, a is the semimajor axis, E_t is the eccentric anomaly at t calculated from the Keplerian orbital parameters and F is a constant:

$$F = \sqrt{\mu}/c^2 = -4.442807309 \cdot 10^{-10} \quad (s/m^{1/2}), \tag{1.139}$$

where μ and c are the Earth gravitational constant and the speed of light, respectively.

Broadcast Group Delays

A sometimes omitted term in the position computation is the satellite broadcast group delay also referred to as Timing/Total Group Delay (TGD), instrumental delay, satellite inter-frequency bias (IFB), differential code bias or hardware bias. It encompasses the time biases in the generation of different signals due to the satellite hardware. Many GNSS, GPS and Galileo among them, provide their clock correction not referring to L1 C/A or E1, but to measurements generated in the receiver by combining two frequencies, L1 and L2 in GPS, and E1 and E5b in Galileo, into the so-called *ionosphere-free* combination. The reason is that the service is formally defined for a dual-frequency user, for a better performance. The clock correction is therefore transmitted to cancel the bias in the dual-frequency combination, and the clock correction for the single-frequency user (e.g. L1 C/A) is not straightforwardly available in the navigation message. As a consequence, in order to navigate with a single-frequency, the receiver needs to subtract the inter-frequency bias to the broadcast clock correction. Therefore, a user navigating with a frequency F_1 will need to apply this additional bias:

$$dt(t)_{F_1} = dt(t)_{F_1,F_2} - B_{F_1,F_2}, \tag{1.140}$$

where B_{F_1,F_2} is also broadcast in the navigation message, or a term allowing to calculate it following the signal specification; and $dt(t)_{F_1,F_2}$ is the term calculated in Eq. (1.137). This is because, as mentioned earlier, the clock corrections of Table 1.4 have been adapted so that these biases cancel when the receiver applies the F_1 and F_2 iono-free combination. For more details on multifrequency navigation and generating ionosphere-free measurements, see Eq. (8.11) and Eq. (8.12) in Chapter 8.

Other Formats and Navigation Data

These parameters and corrections are used nowadays by GPS and most GNSS. However, there are multiple ways to provide the satellite position. For example, GLONASS transmits the satellite position coordinates, velocity and acceleration perturbations, as discussed later in Chapter 3, and Galileo has recently started providing a short ephemeris message in similar, yet reduced format [75].

The navigation message also includes other parameters supporting the receiver. One of them is an ionospheric model (Klobuchar for GPS, NeQuick for Galileo), which corrects around half of the ionospheric error and is outlined later. Another part of the message is the *almanac*. The almanac is a long message by which each satellite transmits the rough (at the few-kilometer error level) position of all the satellites in the constellation, with the purpose of speeding up the acquisition time: If a receiver has a rough approximation of its own position, and that of the satellites, it will not search for satellite PRNs that are not visible. Almanacs are not widely used nowadays, but they are still transmitted for legacy reasons. Other information includes an indicator of the accuracy of the transmitted signal and its parameters, named URA (User Range Accuracy) or SISA (Signal-In-Space Accuracy) for GPS and Galileo, respectively, other satellite flags reporting general satellite and data health or the offset between the GNSS time and other time references such as UTC (Universal Time Coordinated).[39] Galileo will also transmit data authentication, also known as OSNMA (Open Service Navigation Message Authentication) [76].

1.9 Pseudorange Errors

Once the navigation data are decoded, the receiver can determine the coordinates of the satellite position as well as the time at which the signal is transmitted. The receiver can then estimate the exact time of arrival (ToA) of the signal. Based on the time difference between the transmission and reception time, the receiver can compute the distance between the satellite and the receiver by multiplying the time difference by the speed of light. However, this is not the actual distance, but a *pseudodistance*, as it includes also the offset in the receiver clock. The *pseudorange* is therefore the distance between a satellite and a receiver including the receiver clock offset. It is composed by the integer number of codes since a bit transition in the message, that the receiver can time-tag once it is synchronized, and the fractional code delay measurement. The reason we speak of pseudoranges rather than ranges is therefore because the quantity includes a receiver clock offset which is common to all pseudoranges. Together with the satellite position, it is one of the unknowns to be solved in our system of equations, as described later.

The pseudorange also contains, together with the true geometric distance and the receiver clock bias, some additional distance errors related to propagation effects

---

[39] UTC is useful to align the reference time with some standards, such as the National Marine Electronics Association (NMEA) standard, which is used by many applications far beyond the maritime domain.

due to the ionosphere, the troposphere, local effects in the vicinity of the receiver leading to multipath errors and receiver noise. This is expressed in Eq. (1.141) as follows:

$$\rho_m^{(p)} = r^{(p)} + c(dt_u - dt^{(p)}) + I^{(p)} + T^{(p)} + \varepsilon^{(p)}, \tag{1.141}$$

where $\rho_m^{(p)}$ is the *measured* pseudorange for satellite (p), before applying any correction; $r^{(p)}$ is the range, or actual distance between the satellite and the receiver; c is the speed of light; dt_u is the user receiver clock bias; $dt^{(p)}$ is the satellite clock bias $I^{(p)}$ is the range ionospheric error; $T^{(p)}$ is the tropospheric error and $\varepsilon^{(p)}$ contains other unmodelled measurement errors including multipath, interference and noise. These errors are described in the next subsections.

1.9.1 Ionospheric Error

The ionosphere, as the name suggests, is a layer of the atmosphere that contains electrically charged, or *ionized*, particles that affect the propagation of electromagnetic waves. It can be the most significant source of errors in GNSS measurements, up to a hundred meters. The ionospheric variations depend on the solar activity, which oscillates in cycles of approximately 11 years. The last solar maximum was in 2014 when writing this text. At a solar maximum, there is a higher amount of solar spots and flares, and a higher amount of ionized particles arrive to the Earth and remain in the ionosphere. Therefore, the ionosphere is active during the day, but in some cases, ionospheric perturbations occur at or after sunset. The particles are also affected by the Earth's magnetic field, and therefore, activity is higher close to the poles and the geomagnetic equator. Ionospheric activity is measured by the Total Electron Content (TEC), which is usually expressed in TECU (TEC Units), or total number of electrons per square meter.

Ionosphere affects the signals in two different ways. Ionospheric *refraction* induces a group delay to the ranging code modulation and a carrier phase advance of the same magnitude [4, Chapter 5]. Its effect is present everywhere within the atmosphere. In contrast, ionospheric *scintillation* causes rapid signal power fading and phase fluctuations, and it is usually encountered at equatorial and polar latitudes. In this section, we describe approaches to mitigate the ionospheric refraction; scintillation is not addressed here, but the interested reader may refer to [77] for more information. In the remainder of this section, we will refer to the ionospheric refraction effect as a delay, which corresponds to the refractive effect on the pseudorange measurement. The same concepts apply to the carrier phase measurement but with the sign reversed (see Eq. (8.7) for more details).

The ionospheric delay depends on the carrier frequency. The first-order ionospheric delay is proportional to TEC along the propagation path between the receiver and satellite (p) as

$$I^{(p)}(f) = \frac{40.3 \, \text{TEC}^{(p)}}{f^2}. \tag{1.142}$$

Higher order ionospheric delay effects exist, but their contribution on the ionospheric delay is in the order of centimeters so they can be neglected except for high accuracy applications [78].

There are three main approaches to compensate for the ionospheric delay: evaluating an ionosphere model using the parameters transmitted by the satellites, exploiting the frequency dependence of Eq. (1.142) with a dual-frequency receiver and using external correction data. The first method is outlined here. The second one is described in Chapter 8, which covers a dual-frequency receiver, and some more details about the ionosphere. Some references for the third one are provided in Chapter 7.

Broadcast Ionospheric Correction Models

For single-frequency users, the most straightforward way for mitigating ionospheric delays is to apply the correction parameters broadcast as part of the navigation message. GPS and BeiDou support the Klobuchar ionosphere model [79], whereas Galileo supports the NeQuick G algorithm [80]. This model is a variant of the NeQuick electron density model [81], which has specifically been adapted for GNSS. In addition to the ionosphere parameters, broadcast models require the receiver and satellite positions as well as the time of measurement as inputs.

The ionospheric data transmitted by GNSS satellites are based on predictions and not updated in real time. It is also transmitted in very few bits, for example, 41 and 64 bits for Galileo and GPS, respectively. Therefore, the user cannot expect them to compensate for the ionospheric error perfectly. For instance, the GPS ionospheric corrections are estimated to reduce the RMS error by at least 50% [2], whereas the NeQuick G model has been designed to correct 70% or more of RMS ionospheric delays under regular conditions [80]. Since the ionospheric delay in Eq. (1.142) depends only on the TEC value and the carrier frequency, any model to estimate the TEC value can be used with any GNSS constellation. Consequently, Galileo-based NeQuick G corrections can be used with GPS and BeiDou. Similarly, Klobuchar corrections can be applied to Galileo signals instead of NeQuick G.

The two different approaches are illustrated in Figure 1.36. The Klobuchar model takes eight input parameters $\{\alpha_i, \beta_i\}_{i=0...3}$ that characterize the amplitude and period of the vertical ionospheric delay. It models the ionosphere as a thin shell at a height of 350 km above the Earth surface. The correction algorithm determines the point where the LOS ray from the satellite to the receiver pierces the shell, computes the vertical ionospheric delay at the piercing point and adjusts this estimate based on the satellite elevation angle to obtain the slant ionospheric delay. A detailed description of the algorithm is available in [2].

NeQuick G is fundamentally different from the thin-shell Klobuchar model supported by GPS and BeiDou. NeQuick G is a full model of the ionosphere consisting of three layers (E, F1, F2), allowing the evaluation of the local electron density at an arbitrary point. Therefore, the TEC estimate is obtained by numerically integrating the electron density along the path from the satellite to the receiver antenna. In general, the NeQuick G algorithm is computationally much heavier than the Klobuchar, and

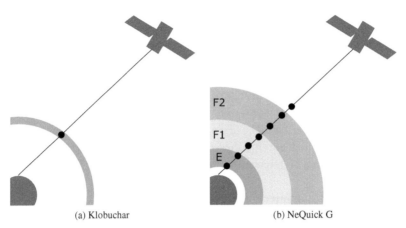

(a) Klobuchar   (b) NeQuick G

Figure 1.36 Principles of ionosphere compensation algorithms (illustrations not to scale). Black dots denote points where the delay is evaluated; note that in (b), the amount and spacing of evaluation points are determined by the numerical integration algorithm.

for this reason, it is recommended to update the TEC estimates at a low rate, such as every 30 seconds [80].

The Galileo navigation message provides three NeQuick G parameters (a_0, a_1, a_2) which are the coefficients of a second-order polynomial to evaluate the effective ionization level. In addition, the receiver must have access to two sets of precomputed data: a grid to compute the modified dip latitude and a set of spherical harmonic coefficients to characterize the monthly median transmission factor and critical frequency of the F2 layer. For details, please refer to the NeQuick G algorithm specification [80].

1.9.2   Tropospheric Error

The troposphere is the lowest layer of the atmosphere. It is between the Earth surface and about 9–10 kilometers high. Unlike the ionosphere, the troposphere is nondispersive, that is, there is no frequency-dependent effect. The temperature, pressure and humidity will, however, alter the velocity of radio waves. The resulting delay highly depends on the distance travelled through the troposphere. Since signals from low-elevation satellites travel a longer distance than a signal from a satellite at zenith, we often model the tropospheric delay using a zenith delay and a mapping function. The zenith tropospheric delay is about 2.4 m, and the tropospheric delay can be as high as 10 m for low-elevation satellites. Applying relatively simple tropospheric models can reduce these errors to below 0.5 m.

The total tropospheric delay in the direction of a satellite can be divided into two parts: the "dry," hydrostatic component, and the "wet", nonhydrostatic component. These two delays are then projected onto the zenith direction using some mapping functions, for example, using the Niell model [82]:

$$\mathrm{SPD}(el) = m_{\mathrm{HD}}(el)\,\mathrm{ZHD} + m_{\mathrm{WD}}(el)\,\mathrm{ZWD}, \qquad (1.143)$$

where SPD(el) is the slant path delay in meters along elevation angle el, ZHD is the zenith hydrostatic delay in meters, ZWD is the zenith wet delay in meters and $m_{HD}(el)$ and $m_{WD}(el)$ are the mapping functions of the hydrostatic and wet path delay along elevation angle el. The model proposed in [82] is currently used as the mapping function in the multi-GNSS SDR receiver accompanying this book. Another tropospheric model commonly used is the UNB3 model [83], used in the SBAS Minimum Operational Performance Standards (MOPS) of the Radio-Technical Commission for Aeronautics (RTCA) [84]. Further information about the troposphere and its models is provided in [85].

1.9.3   Multipath

Multipath is the effect of the reflections of the satellite signal in objects around the receiver. These reflections may be received and tracked by the receiver and the consequence is that our correlation peak is distorted, and this distortion leads to an error in the code delay measurement. The multipath error depends on the amplitude and distance of the reflections. More than one reflection can be received by the receiver, and this is generally the case in reality. The multipath error observed in the measurement depends on many factors: the signal tracked, the amplitude, the distance and the carrier phase of the reflection, the receiver bandwidth, and the correlator spacing. Even if the multipath echoes arrive always later than the original signal, they can lead to a negative delay if the carrier phase is rotated 180° with respect to the original signal. This is usually measured in the multipath error envelope (MEE) plot, treated in more detail in Chapter 10, where an example of MEE is provided (Figure 10.12).

Generally, mitigation techniques are effective in the case of LOS multipath, that is, when both the satellite-to-receiver LOS and the echoes are received. When only the echoes are received, this situation is referred to as non-line of sight (NLOS) multipath, and its effect may be more detrimental in the receiver than LOS multipath, especially if the reflection arrives with a delay of tens, or even hundreds of meters. Further details on state-of-the-art multipath detection and mitigation techniques and their performance can be found in Chapter 10. A low-complexity multipath detection method is also proposed in Chapter 9 for snapshot receivers. For more details on multipath, see [10, Chapter 9].

1.9.4   Interference

GNSS RFI can be defined as unwanted energy that gets into the GNSS receiver, excluding natural causes such as electromagnetic radiation from the sky or the ground, which we consider as noise. Typical GNSS interference can come from multiple sources, such as radars, aviation equipment, jammers, spoofers, GNSS repeaters, solar bursts, energy spillover from signals in other bands, such as TV signals, or even self-interference if the receiver has several RF devices, as it is the case for smartphones. There are multiple ways to classify interference, depending on the interferer's intention (intentional, nonintentional, collateral), whether it's natural (e.g. solar bursts) or man-made, the signal waveform (e.g. pulsed, narrow-band, wideband, chirped, matched spectrum...), or the impact on the receiver (e.g. signal degradation, service denial, or service integrity failure) [86]. For more details, [87] presents a thorough description of interference, including their mathematical representation and countermeasures. In our book and related software, we will not develop specific interference mitigation or detection techniques, although SDRs are powerful platforms for developing and improving such techniques thanks to their flexibility.

1.9.5   Receiver Noise

We refer to receiver noise as all other effects that contribute to the measurement error which are not described before. They include the electromagnetic noise radiated by the sky and the ground that reach the antenna together with the signal, and the thermal noise added by the receiver. The combination of these noise sources are represented by the noise power in Eq. (1.44). The signal filters also add a delay to our measurement. This delay can be considered noise as well, although, if constant, it is estimated as part of the receiver clock bias. If it is frequency-dependent, it is considered as a receiver group delay or hardware bias and can be estimated separately or removed with the ionospheric delay. Other noise effects relate to the sampling and quantization of the signal, described previously in Section 1.4.3. More details on receiver noise components can be found in [12, Chapter 6].

1.9.6   Error Budgets

Once we have introduced all the errors, we present in Table 1.5 the error budgets or contributions to the pseudorange measurement. The purpose of the table is to provide the reader a grasp of the different error contributions, but the error magnitudes can vary significantly depending on many factors. Satellite orbit and clock error contributions are gradually reduced due to the improvements in satellite on-board clocks and ground orbit determination capabilities. Ionosphere and multipath are very variable, as described before, and multipath and noise highly depend on the receiver signal and measurement processing. Note also that the error contribution can be composed of a fixed term (bias) and a random term (variance). Finally, the table illustrates the error

Table 1.5 Approximate average error contributions to the pseudorange measurement. Some of them can significantly vary and depend on many factors.

Source	Typical Error (m)	Comments
Satellite orbit	0.2–0.5	Gradually reduced with GNSS modernization.
Satellite clock	0.2–0.5	Gradually reduced with GNSS modernization.
Ionosphere	2.5–5	Very variable. 5–10 m before navigation message correction.
Troposphere	0.5	Around 3–10 m before receiver model correction.
Multipath	1–2	In open-sky conditions. Very variable and up to around 20 m in degraded conditions or higher if NLOS.
Receiver noise	0.5	Includes receiver hardware noise, filtering, sampling and quantization.

in the *corrected* pseudorange ρ, while the error in the *measured* pseudorange ρ_m from Eq. (1.141) is quantified in the right column, where pertinent. Further considerations on error modeling are provided in section 1.10.4. For a more exhaustive measurement error characterization, see [4, Chapter 5], [10] or [11].

1.10 Computation of Position and Time

The principle of GNSS positioning is simple. By knowing the position of the satellites and the satellite-receiver distance, or range, as the signals travel at the speed of light, the receiver can compute its position. In order to calculate a 3D position, three exact range measurements would be required, but the receiver does not know its own time, so it cannot calculate the exact time of arrival (ToA). The receiver time uncertainty, or bias b, is common to all range measurements, so it is calculated as an additional unknown. This means we need at least four pseudoranges to build a system of equations, as follows:

$$\rho^{(p)} = \sqrt{(x^{(p)} - x)^2 + (y^{(p)} - y)^2 + (z^{(p)} - z)^2} + b + e^{(p)}, \tag{1.144}$$

where the unknonws are in the vector $\mathbf{x} = (x, y, z, b)$, that we call the *state vector*, $(x^{(p)}, y^{(p)}, z^{(p)})$ are satellite p's coordinates, and b is the receiver clock bias, in meters ($b = dt_u c$). When we have four measurements, we can solve the system. Note that $\rho^{(p)}$ is the *corrected* pseudorange after removing the errors we can estimate: ionosphere and satellite clock bias from the navigation message, and troposphere from our model. However, $\rho^{(p)}$ still has an error $e^{(p)}$ that includes everything that has not been corrected yet.

1.10.1 Linearization

The system of equations from Eq. (1.144) when $(p) = 1, \ldots, 4$ is nonlinear, so it has to be linearized. We will explain the linearization process algebraically and then geometrically.

Algebraic Explanation

In order to solve a nonlinear system of equations, it is typical to use the *Newton*, or *Netwon-Raphson*, approximation, which consists of approaching the value of a function at a certain point by its value in the proximity and its derivative multiplied by an increment. A function can be expressed as a series of powers, according to a Taylor series, which also depend on the function derivatives. In our case, the function is the pseudorange $\rho^{(p)}$, which depends on our state vector $\mathbf{x} = (x, y, z, b)$, and the Taylor series is expanded as follows:

$$\rho^{(p)} = \rho^{(p)}(\mathbf{x}_0) + \frac{\partial \rho^{(p)}(\mathbf{x}_0)}{\partial \mathbf{x}} \Delta \mathbf{x} + \frac{1}{2!} \frac{\partial^2 \rho^{(p)}(\mathbf{x}_0)}{\partial^2 \mathbf{x}} \Delta \mathbf{x}^2 + \ldots \tag{1.145}$$

where $\rho^{(p)}(\mathbf{x}_0)$ is the pseudorange between the satellite position and our initial position estimate $\mathbf{x}_{0, (1:3)}$ (note that full initial state vector $\mathbf{x}_0$ includes also the bias in the fourth term). We consider sufficient to take the first derivative, as later we will perform several iterations. Thus, and developing Eq. (1.145) into the partial derivatives on the variables of our four-state vector, we have:

$$\rho^{(p)} \approx \rho^{(p)}(\mathbf{x}_0) + \frac{\partial \rho^{(p)}(x_0)}{\partial x}\Delta x + \frac{\partial \rho^{(p)}(y_0)}{\partial y}\Delta y + \frac{\partial \rho^{(p)}(z_0)}{\partial z}\Delta z + \frac{\partial \rho^{(p)}(b_0)}{\partial b}\Delta b.$$

(1.146)

If we perform the partial derivatives and slightly rearrange Eq. (1.146), we obtain the following:

$$\rho^{(p)} - \rho^{(p)}(\mathbf{x}_0) \approx \frac{x - x^{(p)}}{r^{(p)}}\Delta x + \frac{y - y^{(p)}}{r^{(p)}}\Delta y + \frac{z - z^{(p)}}{r^{(p)}}\Delta z + \Delta b,$$

(1.147)

where $r^{(p)} = r^{(p)}(\mathbf{x}_0) = \sqrt{(x^{(p)} - x_0)^2 + (y^{(p)} - y_0)^2 + (z^{(p)} - z_0)^2}$. The terms applying to the user position coordinates $(\Delta x, \Delta y, \Delta z)$ can be expressed as the unitary vector $\mathbf{e}^{(p)}$ that points from the estimated position solution $\mathbf{x}_0$ to satellite (p) position, $\mathbf{x}^{(p)}$, with opposite sign:

$$\mathbf{e}^{(p)} = \left(\frac{x^{(p)} - x}{r^{(p)}}, \frac{y^{(p)} - y}{r^{(p)}}, \frac{z^{(p)} - z}{r^{(p)}}\right) = \frac{\mathbf{x}^{(p)} - \mathbf{x}_{0,(1:3)}}{|\mathbf{x}^{(p)} - \mathbf{x}_{0,(1:3)}|}.$$

(1.148)

Finally, our system of equations can be expressed in a matrix form as follows:

$$\Delta\rho = \begin{bmatrix} -\mathbf{e}^{(1)} & 1 \\ \cdots & \end{bmatrix} \cdot \Delta\mathbf{x},$$

(1.149)

where $\Delta\mathbf{x} = \mathbf{x} - \mathbf{x}_0$ and $\Delta\rho$ is a vector of differences between the corrected pseudoranges and the theoretical estimation based on the initial position. For satellite (p):

$$\Delta\rho^{(p)} = \rho^{(p)} - \rho^{(p)}(\mathbf{x}_0).$$

(1.150)

Note that, in some other references, $\rho^{(p)}(\mathbf{x}_0)$ is referred as $\hat{\rho}^{(p)}(\mathbf{x}_0)$, or just $\hat{\rho}^{(p)}$, to highlight that it is a *theoretical* estimation of the pseudorange. The linear system in Eq. (1.149) leads to a recursion where the new receiver position is used as initial position $\mathbf{x}_0$ for the next iteration, and so on, until the position improvement converges. Usually, around four iterations are sufficient, and more iterations do not improve the result. We will call the matrix with the unitary vectors as the *geometry matrix*, also known as *observation matrix*, or *design matrix*, and it will be denoted by $\mathbf{A}$:

$$\Delta\rho = \mathbf{A} \cdot \Delta\mathbf{x}.$$

(1.151)

For more information on the linear algebra behind the satellite navigation positioning equations, see [88].

Geometric Explanation

Our geometric interpretation of the positioning problem is depicted in Figure 1.37, again for satellite (p). For our explanation, we reduce the state vector to the position vector (x, y, z). In the figure, we start from an initial position vector $\mathbf{x}_0$ and we want

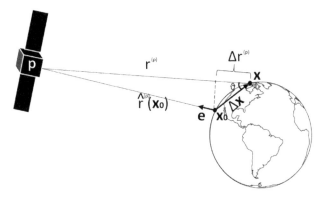

Figure 1.37 Linear relationship between measurements and position.

to obtain vector $\Delta\mathbf{x}$ toward $\mathbf{x}$, the true position, through a linear relationship with our pseudoranges. We define the delta range for satellite (p) as

$$\Delta r^{(p)} = r^{(p)} - \hat{r}^{(p)}(\mathbf{x}_0). \tag{1.152}$$

Then, as shown in Figure 1.37:

$$\Delta r^{(p)} \approx |\Delta\mathbf{x}| \cos\theta, \tag{1.153}$$

which can also be expressed as the dot product of vectors $-\mathbf{e}^{(p)}$ and $\Delta\mathbf{x}$,

$$\Delta r^{(p)} \approx -\mathbf{e}^{(p)} \cdot \Delta\mathbf{x} = |\Delta\mathbf{x}| \cos\theta. \tag{1.154}$$

We use $\approx$ in Eq. (1.153) because of the linearization error, which is reduced at each iteration.

We have now established our linear relationship between measurements and position. We add the bias term b to the expression and obtain the system of equations from Eq. (1.149).

1.10.2 Least Squares and Weighted Least Squares

When we have four pseudorange measurements, we can solve our four unknowns by calculating the inverse of matrix $\mathbf{A}$ for each iteration:

$$\Delta\mathbf{x} = \mathbf{A}^{-1} \cdot \Delta\rho, \tag{1.155}$$

where $(\cdot)^{-1}$ is the inverse matrix operator. However, even after applying the ionospheric and tropospheric models and satellite clock corrections, our measurements still have errors, and these errors will propagate into the position error. In order to minimize this error, we want to use more measurements. When we have more than four measurements, our system is overdetermined, meaning that it has many possible solutions for our state vector. What we want is that the chosen solution minimizes the position error. In order to minimize this position error, we use the *least-squares* estimator, which computes the state vector that minimizes the sum of the squares of the residual measurement errors with respect to our solution, that is, the differences between the measured and theoretical pseudoranges for our solution ($\Delta\rho$ in the last iteration). When applying least squares, our iterative system of equations is expressed as

$$\Delta\mathbf{x} = (\mathbf{A}^T\mathbf{A})^{-1}\mathbf{A}^T\Delta\rho, \qquad (1.156)$$

where $(\cdot)^T$ is the transpose matrix operator. Sometimes the matrix $(\mathbf{A}^T\mathbf{A})^{-1}\mathbf{A}^T$ is referred to as the *pseudo-inverse* matrix.

Some measurements will be of higher quality than others. For example, those from a higher satellite elevation use to have a lower ionospheric and tropospheric delay, less multipath and possibly a higher signal strength due to the satellite and receiver antenna gain patterns. The estimator may *weight* the measurements so that those believed to be better have a higher weight in the position solution. In this case, we use

$$\Delta\mathbf{x} = (\mathbf{A}^T\mathbf{W}\mathbf{A})^{-1}\mathbf{A}^T\mathbf{W}\Delta\rho, \qquad (1.157)$$

where $\mathbf{W}$ is the *weight matrix*, a diagonal ($m \times m$) matrix, where m is the number of measurements and where each coefficient in the diagonal represents the weight. Usually, weights are determined depending on the satellite elevation or C/N_0, but other criteria can be used, such as the satellite type or information in the navigation message, like URA/SISA as explained later in this chapter. If the variances of the satellite errors could be estimated, its weight could be their inverse, so that $W^{(p)} = 1/\sigma_{(p)}^2$ (we will later refer to these variances as UERE^2).

Least squares is one of the simplest methods to estimate the receiver position, but it has the inconvenience that the resulting position is not smoothed in time, as it depends on the instantaneous errors of the measurements. Kalman filters [89] are often used for that purpose. Other standard algorithms use least squares, but over carrier-smoothed pseudoranges such as Hatch filters discussed in [90] or in Chapter 8 (8.18). Also, for the estimation of the receiver velocity based on the range rate measurements, see [4, Chapter 6].

1.10.3   Coordinates and Reference Frames

Our satellite position was calculated in ECEF Cartesian coordinates, so our position solution (x, y, z) is by default referred to ECEF as well. However, users are often interested in the longitude and latitude, or *geographic* coordinates. As shown in Figure 1.38, the transformation between latitude and longitude[40] (l_a, l_o) and (x, y, z) is as follows:

$$x = R\cos(l_a)\cos(l_o)$$
$$y = R\cos(l_a)\sin(l_o) \qquad (1.158)$$
$$z = R\sin(l_a),$$

where R is the position vector module. However, we often want to express our coordinates with respect to a *reference frame*. A reference frame includes not only

---

[40] We avoid the usual notation of φ and λ symbols as both have been used before in this chapter.

latitude and longitude but also height, among other information. For example, GPS uses WGS84. WGS84 defines the specific values of the main constants used in the position computation: the Earth rotation rate $\dot{\Omega}_e$ and the Earth gravitational constant μ. WGS84 also defines an ellipsoid that models the Earth. This way, a position can be expressed in (l_a, l_o, h) where h is the height with respect to the ellipsoid. The ellipsoid is defined by the Earth semimajor axis ($a = 6{,}378{,}137$ m) and the flattening factor ($f = 1/298.257223563$). Then, l_a, l_o and h can be calculated with respect to the WGS84 ellipsoid as follows:

$$l_o = \operatorname{atan}\left(\frac{y}{x}\right) \tag{1.159}$$

$$N_\varphi = \frac{a}{\sqrt{1 - f(2 - f)\sin^2(l_a)}} \tag{1.160}$$

$$l_a = \operatorname{atan}\left(\frac{z}{\sqrt{x^2 + y^2}}\left(1 - \frac{(2 - f)f N_\varphi}{N_\varphi + h}\right)^{-1}\right) \tag{1.161}$$

$$h = \frac{\sqrt{x^2 + y^2}}{\cos(l_a)} - N_\varphi. \tag{1.162}$$

Note that, once again, we have a nonlinear system to solve N_φ, l_a and h. This is usually solved by iteration.

In addition to the simple ellipsoid, WGS84 defines the *geoid* as a more irregular surface through some harmonic corrections. Further information on the WGS84 and all its parameters, which also include gravitational models, magnetic models and datum transformations, can be found in [91].

Topocentric coordinate systems are also relevant for receivers. They express coordinates with respect to the horizon, or a horizontal plane that is tangent to the Earth surface at the user location. They are particularly relevant to express errors and covariances in the horizontal plane and vertical component, as mentioned later. A usual right-handed topocentric coordinate system is ENU (East, North, Up). It is also depicted in Figure 1.38. Another one is NED (North, East Down).

Reference frames, also known as *datums*, can be different for different GNSS. Reference frame parameters are defined by a set of monitor stations precisely located on Earth. Each terrestrial reference frame, like Galileo Terrestrial Reference Frame (GTRF) for Galileo, or WGS84 for GPS, uses its own set of stations. Both GTRF and WGS84 are based on International Terrestrial Reference Frame, produced by the International Earth Rotation and Reference Systems Service (IERS), and their differences are minimal, in the order of centimeters, so they are not relevant for code-based positioning. They may be relevant though for high accuracy centimeter-level positioning, which may require reference frame transformations to avoid this error. More details on reference frames can be found in [11, Chapter 2] and [92, Chapter 2].

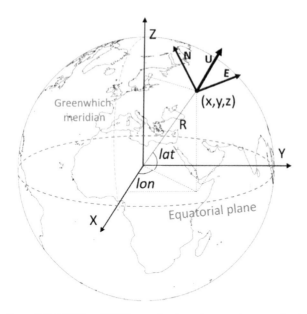

Figure 1.38 ECEF and ENU coordinate systems and conversion to geographic coordinates.

1.10.4   Dilution of Precision, Measurement Residuals and Position Accuracy

Dilution of Precision

Position accuracy depends on two independent factors: satellite geometry and measurement errors. If we have four satellites in view but they are aligned in the sky, our matrix $\mathbf{A}$ will be rank-deficient and no solution will be obtained, or at best our precision would be highly *diluted*. This is measured by the *Dilution of Precision* (DOP), and is reflected in the DOP matrix $\mathbf{D}$:

$$\mathbf{D} = (\mathbf{A}^T\mathbf{A})^{-1} \tag{1.163}$$

$\mathbf{D}$ is a (4×4) matrix where the diagonal terms d_{11} to d_{44} express how good our geometry is. The advantage of DOP is that it gives an indication of the solution precision (or its dillution) without looking at the pseudorange measurements. Depending on our interest, we can look at different terms:

- $\sqrt{d_{11} + d_{22} + d_{33} + d_{44}}$: Geometric Dilution of Precision (GDOP). It expresses the overall impact of geometry in the position and time solution.
- $\sqrt{d_{11} + d_{22} + d_{33}}$: Position Dilution of Precision (PDOP). It expresses the geometry impact in the position solution.
- $\sqrt{d_{44}}$: Time Dilution of Precision (TDOP). It expresses the geometry impact in the time solution.

One limitation of these DOP indicators is that they do not allow expressing DOP in the receiver vertical and horizontal components, which is desirable. For this, we need to express $\mathbf{D}$ in a topocentric coordinate system. We can rotate our DOP matrix and express it in ENU coordinates as follows:

$$\mathbf{D}_{\text{enu},3\times3} = \mathbf{F}_{\text{xyz}}^{\text{enu}T} \cdot \mathbf{D}_{3\times3} \cdot \mathbf{F}_{\text{xyz}}^{\text{enu}}, \tag{1.164}$$

where $\mathbf{D}_{\text{enu},3\times3}$ is the (3×3) DOP matrix rotated to ENU including elements $d_{\text{enu},11}$ to $d_{\text{enu},33}$, and $\mathbf{D}_{3\times3}$ is the (3×3) DOP matrix in XYZ, including elements $d_{\text{enu},11}$ to $d_{\text{enu},33}$. We define here matrix $\mathbf{F}_{\text{xyz}}^{\text{enu}}$, that allows to convert from a Cartesian to an ENU coordinate system, as follows:

$$\mathbf{F}_{\text{xyz}}^{\text{enu}} = \begin{bmatrix} -\sin(l_o) & -\sin(l_a)\cos(l_o) & \cos(l_a)\cos(l_o) \\ \cos(l_o) & -\sin(l_a)\sin(l_o) & \cos(l_a)\sin(l_o) \\ 0 & \cos(l_a) & \sin(l_a) \end{bmatrix}. \tag{1.165}$$

Now we can estimate DOP in the horizontal plane and vertically:

- $\sqrt{d_{\text{enu},11} + d_{\text{enu},22}}$: Horizontal Dilution of Precision (HDOP). Geometry impact in the horizontal components of the position.
- $\sqrt{d_{\text{enu},33}}$: Vertical Dilution of Precision (VDOP). Geometry impact in the vertical component of the position.

VDOP and HDOP are widely used for many applications. For example, an airplane using GNSS for precision approach will be sensitive to VDOP, while a car will be more sensitive to HDOP. Generally, VDOP will be worse than HDOP, as satellites are only in the sky, that is, in the upper side in the vertical dimension. HDOP and VDOP usually range between almost one and two, where lower implies a better geometry. Thanks to multi-GNSS, geometry can be highly improved. Practical examples of HDOP and VDOP values, for different GNSS combinations, are shown in Chapter 7, Table 7.6.

User Equivalent Range Error and Covariance Matrix

The other element that drives position accuracy is the measurement error. In practice, when solving Eq. (1.149) we apply the troposphere, ionosphere and satellite clock corrections to our receiver measurement ρ_m and the satellite orbital errors to the estimated pseudorange $\hat{\rho}$ and $\mathbf{A}$. We may apply multipath mitigation through specific tracking loop discriminators, or smoothing of the pseudorange measurements. Note also that no interference error is explicitly considered as, unlike the others, interference is sporadic or considered under the noise term. With these actions, we reduce the measurement errors as much as we can, but we still want to estimate how big these errors are. The estimations of error contributions introduced in Section 1.9 are aggregated into the *User Equivalent Range Error* (UERE), which represents the standard deviation of the total error:

$$\text{UERE} = \sqrt{\sigma_{\text{orb}}^2 + \sigma_{\text{clk}}^2 + \sigma_{\text{iono}}^2 + \sigma_{\text{tropo}}^2 + \sigma_{\text{mpath}}^2 + \sigma_{\text{noise}}^2}, \tag{1.166}$$

where σ_{orb}, σ_{clk}, σ_{iono}, σ_{tropo}, σ_{mpath}, and σ_{noise} are the standard deviations of error distributions of the satellite orbits and clocks, ionospheric and tropospheric delays, multipath and receiver noise. In Eq. (1.166), we are assuming that the variance of the sum of stochastic variables is the sum of the variances of each variable. This implies that all variables are independent, that is, each error source is uncorrelated with the rest.

Usually, UERE is divided in two components: signal-in-space ranging error (SISRE), including the error of the orbits, clocks and group delays when applied, and user equipment error (UEE), which includes everything else[41]:

$$\text{SISRE} = \sqrt{\sigma_{\text{orb}}^2 + \sigma_{\text{clk}}^2} \tag{1.167}$$

$$\text{UEE} = \sqrt{\sigma_{\text{iono}}^2 + \sigma_{\text{tropo}}^2 + \sigma_{\text{mpath}}^2 + \sigma_{\text{noise}}^2} \tag{1.168}$$

$$\text{UERE} = \sqrt{\text{SISRE}^2 + \text{UEE}^2}. \tag{1.169}$$

The assumption that errors are uncorrelated may not be correct sometimes. Also, if we want to characterize the errors by their standard deviation, we are assuming that they are zero-mean, which may also not be true. Therefore, UERE is not very representative in many cases. In spite of that, it is widely used. For example, it allows us to roughly characterize the horizontal user position error as UERE × HDOP, and the vertical position error as UERE × VDOP. We can also use a generic UERE value to estimate the covariance matrix of our solution that shows the variances of the different components ($\sigma_x^2, \sigma_y^2...$) and their interrelationships, or *covariances*, ($\sigma_{xy}, \sigma_{xz}...$), as

$$\mathbf{COV}_{\text{xyz}} = \text{UERE}^2 \cdot \mathbf{D}. \tag{1.170}$$

Measurement Residuals

So far, DOP and UERE have been described independently from the actual measurements. In order to assess the goodness of our position in relation to the measurement errors, we use the measurement *residuals*. Once we obtain our solution, our residuals vector β is commonly expressed as

$$\beta = \mathbf{A}\mathbf{x} - \Delta\rho. \tag{1.171}$$

The residuals are a good, yet limited, indicator of the position accuracy. Having a vector with low residuals is often an indicator of a good solution accuracy. It means that the measurements "agree" with the computed position. However, this is a good indicator only if the measurement errors are uncorrelated. If, for example, measurements are reflections from a building, the position may be off by much more than the residuals suggest. Therefore, measurement residuals are not a perfect indicator of the user accuracy either.

Measuring and Bounding Position Errors and Integrity

We have presented an overview of how the receiver can estimate the goodness of its position through its geometry (DOP), a-priori errors (UERE) and measurement residuals β. However, in the next chapters, the actual errors are measured with respect to the true position, which is known, instead of using UERE and DOP. When we measure

---

[41] Note that, in this convention, we include all propagation errors as part of the user equipment error. Note also that SISRE is sometimes referred to as signal-in-space error (SISE).

position errors, we have to do it through an unambiguous statistical characterization. For example, while RMS and standard deviation (σ) are used sometimes as synonyms, and the same occurs for 95 percentile and 2σ, there are subtleties that depend on the number of dimensions of the error used and other factors. In the next chapters, the position accuracy is provided through the horizontal RMS error, vertical RMS error and 3D RMS error. For more details on the differences on accuracy indicators, see the excellent analysis in [93]. When the user position is unknown but we want to better characterize our errors beyond an a-priori UERE, we can use the information provided by some GNSS or augmentation systems. For example, GPS and Galileo provide an indicator of the error statistical distribution in the user range accuracy (URA) and signal-in-space accuracy (SISA), respectively. URA bounds the user range error, or URE, which includes the orbit and clock errors[42] by 4.42 times with a probability of $1 - 10^{-5}$ every hour (or 5.73 times with a probability of $1 - 10^{-8}$ every hour, depending on the navigation flags). Similarly, SISA provides the standard deviation (1σ) of the zero-mean Gaussian distribution that bounds the URE. The exact definitions and further details are provided in the official GNSS service definition documents [94] [95].

Measurement errors are at the core of satellite navigation because they drive the position accuracy. Also, by modeling the error statistical distributions correctly, the user can have an estimate of the goodness of its position and the maximum bound of its error with a certain probability. This is important for critical applications, such as airplane precision approaches, which in addition to accuracy, require position *integrity*. Integrity can only be provided effectively with additional monitoring means, either on-board, through external information or combined. The GNSS integrity concept also requires that when the errors cannot be bounded, the receiver is alerted in a few seconds maximum. On-board integrity is usually called receiver autonomous integrity monitoring (RAIM). RAIM is usually based on looking at a priori and/or residuals information available, and detecting and excluding faulty measurements from the solution. Augmentations monitor the satellites and sometimes ionosphere and transmit error bounds and alert the users through satellite or local ground channels, in the case of SBAS or ground-based augmentation systems (GBAS), respectively. GPS III already provides an "integrity flag," that allows to switch from the $4.42, 10^{-5}$ to the $5.73, 10^{-8}$ URA characterization mentioned earlier. New SBAS provide the covariance matrix of each satellite, allowing a better computation of **COV** than that in Eq. (1.170). Other modern integrity systems, such as advanced RAIM (ARAIM), combine error bounds provided from the system with RAIM. Integrity is beyond the scope of this book, but for more details on RAIM, SBAS, GBAS and ARAIM, see [96], [97], [98] and [99] respectively.

---

[42] Note that before we defined SISRE/SISE as the one-sigma or RMS URE.

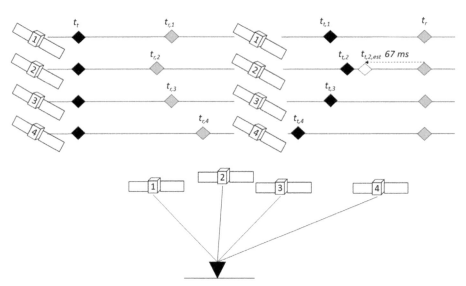

Figure 1.39 Satellite position computation for pseudoranges measured at the same time.

1.10.5   Other Practical Aspects

How to put our PVT engine to work is not so obvious in practical terms, so we will discuss some practicalities in this subsection.

Earth Rotation and Receiver Time Reference

Apart from the satellite clock offset, which can be neglected here, all satellites are synchronized and provide the time of the start of the frame in the transmitted signal t_t, with which the receiver can synchronize, as illustrated in Figure 1.39, left. As the satellites are located at different distances to the receiver, this time tag will be received at different instants $t_{r,1}, t_{r,2}$, etc. However, our receiver measures the pseudoranges at a given instant for *all* satellites, t_r, as shown in the top right part. What is measured is a satellite spreading code that was transmitted by each satellite at a different time, $t_{t,1}, t_{t,2}$, etc. The subtlety here is that, for our range estimation, we need to have the satellite positions in ECEF. To do that, we need to convert from an inertial reference frame, not accounting for the Earth rotation, to ECEF, but conversion to ECEF at $t_{t,2}$ is not the same as conversion to ECEF at $t_{t,3}$, as the Earth has rotated between the two timestamps. Therefore, we have to convert to ECEF separately for each satellite. This can be done through the following rotation matrix:

$$\mathbf{x}^{(p)} = \begin{bmatrix} \cos(\dot{\Omega}_e \hat{\tau}^{(p)}) & \sin(\dot{\Omega}_e \hat{\tau}^{(p)}) & 0 \\ -\sin(\dot{\Omega}_e \hat{\tau}^{(p)}) & \cos(\dot{\Omega}_e \hat{\tau}^{(p)}) & 0 \\ 0 & 0 & 1 \end{bmatrix} \mathbf{x}_{nr}^{(p)}, \qquad (1.172)$$

where $\mathbf{x}^{(p)}$ is the rotated satellite position vector, $\dot{\Omega}_e$ is the Earth's rotation speed, $\hat{\tau}^{(p)}$ is the expected ToA for satellite (p) ($\hat{\tau}^{(p)} = r^{(p)}/c$) and $\mathbf{x}_{nr}^{(p)}$ is the satellite position vector before rotation. This per-satellite rotation is sometimes called *Sagnac effect*.

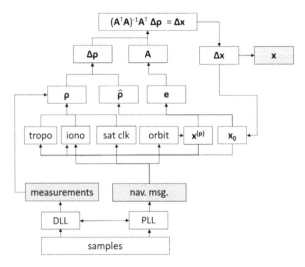

Figure 1.40 Position and time computation process, including inputs (measurements and navigation message), outputs (position and time) and dependencies.

The concept of receiver time reference in an SDR, and more generally, in postprocessing, may not be obvious. Our receivers make an estimation of the reference time based on the GNSS signal. A way to estimate its own time is to assume that the signal of the closest GPS satellite takes, for example, 67 ms, which corresponds to the approximate distance of a GPS satellite at the zenith and use this as an initial time reference that will be refined as part of the position solution. This is also illustrated in Figure 1.39, right. Also, the receiver can estimate the time just with one satellite, with a certain error, by subtracting a rough arrival time estimation from one satellite.

Data Flows and Dependencies

A data flow diagram with the dependencies of the position computation elements is presented in Figure 1.40. The pseudorange measurements are generated from the DLL, after the addition of the integer number of codes to the code delay, and the data bits from the PLL are decoded to form the navigation message. From the navigation message, we obtain the orbits, clocks and ionospheric information (note that signal synchronization is assumed to be achieved and removed from the figure for simplicity). The satellite positions $\mathbf{x}^{(p)}$ at the expected time are computed and used, together with the estimated receiver position in $\mathbf{x_0}$, for the predicted pseudoranges $\hat{\rho}$ and geometry matrix $\mathbf{A}$. $\mathbf{A}$ requires the receiver-to-satellite unitary vector $\mathbf{e}$, also using the estimated receiver position in $\mathbf{x_0}$. Our best initial estimation ($\mathbf{x_0}$ in Figure 1.40) can be the Earth center or the center between sub-satellite points. The measured pseudoranges (ρ_m) are corrected with the iono and tropo models and clock corrections to generate the corrected pseudoranges ρ. From these, we subtract the predicted pseudoranges $\hat{\rho}$ and plug the result $\Delta\rho$ into the navigation equation. The equation is iterated until the final solution, $\mathbf{x}$, is obtained.

References

[1] J. W. Betz, "Binary Offset Carrier Modulations for Radio Navigation," *Navigation: Journal of the Institute of Navigation*, vol. 48, no. 4, pp. 227–246, 2002.

[2] GPS Navstar Joint Program Office, "Navstar GPS Space Segment/Navigation User Segment Interfaces, IS-GPS-200," Revision M, May 21, 2021.

[3] L. Lestarquit, G. Artaud, and J.-L. Issler, "AltBOC for Dummies or Everything You Always Wanted to Know about AltBOC," in *Proceedings of ION GNSS, Savannah, GA*, pp. 961–970, 2008

[4] P. Misra and P. Enge, *Global Positioning System: Signals, Measurements and Performance, Revised Second Edition*. Ganga-Jamuna Press, 2011.

[5] S. M. Kay, *Fundamentals of Statistical Signal Processing, Volume I: Estimation Theory*. Prentice-Hall, 1993.

[6] J. Proakis and M. Salehi, *Communication Systems Engineering*. Prentice-Hall, 2002.

[7] J. W. Betz, "On the Power Spectral Density of GNSS Signals, with Applications," in *Proceedings of ION International Technical Meeting (ITM), San Diego, CA*, pp. 859–871, January 2010.

[8] A. V. Oppenheim, A. S. Willsky, and S. Hamid, *Signals and Systems*. Prentice-Hall, 1996.

[9] S. Ramakrishnan, T. Reid, and P. Enge, "Leveraging the L1 Composite Signal to Enable Autonomous Navigation at Geo and beyond," in *Proceedings of the 26th International Technical Meeting of The Satellite Division of the Institute of Navigation (ION GNSS'2013), Nashville, TN, USA*, pp. 16–20, 2013.

[10] E. D. Kaplan and C. J. Hegarty, *Understanding GPS: Principles and Applications*. Artech House, Inc., 2017.

[11] P. Teunissen and O. Montenbruck, *Springer Handbook of Global Navigation Satellite Systems*. Springer, 2017.

[12] F. van Diggelen, *A-GPS, Assisted GPS, GNSS and SBAS*. Artech House, 2009.

[13] J. W. Betz, *Engineering Satellite-Based Navigation and Timing: Global Navigation Satellite Systems, Signals, and Receivers*. Wiley IEEE Press, 2016.

[14] K. Borre, D. Akos, N. Bertelsen, P. Rinder, and S. H. Jensen, *A Software-Defined GPS and Galileo Receiver, Single Frequency Approach*. Birkhäuser, 2007.

[15] K. J. Quirk and M. Srinivasan, "PN Code Tracking Using Noncommensurate Sampling," *IEEE Transactions on Communications*, vol. 54, no. 10, pp. 1845–1856, October 2006.

[16] V. Tran, N. Shivaramaiah, T. Nguyen, J. Cheong, E. Glennon and A. Dempster, "Generalised Theory on the Effects of Sampling Frequency on GNSS Code Tracking." *Journal of Navigation*, vol. 71, no. 4, 257–280.

[17] D. M. Akos, M. Stockmaster, J. B. Y. Tsui, and J. Caschera, "Direct Bandpass Sampling of Multiple Distinct RF Signals," *IEEE Trans on Communcations*, vol. 47, no. 7, pp. 983–988, July 1999.

[18] A. J. Van Dierendonck, "GPS receivers," in *Part I: GPS Fundamentals* (J. J. Spilker Jr., P. Axelrad, B. W. Parkinson, and P. Enge, eds.), *Global Positioning System: Theory and Applications*, ch. 8, American Institute of Aeronautics and Astronautics, 1996.

[19] A. Lakhzouri, E. S. Lohan, and M. Renfors, "Reduced-Complexity Time-Domain Correlation for Acquisition and Tracking of BOC-Modulated Signal," in *Proceedings of ESA NAVITEC, Noordwijk, The Netherlands*, December 2004.

[20] H. Sorokin, E. S. Lohan, and J. Takala, "Memory-Efficient Time-Domain Correlation for BOC-Modulated Signals," in *Proceedings of ESA NAVITEC, Noordwijk, The Netherlands*, December 2006.

[21] T. Pany, B. Riedl, and J. Winkel, "Efficient GNSS Signal Acquisition with Massive Parallel Algorithms Using GPUs," in *Proceedings of ION Intl. Technical Meeting of the Satellite Division, Portland, OR*, pp. 1889–1895, 2010.

[22] M. L. Psiaki, "Block Acquisition of Weak GPS Signals in a Software Receiver," in *Proceedings of ION GPS*, pp. 2838–2850, *Salt Lake City, UT*, September 2001.

[23] B. Chibout, C. Macabiau, A. C. Escher, L. Ries, J. L. Issler, S. Corrazza, and M. Bousquet, "Comparison of Acquisition Techniques for GNSS Signal Processing in Geostationary Orbit," in *Proceedings of ION National Technical Meeting (NTM), San Diego, CA*, pp. 637–649, 2007.

[24] A. Papoulis, *Probability, Random Variables and Stochastic Processes*. McGraw-Hill, 1991.

[25] D. W. Allan, "Time and Frequency (Time-Domain) Characterization, Estimation and Prediction of Precision Clocks and Oscillators," *IEEE Transactions on Ultrasonics, Ferroelectrics and Frequency Control*, vol. 34, no. 6, pp. 647–654, 1987.

[26] E. Domínguez, A. Pousinho, P. Boto, D. Gómez-Casco, S. Locubiche, G. Seco-Granados, J. A. López-Salcedo, H. Fragner, F. Zangerl, O. Peña, and D. Jimenez-Baños, "Performance Evaluation of High Sensitivity GNSS Techniques in Indoor, Urban and Space Environments," in *Proceedings of ION GNSS+, Portland, OR*, pp. 373–393, September 2016.

[27] G. Corazza and R. Pedone, "Generalized and Average Likelihood Ratio Testing for Post Detection Integration," *IEEE Transactions on Communications*, vol. 55, no. 11, pp. 2159–2171, 2007.

[28] M. Kokkonen and S. Pietilla, "A New Bit Synchronization Method for a GPS Receiver," in *Proceedings of the IEEE Position Location and Navigation Symposium, Palm Springs, CA*, pp. 85–90, 2002.

[29] T. Ren and M. Petovello, "An Analysis of Maximum Likelihood Estimation Method for Bit Synchronization and Decoding of GPS L1 C/A Signals," *Journal on Advances in Signal Processing*, vol. 3, pp. 1–12, 2014.

[30] G. López-Risueño and G. Seco-Granados, "Measurement and Processing of Indoor GPS Signals Using One-Shot Software Receiver," in *Proceedings of ESA NAVITEC, Noordwijk, The Netherlands*, pp. 1–9, 2004.

[31] D. Gómez-Casco, J. A. López-Salcedo, and G. Seco-Granados, "Optimal Fractional Non-coherent Detector for High-Sensitivity GNSS Receivers Robust against Residual Frequency Offset and Unknown Bits," in *Proceedings of IEEE Workshop on Positioning, Navigation and Communications (WPNC), Bremen, Germany*, pp. 1–5, October 2017.

[32] D. Borio and L. L. Presti, "Data and Pilot Combining for Composite GNSS Signal Acquisition," *International Journal of Navigation and Observation*, pp. 1–12, 2008.

[33] B. A. Siddiqui, J. Zhang, M. Z. H. Bhuiyan, and E. S. Lohan, "Joint Data-Pilot Acquisition and Tracking of Galileo E1 Open Service Signal," in *Proceedings of Ubiquitous Positioning Indoor Navigation and Location Based Service, Kirkkonummi, Finland*, pp. 1–7, October 2010.

[34] J. Zhou and C. Liu, "Joint Data-Pilot Acquisition of GPS L1 Civil Signal," in *Proceedings of International Conference on Signal Processing (ICSP), Hangzhou, China*, pp. 1628–1631, October 2014.

[35] S. M. Kay, *Fundamentals of Statistical Signal Processing, Vol 2.: Detection Theory*. Prentice Hall, 1993.

[36] M. K. Simon and M. S. Alouini, *Digital Communication over Fading Channels*. John Wiley & Sons, 2000.

[37] G. J. Povey, "Spread Spectrum PN Code Acquisition Using Hybrid Correlator Architectures," *Wireless Personal Communications*, vol. 8, pp. 151–164, 1998.

[38] B. J. Kang and I. K. Lee, "A Performance Comparison of Code Acquisition Techniques in DS-CDMA System," *Wireless Personal Communications*, vol. 25, pp. 163–176, 2003.

[39] E. S. Lohan, A. Lakhzouri, and M. Renfors, "Selection of the Multiple-Dwell Hybrid-Search Strategy for the Acquisition of Galileo Signals in Fading Channels," in *Proceedings of IEEE Personal and Indoor Mobile Radio Communications (PIMRC)*, vol. 4, *Barcelona, Spain*, pp. 2352–2356, September 2004.

[40] E. S. Lohan, A. Burian, and M. Renfors, "Acquisition of GALILEO Signals in Hybrid Double-Dwell Approaches," in *Proceedings of ESA NAVITEC, Noordwijk, The Netherlands*, December 2004.

[41] W.-H. Sheen and S. Chiou, "Performance of Multiple-Dwell Pseudo-Noise Code Acquisition with IQ Detector on Frequency-Nonselective Multipath Fading Channels," *Wireless Networks*, vol. 5, no. 1, pp. 11–21, 1999.

[42] L. E. Aguado, G. J. Brodin, and J. A. Cooper, "Combined GPS/Galileo Highly-Configurable High-Accuracy Receiver," in *Proceedings of ION GNSS, Long Beach, CA*, September 2004.

[43] A. Konovaltsev, A. Hornbostel, H. Denks, and B. Bandemer, "Acquisition Tradeoffs for Galileo SW Receiver," in *Proceedings of European Navigation Conference, Manchester, UK*, 2006.

[44] A. Polydoros and C. L. Weber, "A Unified Approach to Serial Search Spread Spectrum Code Acquisition—Part II: A Matched-Filter Receiver," *IEEE Transactions on Communications*, vol. 32, pp. 550–560, 1984.

[45] N. M. O'Mahony, *Variable Dwell Time Verification Strategies for CDMA Acquisition with Application to GPS Signals*. PhD thesis, Department of Electrical and Electronic Engineering, University College Cork, Ireland, 2010.

[46] A. Konovaltsev, H. Denks, A. Hornbostel, M. Soellner, and M. Kaindl, "General Approach to Analysis of GNSS Signal Acquisition Performance with Application to GALILEO Signals," in *Proceedings of ESA NAVITEC, Noordwijk, The Netherlands*, 2006.

[47] M. Villanti, C. Palestini, R. Pedone, and G. Corazza, "Robust Code Acquisition in the Presence of BOC Modulation for Future GALILEO Receivers," in *Proceedings of IEE International Conference on Communications (ICC), Glasgow, UK*, pp. 5801–5806, 2007.

[48] W. Suwansantisuk and M. Z. Win, "Multipath Aided Rapid Acquisition: Optimal Search Strategies," *IEEE Transactions on Information Theory*, vol. 53, pp. 174–193, 2007.

[49] G. López-Risueño and G. Seco-Granados, "Measurement and Processing of Indoor GPS Signals Using a One-Shot Software Receiver," in *Proceedings of ESA NAVITEC, Noordwijk, The Netherlands* pp. 1–9, 2004.

[50] D. Jiménez-Baños, N. Blanco-Delgado, G. López-Risueño, G. Seco-Granados, and A. Garcia, "Innovative Techniques for GPS Indoor Positioning Using a Snapshot Receiver," in *Proceedings of ION GNSS, Fort Worth, TX*, pp. 2944–2955, 2006.

[51] A. Polydoros and M. Simon, "Generalized Serial Search Code Acquisition: The Equivalent Circular State Diagram Approach," *IEEE Transactions on Communications*, vol. 32, pp. 1260–1268, 1984.

[52] V. M. Jovanovic, "Analysis of Strategies for Serial-Search Spread-Spectrum Code Acquisition–Direct Approach," *IEEE Transactions on Communications*, vol. 36, pp. 1208–1220, 1998.

[53] A. Burian, E. S. Lohan, and M. Renfors, "BPSK-Like Methods for Hybrid-Search Acquisition of GALILEO Signals," in *Proceedings of ICC 2006, Istanbul, Turkey*, 2006.

[54] E. Pajala, E. S. Lohan, and M. Renfors, "CFAR Detectors for Hybrid-Search Acquisition of GALILEO Signals," in *Proceedings of ENC-GNSS, Munich, Germany*, pp. 1–7, 2005.

[55] W. Zhuang, "Noncoherent Hybrid Parallel PN Code Acquisition for CDMA Mobile Communications," *IEEE Transactions on Vehicle Technology*, vol. 45, pp. 643–656, 1996.

[56] Y. Zheng, "A Software-Based Frequency Domain Parallel Acquisition Algorithm for GPS Signal," in *Proceedings of IEEE International Conference on Anti-Counterfeiting Security and Identification in Communication (ICASID), Chengdu, China*, pp. 298–301, 2010.

[57] C. O'Driscoll, *Performance Analysis of the Parallel Acquisition of Weak GPS Signals*. PhD thesis, University College Cork, 2007.

[58] W. C. Lindsey and C. M. Chie, "A Survey of Digital Phase-Locked Loops," *Proceedings of the IEEE*, vol. 69, no. 4, pp. 410–431, 1981.

[59] G.-C. Hsieh and J. C. Hung, "Phase-Locked Loop Techniques. A Survey," *IEEE Transactions on Industrial Electronics*, vol. 43, no. 6, pp. 609–615, 1996.

[60] F. M. Gardner, *Phaselock Techniques*. John Wiley & Sons, 2005.

[61] U. Mengali and A. D'Andrea, *Synchronization Techniques for Digital Receivers*. Plenum Press, 1997.

[62] B. W. Parkinson and J. J. Spilker, *Global Positioning System: Theory and Applications Volume I*. Applied Mechanics Reviews American Institute of Astronautics, 1996.

[63] Y. J. Morton, van Diggelen, F. S. Jr, P. J. J., S. B. W., Lo, and G. E. Gao, *Position, Navigation, and Timing Technologies in the 21st Century: Integrated Satellite Navigation, Sensor Systems, and Civil Applications*. John Wiley & Sons, 2021.

[64] The US Government, Global Positioning Systems Directorate, *IS-GPS-200*, j ed., 2018.

[65] I. Fernández-Hernández, T. Senni, D. Borio, and G. Vecchione, "High-Parity Vertical Reed-Solomon Codes for Long GNSS High-Accuracy Messages," *Navigation: Journal of the Institute of Navigation*, vol. 67, no. 2, pp. 365–378, 2020.

[66] M. Miya, S. Fujita, Y. Sato, K. Ota, R. Hirokawa, and J. Takiguchi, "Centimeter Level Augmentation Service (clas) in Japanese Quasi-zenith Satellite System, Its User Interface, Detailed Design, and Plan," in *Proceedings of the 29th International Technical Meeting of the Satellite Division of the Institute of Navigation (ION GNSS+ 2016), Portland, OR*, pp. 2864–2869, 2016.

[67] A. Viterbi, "Error Bounds for Convolutional Codes and an Asymptotically Optimum Decoding Algorithm," *IEEE Transactions on Information Theory*, vol. 13, no. 2, pp. 260–269, 1967.

[68] G. D. Forney, "The Viterbi Algorithm," *Proceedings of the IEEE*, vol. 61, no. 3, pp. 268–278, 1973.

[69] A. Goldsmith, *Wireless Communications*. Cambridge University Press, 2005.

[70] O. Montenbruck, E. Gill, and F. Lutze, "Satellite Orbits: Models, Methods, and Applications," *Applied Mechanics Reviews*, vol. 55, no. 2, pp. B27–B28, 2002.

[71] European Union, "European GNSS (Galileo) Open Service, Signal-in-Space Interface Control Document, Issue 2.0," January 2021.

[72] China Satellite Navigation Office, "BeiDou Navigation Satellite System, Signal in Space Interface Control Document, Open Service Signal B1I," February 2019. version 3.0.

[73] Russian Institute of Space Device Engineering, "Global Navigation Satellite System GLONASS, Interface Control Document, Navigational Radiosignal in Bands L1, L2," 2008. Edition 5.1.

[74] G. Tobías, A. García, C. García, M. Laínez, P. Navarro, and I. Rodríguez, "Advanced GNSS Algorithms and Services Based on Highly-Stable On-board Clocks," in *Proceedings of the 28th International Technical Meeting of the Satellite Division of the Institute of Navigation (ION GNSS+ 2015), Tampa, FL*, pp. 2801–2808, 2015.

[75] M. Paonni, M. Anghileri, T. Burger, L. Ries, S. Schlötzer, B. Schotsch, M. Ouedraogo, S. Damy, E. Chatre, M. Jeannot, J. Godet, and D. Hayes, "Improving the Performance of Galileo E1-OS by Optimizing the I/NAV Navigation Message," in *Proceedings of the 32nd International Technical Meeting of the Satellite Division of the Institute of Navigation (ION GNSS+ 2019), Miami, FL, USA*, pp. 1134–1146, 2019.

[76] I. Fernández-Hernández, V. Rijmen, G. Seco-Granados, J. Simon, I. Rodríguez, and J. D. Calle, "A Navigation Message Authentication Proposal for the Galileo Open Service," *Navigation: Journal of the Institute of Navigation*, vol. 63, no. 1, pp. 85–102, 2016.

[77] P. M. J. Kintner, T. Humphreys, and J. Hinks, "GNSS and Ionospheric Scintillation: How to Survive the Next Solar Maximum," *Inside GNSS*, pp. 22–30, July/August 2009.

[78] A. Aragon-Angel, M. Hernandez-Pajares, P. Defraigne, N. Bergeot, and R. Prieto-Cerdeira, "Modelling and Assessing Ionospheric Higher Order Terms for GNSS Signals," in *Proceedings of ION GNSS+, Tampa, FL*, pp. 3511–3524, September 2015.

[79] J. A. Klobuchar, "Ionospheric Time-Delay Algorithm for Single-Frequency GPS Users," *IEEE Transactions on Aerospace and Electronic Systems*, vol. 23, no. 3, pp. 325–331, 1987.

[80] European Commission, "European GNSS (Galileo) Open Service–Ionospheric Correction Algorithm for Galileo Single Frequency Users," Tech. Rep., September 2016.

[81] B. Nava, P. Coïsson, and S. M. Radicella, "A New Version of the NeQuick Ionosphere Electron Density Model," *Journal of Atmospheric and Solar-Terrestrial Physics*, vol. 70, no. 15, pp. 1856–1862, 2008.

[82] A. E. Niell, "Global Mapping Functions for the Atmosphere Delay at Radio Wavelengths," *Journal of Geophysical Research*, vol. B2, no. 101, pp. 3227–3246, 1996.

[83] P. Collins, R. Langley, and J. LaMance, "Limiting Factors in Tropospheric Propagation Delay Error Modelling for GPS Airborne Navigation," in *Proceedings of the 52nd Annual Meeting of The Institute of Navigation*, vol. 3, 1996.

[84] RTCA, "Minimum Operational Performance Standards for Global Positioning System/Wide Area Augmentation System Airborne Equipment," *RTCA/DO-229D*, 2006.

[85] T. Hobiger and N. Jakowski, "Atmospheric Signal Propagation," in *Springer Handbook of Global Navigation Satellite Systems*, Springer, pp. 165–193, 2017.

[86] I. Fernández-Hernández, T. Walter, K. Alexander, B. Clark, E. Châtre, C. Hegarty, M. Appel, and M. Meurer, "Increasing International Civil Aviation Resilience: A Proposal for Nomenclature, Categorization and Treatment of New Interference Threats," in *Proceedings ION International Technical Meeting (ITM), Reston, VA*, pp. 389–407, January 2019.

[87] F. Dovis, *GNSS Interference Threats and Countermeasures*. Artech House, 2015.

[88] K. Borre and G. Strang, *Algorithms for Global Positioning*. Cambridge University Press, 2012.

[89] M. S. Grewal, "Kalman Filtering," *International Encyclopedia of Statistical Science*. Springer, pp. 705–708, 2011.

[90] R. Hatch, "The Synergism of GPS Code and Carrier Measurements," in *Proceedings of International Geodetic Symposium on Satellite Doppler Positioning, Las Cruces, NM*, pp. 1213–1231, 1983.

[91] Defense Mapping Agency, "World Geodetic System 1984: Its Definition and Relationships with Local Geodetic Systems," Tech. Rep., 2014.

[92] B. Hofmann-Wellenhof, H. Lichtenegger, and E. Wasle, *GNSS–Global Navigation Satellite Systems: GPS, GLONASS, Galileo, and More.* Springer Science & Business Media, 2007.

[93] F. van Diggelen, "Update: GPS accuracy: Lies, Damn Lies, and Statistics," *GPS WORLD*, p. January, 2007.

[94] GPS Navstar Joint Program Office, "Global Positioning System Standard Positioning Service Performance Standard," 5th Edition, April 2020.

[95] European Commission, "Galileo Open Service - Service Definition Document", Issue 1.2, November 2021.

[96] R. G. Brown, "A Baseline GPS RAIM Scheme and a Note on the Equivalence of Three RAIM Methods," *Navigation*, vol. 39, no. 3, pp. 301–316, 1992.

[97] P. Enge, T. Walter, S. Pullen, C. Kee, Y.-C. Chao, and Y.-J. Tsai, "Wide Area Augmentation of the Global Positioning System," in *Proceedings of the IEEE*, vol. 84, no. 8, pp. 1063–1088, 1996.

[98] S. Pullen, T. Walter, and P. Enge, "1.5 System Overview, Recent Developments, and Future Outlook for WAAS and LAAS," 2002.

[99] J. Blanch, T. Walter, P. Enge, S. Wallner, F. Amarillo Fernandez, R. Dellago, R. Ioannides, I. Fernandez Hernandez, B. Belabbas, A. Spletter, *et al.*, "Critical Elements for a Multi-constellation Advanced Raim," *Navigation: Journal of the Institute of Navigation*, vol. 60, no. 1, pp. 53–69, 2013.

2 GPS L1 C/A Receiver Processing

Kai Borre, M. Zahidul H. Bhuiyan, Stefan Söderholm, Heidi Kuusniemi, Ignacio Fernández-Hernández and José A. López-Salcedo

2.1 Introduction

The GPS started to be developed by the US Department of Defense (DoD) in the 1970s, but it was not until 1995 that it was declared fully operational [1]. At that time it was the very first operational GNSS, and still today it is the one most widely adopted. The system design was primarily driven by the military need of having a positioning system that could operate anytime, anywhere on Earth. The system provides a Precise Positioning Service (PPS) to authorized users through the so-called precision (P) signal, which is broadcast simultaneously at L1 and L2 frequencies. This allows ionospheric effects to be compensated at the receiver side by implementing dual-frequency corrections. The system also provides a Standard Positioning Service (SPS) to civil users, which was originally broadcast in the L1 frequency. Three more civil signals were added in subsequent generations of modernized GPS satellites: the L2C signal broadcast at L2 by block IIR-M satellites in 2005 [2]; the L5 signal added to block IIF satellites in 2010 [3] and finally, the L1C signal broadcast at L1 by the most recent set of modernized GPS satellites, block III, starting in 2018 [4].

Among all signals being broadcast by GPS satellites, this chapter will focus only on the L1 Coarse / Acquisition (C/A) signal for several reasons. First, it is a simple signal that facilitates the understanding of the GNSS signal processing principles described in Chapter 1. Second, it was for some decades the only operational GNSS signal available to civil users. This led to many years of leadership and competitive advantage with respect to the rest of GNSS signals that years, which only decades later have also become operational, such as GLONASS and Galileo. Finally, its properties, in particular its spreading codes, have proven near optimal for consumer applications [5]. It is for these reasons that all of the existing GNSS-based devices available today share the common feature of being all compatible, at least, with the GPS L1 C/A signal. This makes the study of the GPS L1 C/A signal and the related receiver processing a good and long-range starting point for any GNSS newcomer. In the following chapters, we will often refer to the GPS L1 C/A case either to make a parallelism with what conventional GPS receivers do or to highlight some differences worth being pointed out. In any case, understanding the rationale behind GPS L1 C/A signal processing becomes a cornerstone for understanding the rest of GNSS and the operation of the software companion of the present book.

2.2 GPS L1 C/A Signal Characteristics

2.2.1 Signal Structure

The structure of the C/A signal was driven to overcome the difficulties in processing the precision signal, which is an SS signal based on a very long spreading code sequence that repeats itself within a one-week period. Such a long period of time poses many difficulties for its acquisition, since one full week is needed to observe the whole code. This means that a receiver that is not already synchronized does not know what portion of the precision signal has to generate in order to correlate with the received signal. The C/A signal was designed to circumvent this limitation by providing a short and easy-to-acquire signal that could assist the subsequent acquisition of the Precision code. The latter, referred to as the P-code, is publicly available in the GPS ICD [2] so it has the additional drawback of potentially being spoofed by a third party, thus compromising the provision of the precision service. To prevent this weakness, an encrypted version of the P-code was made available to authorized users only. This alternative code is referred to as the Y-code and can be transmitted instead of the P-code whenever the anti-spoof (AS) mode of GPS is activated, a situation that has been taking place since 1994. Because of this dual behavior, the resulting signal is referred to as the P(Y) signal. It is broadcast on both L1 and L2 frequencies so that ionospheric effects can be mitigated using dual-frequency corrections. Finally, it is interesting to note that even though access to the P(Y) signal is only possible to authorized users, some benefits of this signal can be exploited anyway without the need of explicitly acquiring the signal. These techniques are commonly known as *codeless*, *semi-codeless* or *quasi-codeless* techniques [6] [7, Ch. 4], and they are based on correlating the L1 and L2 signals with one another.

In summary, the C/A signal originally came up as an auxiliary signal to assist the acquisition of the P(Y) signal. Its nominal power is also 3 dB higher than the one for the P(Y) signal (−158.5 versus −161.5 dBW), which further helps in this task. The C/A spreading code, similarly to the P-code, is available in the GPS ICD and thus can be used by any user.

Both the P(Y) and C/A signals are accommodated into the same L1 carrier by modulating both signals in phase quadrature with each other, similarly to Eq. (1.2). According to the GPS ICD, the P(Y) signal is placed on the in-phase component, while the C/A signal is placed on the quadrature component. This is actually a mere convention because the received signal will always be affected by an unknown phase rotation that will cause these signals to appear somewhere in between the in-phase and the quadrature components. Counteracting this effect and properly placing the in-phase and quadrature components is a task done by the tracking stage of the receiver. Since the C/A signal is the one of interest for civil users and the P(Y) signal is often disregarded, it is customary to express the signal model of the GPS L1 signal in such a way that the C/A signal appears in the in-phase component, instead. This leads to the following signal model for the composite L1 signal[1]:

$$s_{L1}(t) = \sqrt{2P_{C/A}}\, s_{C/A}(t) \cos\left(2\pi F_{L1}t + \theta_{L1}\right) + \sqrt{2P_{P(Y)}}\, s_{P(Y)}(t) \sin\left(2\pi F_{L1}t + \theta_{L1}\right),$$

$$(2.1)$$

with RF power $P_{C/A}$ and $P_{P(Y)}$ for the C/A and P(Y) components.

The inner structure of the C/A signal component $s_{C/A}(t)$ follows the data-modulated SS signal model already introduced in Eqs. (1.7)–(1.9). A summary of the main signal parameters is shown in Table 2.1 following the same conventions defined in Section 1.3.1. A primary spreading code with length 1,023 chips and duration 1 ms is used, thus resulting in a chip rate of $R_c = 1.023$ MHz. The chips of the primary spreading code are shaped using a rectangular pulse[2] with the same duration as the chip period, given by $T_c = 1/R_c = 997.5$ ns $\approx 1\,\mu$s. One chip corresponds to a distance of 300.54 m when propagated at the speed of light.

The resulting stream of chip amplitude-modulated pulses becomes the spreading code signal in Eq. (1.9). No overlay or secondary code is present, meaning that the primary code is continuously repeated in the C/A signal. The navigation message is conveyed by BPSK-modulating the data bits at a rate of 50 bps, thus resulting in a bit duration of 20 ms. It is for this reason that the value $N_r = 20$ has been set in Table 2.1.

Table 2.1 Parameters of the general signal model in Eq. (1.13) particularized for the GPS L1 C/A signal.

		GPS L1 C/A
Multiplexing technique		CDMA
Center frequency		1575.42 MHz
Primary (spreading) code type	$\{v[m]\}_{m=0}^{N_c-1}$	Gold Code
Primary code length	N_c	1,023
Primary code duration	T_{code}	1 ms
Primary code chip rate	R_c	1.023 MHz
Primary code chip duration	$T_c = 1/R_c$	977.5 ns $\approx 1\,\mu$s $\equiv 300$ meters
Primary code chip pulse shape	$p_{T_c}(t)$	$\Pi\left(\frac{t-T_c/2}{T_c}\right)$
Secondary code type	$\{u[k]\}_{k=0}^{N_r-1}$	1 (constant)
Secondary code length	N_r	20
Data modulation		BPSK
Symbol rate	R_d	50 sps
Symbol duration	T_d	20 ms
Minimum receiver bandwidth requirement		2 MHz
Pilot channel available		No
Error detection method		Extended (32,26) Hamming code
Error correction method		None

[1] In this expression, the sin() P(Y) component appears added to the cos() component of the C/A. This is equivalent to placing the P(Y) as a cos() component and the C/A as a -sin(), as per our general signal model in Eq. (1.2).

[2] Filtered at 20.46-MHz bandwidth at the satellite transponder.

Within this period, one can therefore expect the repetition of 20 primary codes with the same amplitude, which allows a maximum coherent integration length of 20 ms. Beyond this time, the presence of data-modulating bits may cause the next primary codes to be sign-reversed with respect to ones in the previous bit period. This would cause the overall coherent integration to add destructively beyond one bit period, thus resulting in a significant degradation on the output signal-to-noise ratio, as discussed in Section 1.6.

2.2.2 Spreading Code

The signal transmitted from each GNSS satellite has its own pseudorandom spreading code, which for the case of GPS L1 C/A is based on the family of Gold codes, described by Robert Gold in [8]. Gold codes are obtained as the sum of two maximum-length or so-called m-sequences, which can easily be generated in hardware using a linear feedback shift register (LFSR). For an LFSR with m stages, there are a total of $N_c = 2^m - 1$ possible m-sequences that can be generated, all composed of N_c binary values referred to as *chips*. Gold codes in GPS L1 C/A signals use m-sequences with $m = 10$, resulting in spreading codes with length $N_c = 1,023$ chips.

As it was mentioned previously, spreading codes in GNSS are nearly orthogonal providing a very small residual in their cross-correlation and outside of the main lag of their autocorrelation. One of the key features of m-sequences is that such residual is constant and bounded, which allows having a prior knowledge on the potential degradation that will be incurred. Actually, m-sequences are all composed of 2^{m-1} ones and $2^{m-1} - 1$ zeros, which are later on converted to ± 1 values to result in a zero-mean spreading sequence. Despite having a bounded cross-correlation, the drawback of m-sequences is that not all output sequences for a given m exhibit good-enough autocorrelation properties. This is the reason why just a small subset of m-sequences are selected for generating the Gold codes to be used by GPS L1 C/A, as described in [2]. For this subset, the cross-correlation of the resulting Gold codes is also bounded and has only three possible output values:

$$R_{v^{(p)}v^{(q)}}[k] = \sum_{n=0}^{N_c-1} v^{(p)}[m]v^{(q)}[m-k]_{N_c} = \begin{cases} -1 \\ -\alpha_m \\ \alpha_m - 2 \end{cases} \tag{2.2}$$

with $\alpha_m = 1 + 2^{\lfloor \frac{m+2}{2} \rfloor}$ and m the number of stages of the constituent m-sequence. For GPS L1 C/A we have $m = 10$ and thus $\alpha_m = 65$. Regarding the autocorrelation of Gold codes, their output values are also discrete and given by

$$R_{v^{(p)}}[k] = \sum_{n=0}^{N_c-1} v^{(p)}[m]v^{(p)}[m-k]_{N_c} = \begin{cases} N_c & k = 0 \\ -1 & k \neq 0 \\ -\alpha_m & k \neq 0 \\ \alpha_m - 2 & k \neq 0 \end{cases}. \tag{2.3}$$

The power ratio between the maximum peak of the autocorrelation and the second largest peak (in magnitude) results in $20\log_{10}(1,023/65) \approx 24$ dB for GPS L1 C/A Gold codes.

2.2.3   Power Spectral Density

According to the general result in Eq. (1.35), the spectrum of a GNSS signal is determined by the product of three different contributions: the spectrum of the chip pulse, $|P(f)|^2$; the spectrum of the primary spreading code, $|V(f)|^2$; and the spectrum of the secondary code, $|U(f)|^2$. Regarding the GPS L1 C/A signal, its primary spreading code sequence $v[m]$ is long enough, so that it can be fairly approximated by a random sequence and thus by a constant spectrum, $|V(f)|^2 \approx 1$. The secondary code sequence $u[k]$ is actually constant, composed of $N_r = 20$ consecutive ones with a time separation given by the primary code period, T_{code}. The corresponding spectral representation is given by Eq. (1.36), which results a sinc shape spectrum having nulls at multiples of the symbol rate (i.e. $1/N_r T_{code} = 1/T_d = 50$ Hz). Replicas of this sinc spectrum appear at multiples of the symbol rate (i.e. every $1/T_{code} = 1$ kHz) in the frequency domain. Finally, the spectrum of the rectangular chip pulse $|P(f)|^2$ is given by the well-known sinc function in Eq. (1.25) with zeros at multiples of the chip rate (i.e. $1/T_c = 1.023$ MHz).

So putting all these terms together, we can see that the envelope of the overall GPS L1 C/A PSD is dominated by the chip pulse spectrum, which is the contribution having the widest spectrum. It has a sinc shape with nulls at multiples of 1.023 MHz, and it actually coincides with the simple PSD for GNSS signals introduced in Eq. (1.26) for arbitrarily long random codes. However, while the spreading code of GPS L1 C/A can fairly be assumed to be long enough, the secondary code it is not. Actually, it is not even a random code. It is a repetition of $N_r = 20$ ones. And because of this repetitive structure of the secondary code, the sinc-shaped envelope of the GPS L1 C/A PSD is filled with a periodic repetition of extremely narrow sinc-like shapes. They have nulls at multiples of just $1/T_d = 50$ Hz, and they are repeated every $1/T_{code} = 1$ kHz. The resulting PSD is then given by

$$S_x(f) = T_c \text{sinc}^2 (fT_c) \frac{1}{N_r} \left| \frac{\sin(\pi f N_r T_{code})}{\sin(\pi f T_{code})} \right|^2. \tag{2.4}$$

Note that since the code rate $1/T_{code} = 1$ kHz is much smaller than the chip rate $1/T_c = 1.023$ MHz, within one lobe of the sinc-shaped envelope of the PSD there are thousand narrow sinc waveforms due to the spectral content of the all-ones secondary code. In practice, these narrow sincs are often ignored and the representation of the PSD for GPS L1 C/A is roughly approximated by its overall envelope, which is given by the sinc-shaped spectrum of the rectangular chip pulse. This leads to the well-known expression already introduced in Eq. (1.26):

$$S_x(f) \approx T_c \text{sinc}^2 (fT_c), \tag{2.5}$$

which is actually the result shown in Figure 2.1.

As it can be seen, most of the GPS L1 C/A power is concentrated at the central main lobe of ~2 MHz bandwidth, which corresponds to twice the chip rate. The minimum received signal power is guaranteed to be at least -158.5 dBW, which is actually below the thermal noise floor of a conventional receiver. This leads to the situation shown in Figure 2.1 where the PSD of the GPS L1 C/A is actually below the noise floor and

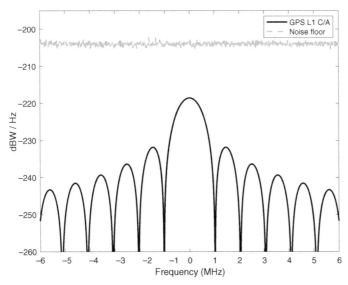

Figure 2.1 Depiction of the power spectral density for the GPS L1 C/A signal, received at a minimum power level of −158.5 dBW, and thermal noise. The center frequency is 1575.42 MHz.

thus cannot be perceived at first glance when observing the overall PSD. In spread spectrum terminology, this feature is often referred to as a signal with LPI.

2.3 Acquisition

The acquisition of GPS L1 C/A signals follows the same general principle already described in Section 1.6, which involves the correlation of the received signal with a local replica of the satellite under analysis. The correlation must be repeated multiple times following a time–frequency search in order to determine the tentative values for the code delay and frequency shift that are closer to the true values in the received signal. This allows the local replica to be aligned with the incoming signal and thus carry out the despreading operation successfully, obtaining a correlation peak well above the noise floor.

Among the different acquisition strategies, parallel acquisition described in Section 1.6.3 is the one most widely adopted in GNSS software receiver implementations such as the multi-GNSS software accompanying this book. The key feature of parallel acquisition is that the correlation between the received signal and the local replica is implemented in the frequency domain by making extensive use of the FFT. First, the discrete-time replica of the GPS L1 C/A spreading code signal is generated at the software receiver. The time-domain samples are converted into the frequency domain by means of the FFT (see Section 9.2.9 for further considerations on the FFT implementation), and then the FFT output is complex conjugated as described in Section 1.6.3. Next, a snapshot of samples from the received signal is taken, the frequency shift is tentatively removed by multiplying with a local carrier at tentative frequency v' and the result is transformed as well into the frequency domain using the FFT. The complex conjugate of the local replica FFT and the FFT of the frequency shift-compensated received samples are both multiplied. The result is converted back into the time domain using the IFFT, thus providing the correlation samples $R_C(\tau', v')$ for $\tau' = 0, 1, \ldots, N_{scode} - 1$ in Eq. (1.65). The resulting samples are stored into the column of the acquisition matrix schematically shown in Figure 1.26(b), which can be non-coherently accumulated using the result obtained from repeating the steps above with consecutive snapshots of received samples. This would lead to the noncoherent correlation in $R_{NC}(\tau', v')$ in Eq. (1.75), which helps in improving the receiver sensitivity when limitations in the coherent integration time are faced.

As described in the search step of Section 1.6, the resolution of the time-domain samples is typically half the one-sided width of the correlation peak. For GPS L1 C/A, this peak width is actually one chip, so the time resolution should then be at least half chip, $\Delta\tau_t'/T_c = 0.5$ chip. This means that the sampling frequency should be at least twice the chip rate. The maximum time delay error would be half of the bin width, thus leading to $\tau_{t,\epsilon,\max}/T_c = 0.25$ chip. According to Figure 1.25(a), this incurs in ~2 dB loss when determining the maximum correlation peak at the acquisition stage.

Regarding the resolution in the frequency domain, the frequency bin width $\Delta v'$ is linked to the coherent integration time as already indicated in Eq. (1.86). When the coherent integration time is set to one code period, as often done in practice, we have $T = T_{code} = 1$ ms in GPS L1 C/A thus leading to a frequency bin width $\Delta v_t' = 500$ Hz. The maximum frequency shift error that is incurred is half of the frequency bin width, which in this example would become $v_{t,\epsilon,\max} = 250$ Hz. This incurs in ~1 dB loss as shown in Figure 1.25(b), on top of the ~2 dB loss already incurred from the time delay search. In total, there would be a maximum of ~3 dB loss at the time–frequency search of the acquisition stage using the aforementioned configuration for GPS L1 C/A (i.e. $\Delta\tau_t'/T_c = 0.5$ chip and $\Delta v_t' = 500$ Hz). This loss accounts for the fact that neither the time delay nor the frequency shift is perfectly compensated at this stage, so some errors are inevitably incurred. These losses can be further reduced by narrowing the time and frequency bin widths at the time–frequency search. This of course comes at the expense of a higher computational load because we would be going through the same search space using smaller steps.

We will now show an example of GPS L1 C/A signal acquisition using the multi-GNSS software receiver described in Chapters 7 and 12. The skyplot for this acquisition data set is shown in Figure 2.2, the acquisition results are shown in Figure 2.3 and the acquisition values are mentioned in Table 2.2. A total of $T = 5$ ms coherent integration and $N_I = 5$ blocks of noncoherent integration are configured. The software automatically sets the resolution of the frequency search to the recommended value $\Delta v_t' = 1/2T$ already discussed in Eq. (1.86), which for this example amounts at $\Delta v_t' = 100$ Hz. The frequency search space then spans along a set of $N_{v'} = 121$ frequency bins according to Eq. (1.89), using the maximum frequency shift of 6000 Hz considered in Table 2.2, which is due to the satellite movement and the clock offset of the RF front end used to gather these samples.

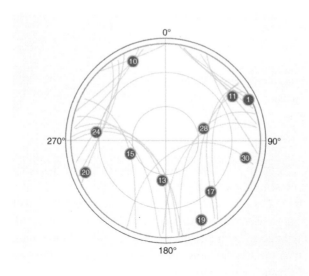

Figure 2.2 Skyplot of GPS satellites on February 21, 2018, 13:10 UTC time at Finnish Geospatial Research Institute with an elevation cutoff angle of 10°, corresponding to the results in Figure 2.3. This figure is obtained with Trimble's online GNSS planning tool.

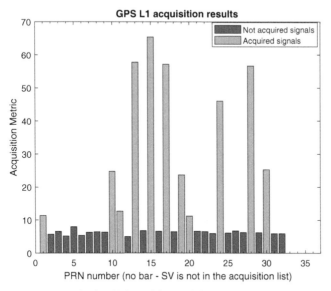

Figure 2.3 Results for GPS L1 C/A acquisition using the companion multi-GNSS software receiver.

The total integration time amounts to $T_{tot} = 25$ ms, which is larger than the typical 10 or 20 ms used in a conventional receiver in order to ensure that all the visible GPS satellites were acquired successfully. The integration parameters and the acquisition threshold were set such that low elevation satellites could also be acquired for tracking. It can be seen from Figure 2.3 that 11 GPS satellites are acquired with PRN numbers {1, 10, 11, 13, 15, 17, 19, 20, 24, 28 and 30}. The results show the value of the

Table 2.2 GPS signal acquisition parameter settings used to obtain the results presented herein.

Parameter	Value	Unit
Coherent integration time	5	ms
Noncoherent block number	5	–
Frequency search range	[± 6000]	Hz
Acquisition threshold	9	–

acquisition metric $R_{NC}(\tau', \nu')$ that is compared with a threshold to determine whether a given satellite is present or not, as indicated in Eq. (1.90).

Looking at both Figures 2.2 and 2.3, one can see that the satellites with the highest elevation angle are the ones received with the highest signal strength, as those at lower elevations are more affected by impairments and may have a lower receiver antenna gain. They have thus the largest acquisition metric. These are satellites with PRN numbers $13, 15, 17, 28$ and 24 to a lower extent.

2.4   Tracking

The implementation of the tracking stage for GPS L1 C/A must bear in mind that BPSK data-modulated symbols are transmitted in this signal component, so noncoherent carrier and code discriminators must be used, as discussed in Section 1.7. An example of the tracking of this signal is shown in Figure 2.4 for the GPS satellite with PRN 17, and for a piece of 100 seconds of the data set already used for Figures 2.2 and 2.3, which is also used later in this chapter for position computation. The results have been obtained with the companion multi-GNSS software receiver, which first starts tracking with an FLL to cope with the large dynamics due to the lack of initial synchronization between the received signal and the local carrier replica. Once the carrier is stabilized and the error is reasonably small, it switches to a PLL in order to precisely track the carrier phase and compensate for any phase offset. Both the FLL and PLL discriminators are based on the atan function, which removes the 180° phase jumps due to the BPSK symbols in the GPS L1 C/A signal. For the same reason, code tracking must be implemented using a DLL with a noncoherent discriminator, in this case the NNEML presented in Section 1.7. The NNEML DLL discriminator is used with a correlator spacing of $\delta = 0.1$ chips between early and prompt correlator. The DLL loop bandwidth is set to 1 Hz, while the PLL loop bandwidth is set to 15 Hz.[3]

The main GPS L1 C/A signal tracking parameters used in the experiment are presented in Table 2.3.[4]

---

[3] Note that the multi-GNSS FGI-GSRx is designed in such a way that the user/developer could flexibly configure each GNSS at different receiver processing stages depending on the requirements set by the signal characteristics of that individual system.

[4] PULL_IN, COARSE_TRACKING and FINE_TRACKING regions are defined in Chapter 7.

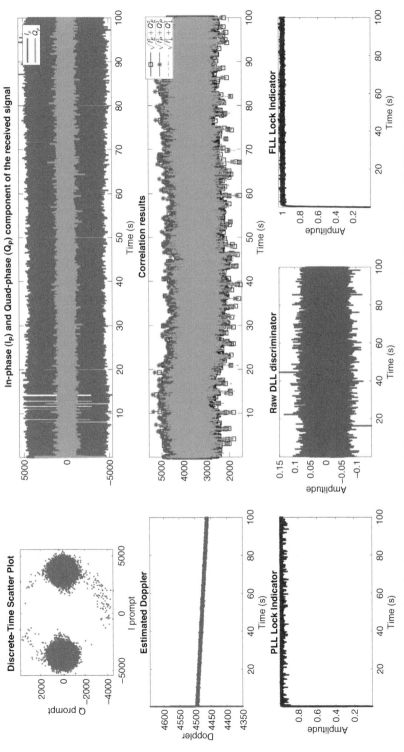

Figure 2.4 Tracking results provided by the companion multi-GNSS software receiver when tracking the GPS satellite with PRN 17 from the data set under analysis.

Table 2.3 GPS signal tracking parameter settings used to obtain the results presented herein.

Parameter	Value	Unit
Correlator spacing	0.1	chips
Number of tracking states	3	–
Tracking states	[PULL_IN; COARSE_TRACKING; FINE_TRACKING]	–
DLL loop filter bandwidth	[1; 1; 1]	Hz
FLL loop filter bandwidth	[200; 100; 5]	Hz
PLL loop filter bandwidth	[15; 15; 10]	Hz

The results in Figure 2.4 show many output plots corresponding to different indicators that help in understanding how the tracking stage is evolving. First of all, the plot on the upper left hand side is showing the complex representation of the prompt correlation output samples (I_P and Q_P) from the DLL.[5] Once the carrier phase is properly compensated, these points correspond to a BPSK constellation with two antipodal clouds of points lying over the real (in-phase) axis, where the positive and negative I_P imply a +1 or −1, respectively, in the navigation bit being transmitted. When the carrier phase is not perfectly aligned, especially at the beginning of the tracking phase, we can see lower I_P and higher Q_P absolute values, which is observed as a *phase rotation*.

The plot on the upper right-hand side represents the time evolution of the real (in-phase) and imaginary (quadrature) components of the DLL prompt correlator output (I_P and Q_P). As it can be seen, most of the power lies on the in-phase component, in line with what is observed in the aforementioned constellation plot. The quadrature component just contains the contribution due to the thermal noise and becomes stable indicating that the constellation has been properly aligned and the phase offset has been properly compensated almost from the very beginning of the observation interval.

The middle plots in Figure 2.4 show the estimated Doppler (left), in hertz, and the power at the DLL correlation results for the early, late and prompt correlators. Despite the noise, we can see that the prompt correlator, as expected, provides a slightly higher correlation output than that from the early and late correlators.

The bottom plots show the PLL lock indicator (left), the DLL discriminator output (middle) and the FLL lock indicator (right). Both PLL and FLL lock indicators provide information on whether the carrier has been properly compensated or not. This is done in the software receiver by measuring the power difference between the in-phase and the quadrature components of the prompt correlator output values (I_P and Q_P). The result is normalized so that when all power is on the in-phase component, the lock indicator outputs 1, while when the same power is on both the in-phase and the quadrature components (as it happens when the phase is not tracked and the constellation points move circularly in the complex plane), the lock indicator outputs 0.

---

[5] Also called *constellation diagram*, which represents PSK and quadrature amplitude-modulated signals, as the ones under study.

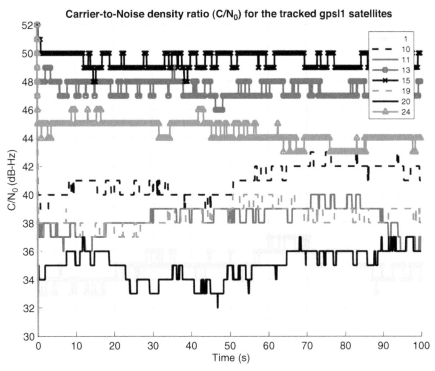

Figure 2.5 Estimated carrier-to-noise density ratio (C/N_0) of the tracked GPS satellites by the multi-GNSS software receiver for the data set under analysis.

Regarding the DLL discriminator output, it outputs 0 when the correlation peak is tracked, and when it deviates in any direction, the DLL adjusts itself to bring it back to 0. The raw discriminator result, that is without filtering, is what we observe in the bottom-middle plot.

In our example, the estimated C/N_0 via NWPR-based method for this data set is shown in Figure 2.5. As can be seen, the largest C/N_0 is obtained for satellites with PRN 13, 15, 17 and 28, which were the ones already providing the largest mark in the acquisition metric shown in Figure 2.3, and the ones with higher elevation with respect to the user's position. They benefit from potentially higher antenna gains, propagate through a thinner layer of the atmosphere, and thus suffer from less propagation losses compared to satellites with lower elevation angles, and also may suffer less multipath.

2.5 Computation of Position and Time

As soon as the signals are tracked properly, the C/A code and carrier wave can be removed from the signal, leaving the navigation data bits. The value of a data bit is found by integration over a navigation bit period of $T_d = 20$ ms. Up to this point, the receiver is able to decode bits from the prompt correlator output and compute time-delay measurements from the code delay of the DLL. However, it is not yet able to compute the actual range between the satellite and the receiver. The reason is the code delay integer ambiguity: The DLL outputs a code delay measurement that is a point in time within a 1 ms GPS L1 C/A code. In order to calculate the full pseudorange, the receiver needs to know the integer number of 1-ms codes to be added to the code delay measurement, as described in Section 1.8.1. In this way, the receiver can determine the full pseudorange, by comparing the GPS time with the receiver time, with some subtleties described in Section 1.10. Once the receiver is synchronized with the signal, it can also recognize the GPS L1 C/A message structure and decode the satellite positions and clock offsets, required to compute its own position. Finally, the receiver position is computed from the estimated pseudoranges, as described in Section 1.10.

The GPST is the reference time provided by the GPS constellation. The GPS signals are referred to GPST, and therefore, the receiver position calculated on GPS signals is referred to an instant in GPST too. GPST counts in weeks and seconds of week starting on January 6, 1980. Each week has its own number. Time within a week is counted in seconds, starting from midnight between Saturday and Sunday (day 1 of the week). Regarding the UTC, it goes at a different rate that is connected to the actual speed of the rotation of the Earth. This leads to what we call *leap seconds*, which are additional seconds added to UTC (i.e. UTC counts one additional second) in June 30 or December 31 of some years, depending on the Earth's rotation. At the time of writing this section (first half of 2020), 18 seconds have to be added to UTC in order to get GPST. At the time GPS was declared operational (July 1995), there were only 8 leap seconds. Regarding the coordinate system, GPS uses WGS84 (World Geodetic System 1984) ECEF reference frame.

The position solution was computed at 10 Hz rate and according to other parameters presented in Table 2.4. Finally, the position error variations obtained with the multi-GNSS software receiver are shown in Figure 2.6 in ENU frame for GPS-only position fix. The corresponding position error statistics are presented in Table 2.5. The error statistics were computed for a standalone code-based position solution. The broadcast Klobuchar ionospheric model parameters are utilized to calculate the ionospheric corrections. Note that, as the Klobuchar parameters may or may not be transmitted within the 100 seconds of data, the FGI-GSRx receiver utilizes some predefined "default" parameter values for computing the ionospheric error. The user would prefer either to use the "default" values or to obtain those values from any other external sources (e.g., from RINEX navigation file of any GNSS station with matching time duration).

Table 2.4 GPS L1 C/A navigation parameter settings used to obtain the results presented herein.

Parameter	Value	Unit
Navigation solution period	100	ms
Number of smoothing samples	10	–
Elevation mask	5	degree
C/N_0 mask	30	dB-Hz

Table 2.5 Position error statistics with respect to true position.

GNSS Signal	Horizontal				Vertical				3D
	RMS (m)	Mean (m)	Max (m)	Mean HDOP	RMS (m)	Mean (m)	Max (m)	Mean VDOP	RMS (m)
GPS	1.76	1.43	4.34	1.17	1.61	1.30	4.63	0.91	2.38

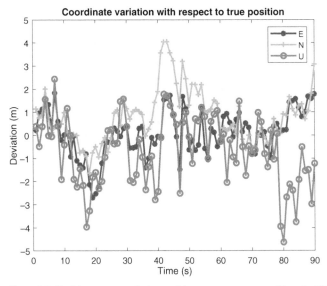

Figure 2.6 Position error variations with respect to true position, in ENU frame, obtained by the multi-GNSS software receiver for the data set under analysis.

If the Klobuchar ionospheric coefficients are not properly obtained and assigned, the receiver's position accuracy will be compromised. The position error statistics were computed with respect to the true known position. The 3D RMS error for GPS L1 is 2.38 m.

2.6 Front End and Other Practical Information

In this final section of this chapter, we provide practical details on the front end used, the data collection, and the MATLAB configuration and functions used to reproduce the results. The same approach is followed in the following chapters.

2.6.1 Front End

The *Stereo* dual-frequency front end from NSL [9] is used herein to capture the live GNSS signals that are processed by the companion multi-GNSS software receiver. The Stereo front end actually embeds two different front ends. The first one is based on the Maxim MAX2769B chipset, and it can process signals between 1550 MHz

Table 2.6 Configuration of the two front ends included in the NSL Stereo v2 for gathering the GNSS samples used in the examples of this book.

	Front end 1	Front end 2
Chipset	MAX2769B	MAX2112
Bandwidth (MHz)	4.2	10
Sampling frequency (MHz)	26	26
Intermediate frequency (MHz)	6.39	0
Number of quantization bits	2	3
Down-converted signal type	Real	Complex

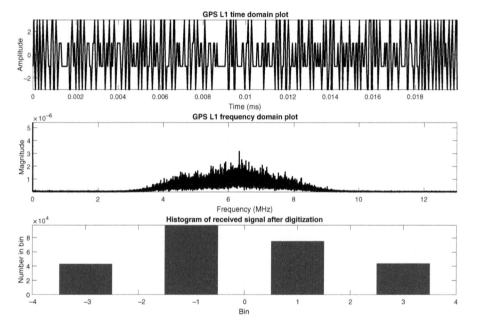

Figure 2.7 GPS L1 time-domain plot (up), signal spectrum (middle) and bin distribution of the digitized IF samples (down) for the data set under analysis.

and 1610 MHz. The second one is based on the Maxim MAX2112 chipset that can be tuned between 900 MHz and 2000 MHz to receive all GNSS frequencies in the L band, and it implements direct conversion to IQ. For the results shown in this chapter, only the MAX2769B is used to receive GPS L1 C/A signals at 1575.42 MHz with the configuration indicated in Table 2.6.

Figure 2.7 shows some of the features of the time-domain signal obtained at the front-end output. The upper plot shows the time-domain samples where it can be seen that the amplitude is discretized to four values only, in line with the use of 2 bits for the signal quantization at the ADC. This can also be seen in the bottom plot of Figure 2.7 where the histogram of the time-domain samples is shown. Since the GNSS signal is actually below the noise floor, the samples that one can see in the upper plot are mostly dominated by the thermal noise. As such, the time-domain samples are Gaussian distributed and this is reflected in the histogram shown in the bottom plot of Figure 2.7, which resembles a four-sample version of Gaussian distribution. The middle plot in Figure 2.7 shows the spectrum of the signal at the front-end output. As can be seen, the spectrum is centered at the intermediate frequency of 6.39 MHz and it is shaped by the frequency response of the front-end bandpass filter whose (two-sided) bandwidth, according to the MAX 2769B specifications, is about 4.2 MHz.

2.6.2   Data Collection

The GNSS data processed herein were collected on February 21, 2018, at around 13:12 UTC time at a static position with a roof antenna in Finnish Geospatial Research Institute, Kirkkonummi, Finland. The data were collected for about 100 seconds. The skyplot for GPS constellation at the time of data collection was shown in Figure 2.2. There were ten GPS satellites (i.e. PRNs 10, 11, 13, 15, 17, 19, 20, 24, 28 and 30) visible at the moment of data collection. The FGI-GSRx receiver was able to acquire, track and compute a navigation solution with all the visible GPS satellites.

2.6.3   MATLAB Configuration and Functions

Here, we will discuss how to utilize the supplied FGI-GSRx GPS MATLAB software receiver in order to obtain the results presented in this chapter. For the ease of understanding, the users of FGI-GSRx Multi-GNSS software receiver are provided with parameter files for each GNSS constellation (i.e. for each GNSS chapter) separately. For example, the parameter file name for GPS L1 receiver is "default_param_GPSL1_Chapter2.txt." The user needs to pass this parameter file with the main function in the MATLAB command prompt as follows:

> > gsrx("default_param_GPSL1_Chapter2.txt")

It can be seen in the parameter file that the enabled GNSS signal is chosen as "gpsl1" in this particular case. The user has options to offer two data sources: (i) MATLAB *.mat data source and (ii) Raw IQ data sources collected from a specific GNSS front end. In case of (i), the FGI-GSRx will check whether the GNSS acquisition and tracking were carried out at an earlier instance and the FGI-GSRx processed data were saved in the specified *.mat file. In that case, FGI-GSRx skips carrying out acquisition and tracking again and it directly attempts to decode the navigation message and offers navigation solution with the already processed tracked data. In the case of (ii), the FGI-GSRx will process the raw IQ data provided in the specified file name, and the user must have to specify a minimum set of IQ data configuration parameters in the parameter file. An example of such configuration parameters is given in Table 2.6. In order for the receiver to work properly, the user should take extra care to specify the following front-end parameters: sampling frequency in hertz, sample size in bits (i.e. number of bits in one sample), intermediate frequency, down-converted signal type (real or complex) and IQ data orientation (i.e. in case of a complex signal type, the IQ data can be as it is like I, Q or it can be swapped like Q, I).

Regarding acquisition, the user has the option to select the required number of coherent and noncoherent integration combination for a particular signal acquisition.

The received signal will be retrieved for correlation depending on the selection of these acquisition parameters (i.e. choice of coherent and noncoherent integration parameters). The user has the option to enable the variable "plotAcquisition" as "true" or "false" in the parameter setting file. If the variable is set as "true," the "plotAcquisition.m" function generates the acquisition metric for all the searched PRNs as can be seen in Figure 2.3. It also generates a two-dimensional signal correlation plot for all the successful PRNs that are above the acquisition threshold.

Regarding tracking, apart from Table 2.3, there are a good number of other relevant tracking parameters that can be directly found in the parameter file named "default_param_GPSL1_Chapter2.txt." The system-specific signal parameters are usually known from the Interface Control Document (ICD), and the users could find those parameters in the function called "getSystemParameters.m." Figures 2.4 and 2.5 can be obtained by calling the function "plotTracking.m." The user can choose to enable the variable "plotTracking" as "true" of "false" in the parameter setting file. If the variable is set as "true", the "plotTracking" function produces one plot for each specific tracking channel showing the performance of different key tracking parameters, as can be seen in Figure 2.4. It also generates a plot for each single constellation (in this case, GPS L1 C/A) presenting the signal strength in terms of C/N_0 as seen in Figure 2.5.

As shown in Table 2.4, the user can choose the satellite elevation mask and the C/N_0 mask. The navigation engine then decides on the satellites which pass these two criteria. Navigation solution was computed after every 100-ms interval with a 10-Hz navigation update rate. The navigation solution was further smoothed with 10 navigation solution samples per epoch. The user can also change this value depending on the requirement of the underlying GNSS application. Figure 2.6 was generated within the function "calcStatistics.m" for a standalone static GNSS user. Statistical calculations in terms of accuracy and availability can be found in this function.

References

[1] B. W. Parkinson, T. Stansell, R. Beard, and K. Gromov, "A History of Satellite Navigation," *Navigation: Journal of the Institute of Navigation*, vol. 1, no. 42, pp. 109–164, 1995.

[2] GPS Navstar Joint Program Office, "Navstar GPS Space Segment/Navigation User Segment Interfaces, IS-GPS-200," Revision M, May 21, 2021.

[3] GPS Navstar Joint Program Office, "Navstar GPS Space Segment/Navigation User Segment Interfaces, IS-GPS-705," Revision H, May 21, 2021.

[4] GPS Navstar Joint Program Office, "Navstar GPS Space Segment/Navigation User Segment Interfaces, IS-GPS-800," Revision H, May 21, 2021.

[5] F. Van Diggelen, "Who's Your Daddy? Why GPS Will Continue to Dominate Consumer GNSS," *Inside GNSS*, vol. 9, no. 2, pp. 30–41, 2014.

[6] J. Ashjaee and R. Lorenz, "Precise GPS Surveying after Y-code," in *Proceedings of the 5th International Technical Meeting of the Satellite Division of the Institute of Navigation (ION GPS 1992), Albuquerque, NM*, pp. 657–659, 1992.

[7] B. Hofmann-Wellenhof, H. Lichtenegger, and E. Wasle, *GNSS–Global Navigation Satellite Systems: GPS, GLONASS, Galileo, and More.* Springer Science & Business Media, 2007.

[8] R. Gold, "Optimal Binary Sequences for Spread Spectrum Multiplexing," *IEEE Transactions on Information Theory*, vol. 13, pp. 619–621, October 1967.

[9] Nottingham Scientific Limited (NSL), "Stereo Front-End Datasheet." www.nsl.eu.com/datasheets/stereo.pdf. Accessed: May 25, 2019.

3 GLONASS L1OF Receiver Processing

M. Zahidul H. Bhuiyan, Salomon Honkala, Stefan Söderholm
and Heidi Kuusniemi

3.1 Introduction

The GLONASS navigation satellite system began its development in the 1970s as a military system. The first satellite was launched in 1982, and a full constellation of 24 satellites was reached in 1995. For many years, it was not possible to maintain the full GLONASS satellite constellation as the older satellites could not be replaced due to lack of funding. In 2001, however, the Russian government initiated a 10-year program to rebuild the system and a full constellation of 24 satellites was restored in 2011 [1], which is completely operational nowadays. The first generation of GLONASS satellites (also known as Uragan) from blocks I, II and III was gradually replaced since 2003 by the second generation of GLONASS-M satellites. While GLONASS was originally designed as an FDMA system with each satellite broadcasting at a different carrier frequency [2], the latest GLONASS-M satellites incorporate a CDMA signal, thus bringing GLONASS to the same transmission scheme as all the rest of GNSS [3]. The additional CDMA signal is also present in the current GLONASS-K satellites being launched since 2012, and it will continue in the future GLONASS-K2 satellites to be launched in 2022 and the high-orbit GLONASS-B satellites planned for 2023 [4]. The latter are actually intended to improve the signal availability of users in urban areas, who typically experience difficulties in receiving GNSS signals for low elevation angles.

Similarly to GPS, GLONASS provides an open and a restricted positioning service. The open service was traditionally referred to as "ST," which stands for *Standardtnaya Tochnost* or Standard Precision, while the restricted service was referred to as "VT," which stands for *Visokaya Tochnost* or High Precision. Both services are provided through spread spectrum signals in quadrature one with each other, using a dedicated center frequency specific of each GLONASS satellite according to a FDMA transmission scheme. Both signals are simultaneously transmitted in the GLONASS L1 and the GLONASS L2 frequency bands, which differ from the conventional L1 and L2 bands used in GPS. In particular, the GLONASS L1 band covers the upper part of the L band with a center frequency of 1602.0 MHz, instead of the 1575.42 MHz used in GPS L1. In turn, the GLONASS L2 band is centered at 1246.0 MHz instead of the 1227.6 MHz used in GPS L2. The ST and VT nomenclature as referred to in [2] has evolved to the so-called open service and restricted service, respectively, with the advent of a new CDMA signal transmitted in the GLONASS L3 band centered at 1202.025 MHz, in-between of the GPS L5 and GPS L2 frequency bands. This new signal is described in a recently issued ICD [3] where the GLONASS signals appear to be renamed using a standardized code composed of the frequency band, the type of service and the type

Table 3.1 Summary of the available GLONASS signals as per [3]. For FDMA signals, the lower carrier frequency, frequency interval and higher carrier frequency are provided.

Service	Freq. band	Access type	Center freqs. (MHz)	Signal name
Open	L1	FDMA	1598.0625 : 0.5625 : 1605.375	L1OF
Open	L1	CDMA	1600.995	L1OC
Open	L2	FDMA	1242.9375 : 0.4375 : 1248.625	L2OF
Open	L2	CDMA	1248.060	L2OC
Open	L3	CDMA	1202.025	L3OC
Secure	L1	FDMA	1598.0625 : 0.5625 : 1605.375	L1SF
Secure	L1	CDMA	1600.995	L1SC
Secure	L2	FDMA	1242.9375 : 0.4375 : 1248.625	L2SF
Secure	L2	CDMA	1248.060	L2SC

of multiple access scheme. For instance, an FDMA signal conveying the secure service in the L1 band is referred to as L1SF, while a CDMA signal conveying the open service in the L3 band is referred to as L3OC. A summary of the different GLONASS signals is shown in Table 3.1.

In this chapter, we will concentrate on the L1OF (L1 open service FDMA) signal as an example of the FDMA-based scheme that makes GLONASS a different GNSS system, thus deserving some special attention. It is also the signal processed in the multi-GNSS software receiver accompanying this book.

3.2 GLONASS L1OF Signal Characteristics

3.2.1 Signal Structure

The GLONASS L1OF signal follows the same inner structure already described in Section 1.3.1 composed of a spreading or primary code, a secondary code and a navigation message. The difference compared to other GNSS systems is that each satellite is transmitting the same spreading code but using different carrier frequencies, as indicated in Table 3.1. The spreading code is chosen from an m-sequence of length $N_c = 511$ with a period of $T_{code} = 1$ ms and thus a chip rate of $R_c = 511$ kbps. The navigation message is transmitted using BPSK modulation at a bit rate of $R_d = 50$ bps so that within a bit period, there are $N_r = 20$ consecutive spreading codes.

Unlike GPS L1 C/A, these consecutive spreading codes are amplitude modulated by a secondary code also known as a *meander* sequence (as referred to in [2]), which is thus a sequence with length $N_r = 20$. The GLONASS meander is designed to impose a Manchester encoding over the bit period. That is, the bit period is divided into two halves, and each one is modulated by a signed-reversed amplitude with respect to the other. The result is equivalent to using a secondary code sequence of length N_r where a constant amplitude is repeated for the first $N_r/2$ values, and then, it is sign-reversed and repeated again for the remaining $N_r/2$ values of the sequence. In practice, this $\{+1, -1\}$ sequence introduces a BOC-like shaping at bit level that helps in reducing the spectral impact of the periodic repetition of the primary spreading code. The effect of such repetition would lead to the comb-like spectral shape due already introduced in Eq. (1.36) for GPS L1 C/A and appearing in its PSD in Eq. (2.4). In GLONASS, however, the introduction of the meander sequence helps in smoothing the PSD and improving the robustness in front of narrowband interferences, as it is always the case when using nonconstant secondary codes. The interested reader may find a discussion on the impact of secondary codes and Manchester encoding onto the PSD of GNSS signals in [5].

FDMA is certainly the key distinctive feature of GLONASS. It is based on assigning a different carrier frequency to each satellite transmitting in either the L1 or L2 frequency bands, which are divided into 14 channels whose central frequency is obtained from the following formulas:

$$f_{k,L1} = f_{0,L1} + k\Delta f_{L1}, \quad \Delta f_{L1} = 562.5 \text{ (kHz)} \tag{3.1}$$

$$f_{k,L2} = f_{0,L2} + k\Delta f_{L2}, \quad \Delta f_{L2} = 437.5 \text{ (kHz)}, \tag{3.2}$$

where $f_{0,L1}$ and $f_{0,L2}$ are the center frequencies for each band, Δf_{L1} and Δf_{L2} are the separation between the channels for L1 and L2, respectively, and k is the number of the channel. The channels are numbered for $k = \{-7 \ldots 6\}$. Each satellite is allocated a pair of carrier frequencies, one in L1 and another one in L2, according to its channel number. As there are 24 satellites but only 14 frequency channels, satellites in opposite orbital slots (antipodal satellites) can transmit on the same frequency channel thus allowing for some frequency reuse.

The minimum received signal power is specified for a satellite observed at a 5° elevation angle using a 3 dBi linearly polarized antenna, and it results in −161 dBW for the L1 band and −167 dBW for the L2 band. At higher elevation angles, the power level can be higher but is not expected to be more than −155.2 dBW [2].

3.2.2   Spreading Code

The main difference of the legacy GLONASS signal compared to other GNSS signals is the FDMA multiple access method. All GNSS except for GLONASS use CDMA for the inherent advantages it offers as compared to FDMA. In a CDMA system, all satellites transmit on the same carrier frequency but using different spreading codes that have such cross-correlation properties that the satellite signals can be distinguished from each other [6]. In FDMA, all satellites have their separate frequency channels and can therefore use the same spreading code.

In GPS L1 C/A, Gold codes were shown in previous chapters to be generated from the product of the output of two maximum length 10-stage shift registers driven by a 1.023-MHz clock, resulting in code sequences with 1,023 chips. Unlike GPS L1 C/A, the GLONASS spreading code is a maximum length 9-stage shift register sequence driven by a 0.511-MHz clock, which generates a code sequence with 511 chips as indicated in Table 3.2. The chips are BPSK modulated with rectangular pulses, and thus, the PSD for a single GLONASS signal has a similar shape to that of a GPS L1 C/A signal, but with a null-to-null main lobe of ∼1-MHz bandwidth compared to the ∼2-MHz bandwidth in GPS L1 C/A. Because of the FDMA scheme, a wider

Table 3.2 Parameters of the general signal model in Eq. (1.13) particularized for the GLONASS L1 signal and compared to those of the GPS L1 C/A signal.

		GPS L1 C/A	GLONASS L1OF
Multiplexing technique		CDMA	FDMA
Center frequency		1575.42 MHz	1602 MHz
Primary (spreading) code type	$\{v[m]\}_{m=0}^{N_c-1}$	Gold Code	m-sequence
Primary code length	N_c	1,023	511
Primary code duration	T_{code}	1 ms	1 ms
Primary code chip rate	R_c	1.023 MHz	0.511 MHz
Primary code chip duration	$T_c = 1/R_c$	0.977 μs	1.957 μs
Primary code chip pulse shape	$p_{T_c}(t)$	$\Pi\left(\frac{t-T_c/2}{T_c}\right)$	
Secondary code type	$\{u[k]\}_{k=0}^{N_r-1}$	1	$1, 0 \le k \le \frac{N_r}{2} - 1$ $-1, \frac{N_r}{2} \le k \le N_r - 1$
Secondary code length	N_r	20	20
Data modulation		BPSK	BPSK
Symbol rate	R_d	50 sps	50 sps
Symbol duration	T_d	20 ms	20 ms
Minimum receiver bandwidth requirement		~2 MHz	~8 MHz
Pilot channel available		No	No
Error detection method		Extended (32,26) Hamming code	Extended (85,81) Hamming code
Error correction method		None	None
Coordinate system		WGS84	PZ-90.11

bandwidth is needed in order to gather all satellite channels at the same time. The minimum receiver bandwidth is on the order of 8 MHz compared to the 2 MHz for a GPS L1 C/A receiver.

3.2.3   Comparison with GPS

GLONASS satellites describe almost-circular orbits at a height of 19,100 km, the lowest of all GNSS. The inclination angle of GLONASS satellites is 64.8° compared to 55° for GPS. For that reason, users in higher latitude areas may sometime observe better GLONASS-derived DOP than users of GPS. GLONASS transmits ephemeris information in the form a set of satellite location, velocity and acceleration vectors in three-dimensional Cartesian coordinates. Therefore, calculating GLONASS orbital positions is significantly different from other current GNSS, which all use modified Keplerian elliptical orbital parameters in their ephemeris messages. GLONASS does not offer any ionospheric correction parameters in the navigation message. However, a multiconstellation receiver can apply the ionospheric transmitted by other systems also to GLONASS measurements.

GLONASS time is tied to the UTC time reference. As mentioned in Chapter 2, UTC implements leap second corrections. Leap seconds are added to UTC usually every 1–1.5 years according to notification by the IERS. Each leap second adds one second to the offset between GLONASS time and GPS time. GPS, Galileo and BeiDou use continuous time scales, which do not implement leap seconds. There is also a three-hour difference from UTC in GLONASS time, which is another unique property of GLONASS. The other systems relate their time scales to UTC with a zero-hour offset, which corresponds to the prime meridian located at Greenwich, England.

A summarized comparison of some key characteristics between GPS L1 C/A and GLONASS L1 signals is presented in Table 3.2. With the aforementioned differences, the spreading gain is 3 dB lower for GLONASS compared to GPS, since GLONASS spreading codes are approximately half the length of GPS L1 C/A codes. Also, the bandwidth requirement has to cope with the broader bandwidth required by the 14 different carriers of the FDMA scheme. The coordinate reference frames used in GNSSs, such as PZ-90.11, WGS84 and China Geodetic Coordinate System 2000 (CGCS 2000), are different realizations of the same terrestrial reference system. The reference frames are therefore not identical. The transformation between PZ-90.11 and the WGS84 has been determined to have a linear origin shift of some centimeters [7].

3.3   Acquisition

Once tuned to the specific carrier frequency of the satellite of interest, the GLONASS signal become a single CDMA signal that can thus be processed in the same manner as any other GNSS signal. It is for this reason that the signal acquisition already described in Chapter 2 for GPS L1 C/A is applicable to GLONASS signals as well (once tuned to the proper channel). In the companion multi-GNSS software receiver, it is assumed a Doppler shift within the range of ±5 kHz due to the satellite-to-receiver velocity. Moreover, the receiver oscillator offset can also introduce some Doppler effect at a rate of ~ 1.5 kHz, or 1 ppm (part per million) at the L1 frequency band. Therefore, it is safe to consider an overall a maximum frequency shift of $v_{t,max} = 6$ kHz from the down-converted IF for a static receiver, noting that each satellite has a different carrier frequency.

Regarding the rest of acquisition parameters, it has to be taken into account that a GLONASS receiver usually requires some extra integration time for the acquisition process as compared to GPS for similar performance. This is because the spreading code length for GLONASS is half of the code length of GPS. Selecting longer coherent and noncoherent integration time during acquisition can improve acquisition sensitivity. In GPS L1 C/A, it was possible to set a coherent integration time of $T = 20$ ms, but in GLONASS, one should be cautious because there is a possibility of sign transition every 10 ms due to the presence of the 10 ms meander code. Thus, a coherent integration time $T > 10$ ms could cause loss of signal power as the signal may partly cancel itself out. Therefore, a careful selection of appropriate values for coherent and noncoherent summation is required for successful acquisition of

Table 3.3 GLONASS signal acquisition parameter settings used to obtain the results presented herein.

Parameter	Value	Unit
Coherent integration time	8	ms
Noncoherent block number	2	–
Frequency search range	[± 6000]	Hz
Acquisition threshold	9	–

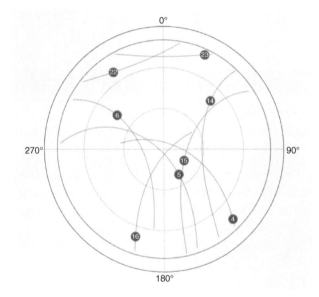

Figure 3.1 Skyplot of GLONASS satellites on February 19, 2018, 11:20 AM UTC time at Finnish Geospatial Research Institute with an elevation cut-off angle of 10°, corresponding to the results in Figure 3.2. This figure is obtained with Trimble's online GNSS planning tool.

visible GLONASS satellites. For the experiment to be shown next, it was chosen a coherent integration time of $T = 8$ ms as indicated in Table 3.3, which incurs in just 1.3-dB average losses due to the presence of sign transitions as per Eq. (1.79). Two consecutive blocks are noncoherently integrated, that is, $N_I = 2$, thus leading to a total integration time $T_{tot} = 16$ ms. Similarly to what already described for GPS L1 C/A, the companion software automatically sets the resolution of the frequency search to the recommended value $\Delta v_t' = 1/2T$ already discussed in Eq. (1.86), which for this example becomes $\Delta v_t' = 62.5$ Hz. The frequency search space then spans along a set of $N_{v'} = 193$ frequency bins according to Eq. (1.89), using the maximum frequency shift of 6 kHz in Table 3.3.

Similarly to Chapter 2, experimental results are shown for the acquisition of the GLONASS L1OF signal using the multi-GNSS software receiver accompanying this book. The skyplot for this acquisition data set is shown in Figure 3.1, the acquisition results are shown in Figure 3.2, and the acquisition values used to obtain these results

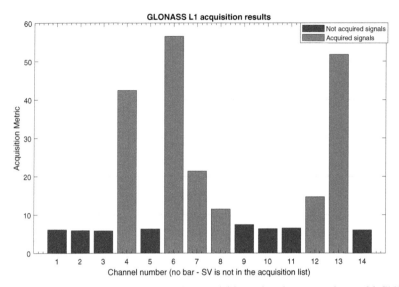

Figure 3.2 Results for GLONASS L1OF acquisition using the companion multi-GNSS software receiver.

are mentioned in Table 3.3. Six GLONASS satellites are acquired for frequency channels 4, 6, 7, 8, 12 and 13. As the PRN information is not available from GLONASS ephemeris, the receiver numbers the satellites based on the estimated frequency slots.

3.4   Tracking

The GLONASS signal processing shares some similarities with GPS L1 C/A because both systems are using spreading codes and BPSK modulation. From a tracking point-of-view, the only major difference between GPS and GLONASS L1 signals is the presence of meander sequence in the GLONASS data bits. Due to the presence of this sequence, the meander-modulated data bit transition for GLONASS L1 is occurs every 10 ms, as compared to the 20 ms transitions of GPS L1 signal. Therefore, most of the GPS tracking functionalities are directly applicable to the GLONASS L1 signal tracking, respecting the fact that the sign transition duration granularity has now changed from 20 ms to 10 ms.

The signal tracking has been carried out on all the acquired channels before starting the data demodulation. The remaining carrier mismatch on the baseband signal provided by the front end is compensated by the carrier tracking loop. Once the Doppler and other remaining frequency errors have been removed, the baseband signal is then correlated with the local replica code by the code tracking loop. The GLONASS tracking functionalities are called at every 1 ms, and the early, prompt and late correlator outputs are assigned into vector data structures for later processing.

Because of the presence of the meander sequence on top of the data bits, the bit transition usually occurs relatively quickly, and therefore, the four-quadrant arctangent discriminator FLL [i.e. as in Eq. (1.102), using the atan2] may not be an ideal choice

Table 3.4 GLONASS signal tracking parameter settings used to obtain the results presented herein.

Parameter	Value	Unit
Correlator spacing	0.25	chips
Number of tracking states	3	–
Tracking states	[PULL_IN; COARSE_TRACKING; FINE_TRACKING]	–
DLL loop filter bandwidth	[1; 1; 1]	Hz
FLL loop filter bandwidth	[200; 100; 5]	Hz
PLL loop filter bandwidth	[15; 15; 10]	Hz

for GLONASS signal tracking. In view of this particular situation, a two-quadrant arctangent discriminator is implemented in the FLL of the software receiver (i.e. using the atan). The implemented two-quadrant arctangent discriminator is insensitive to data bit transition, as discussed in Chapter 1, but it has a reduced tolerance to the frequency uncertainty coming from the acquisition stage. It was shown in [8] that the frequency uncertainty tolerance is reduced by half, as compared to the conventional four-quadrant arctangent discriminator. This restriction in frequency uncertainty can be overcome by proper selection of coherent integration time and therefore frequency bin size, at the acquisition stage. The signal tracking, which starts with the FLL tracking, is then switched to a Costas PLL once the FLL is locked. Results for the GLONASS L1OF carrier tracking are shown in Figure 3.3 along with the results for code tracking. As it can be seen, the plots are quite similar to those obtained for GPS L1 C/A in Figure 2.4. The plot in the upper left corner shows the complex samples at the prompt correlator output. One can see the shape of the BPSK constellation once the carrier has properly been compensated and the samples stop from moving around along a circle. This effect can still be seen and corresponds to those samples from time instants when carrier tracking was still converging. This is also shown in the first one or two seconds in the upper right plot, where one can see that the imaginary part of the prompt correlator output (Q_P) was still containing a significant energy. After these very initial seconds, the signal is properly placed on the real part of the correlator output and on the imaginary part we only have the noise contribution. The power balance between the early, late and prompt correlators is shown on the middle-right side plot. A DLL with a narrow correlator discriminator [9] was used with early and late correlator spacing of $\delta = 0.25$ chips. It is a bit difficult to see because the lines for each correlator are superimposed in this plot, but one can intuitively see that the highest energy lies on the prompt correlator while a lower energy lies on the early and late correlators, with the same contribution on each of them. The main GLONASS L1 signal tracking parameters for these results are presented in Table 3.4.

Once the GLONASS receiver keeps tracking the carrier phase and the code offset of the incoming signal, the next phase is to detect the bit boundary and then to wipe off the meander sequence. Algorithms for the bit boundary detection can be found in [6, 10]. The histogram method, for instance, senses the bit sign changes and keeps

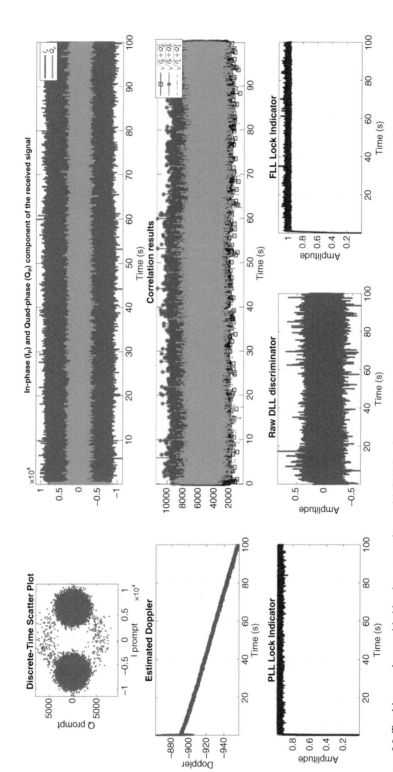

Figure 3.3 Tracking results provided by the companion multi-GNSS software receiver when tracking the GLONASS L1OF satellite at channel 7 from the data set under analysis.

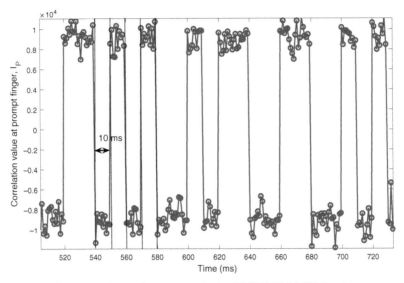

Figure 3.4 Presence of meander sequence in the GLONASS L1OF signal.

a statistic of their position. The presence of meander sequence on top of GLONASS data bits (20 ms long) is shown in Figure 3.4.

3.5   Computation of Position and Time

The navigation message includes the so-called *immediate* (satellite-related) and *non-immediate* (almanac) data transmitted at 50 bits per second. The GLONASS navigation message is generated as continuously repeating super-frames, which consist of frames, and the latter in turn consists of strings. Within each super-frame, a total content of nonimmediate data (almanac for 24 satellites) and the immediate data (for the satellite being tracked) are transmitted. Each super-frame consists of five frames and has a duration of 2.5 minutes. Within each frame, the total content of immediate data for the satellite being tracked and a part of nonimmediate data are transmitted. Each frame consists of 15 strings, each with duration of two seconds. Therefore, the duration of a frame is 30 seconds, like for GPS L1 C/A. Data in strings 1–4 contain the immediate data related to the satellite transmitting the navigation message. Strings 6–15 contain nonimmediate data for all 24 satellites.

After removing the meander code, GLONASS data bits can be obtained from the 50-Hz bit stream. Correlation with the predefined time marker, or preamble, is used to search for the time mark sequence. Once the time mark is found, the string is then decoded according to the ICD [2], and the GLONASS ephemeris, time correction and other satellite data are stored for later use in the navigation processing.

The positioning solution is computed by the navigation block of the companion multi-GNSS software receiver using pseudorange observables and satellite positions

Table 3.5 GLONASS navigation parameter settings used to obtain the results presented herein.

Parameter	Value	Unit
Navigation solution period	100	ms
Number of smoothing samples	10	–
Elevation mask	5	degree
C/N_0 mask	30	dB-Hz

Table 3.6 Position error statistics with respect to true position.

GNSS Signal	Horizontal				Vertical				3D
	RMS (m)	Mean (m)	Max (m)	Mean HDOP	RMS (m)	Mean (m)	Max (m)	Mean VDOP	RMS (m)
GLONASS	8.24	4.86	11.28	2.28	9.71	9.48	27.92	1.54	12.74

produced by the system-specific functions. Observables are corrected by the satellite clock corrections. As the GLONASS message does not transmit atmospheric corrections, if applied, they need to be taken from another source (e.g. another GNSS). The satellite clock corrections in GLONASS and GPS have different sign conventions, which should be taken into account in case of a multi-GNSS solution. In the multi-GNSS software receiver, the GLONASS clock corrections are stored with the sign of the parameter inverted, so they can be used like the GPS corrections.

GLONASS navigation message provides, instead of Keplerian parameters like Galileo or GPS, position, velocity and acceleration vectors, already in an ECEF reference frame, PZ-90.02, at a reference time, of the satellite antenna phase center. The receiver therefore applies, instead of the Keplerian equations, Newton's fundamental orbital differential equation described in Chapter 1 Eq. (1.120), with some linear and harmonic corrections to cope with orbital perturbations, as per [2].

Once the pseudorange measurements and satellite positions are available for at least four satellites, a navigation solution is computed. The positioning calculation currently uses an epoch-by-epoch least-squares method. The main navigation configuration parameters are presented in Table 3.5. A navigation solution was computed after every 100-ms interval with 10-Hz navigation update rate. The navigation solution was further smoothed with 10 navigation solution samples per epoch.

The position error variations in ENU frame are shown in Figure 3.5. The position error statistics are shown in Table 3.6, which were computed for a standalone GLONASS-only code-based position solution. The GPS broadcast Klobuchar ionospheric model parameters at the time of the fix are utilized to calculate the ionospheric corrections. The position error statistics were computed with respect to the true known position. The obtained 3D RMS errors for GLONASS were 12.74 m.

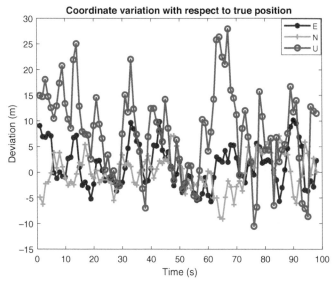

Figure 3.5 Position error variations with respect to true position, in ENU frame, obtained by the multi-GNSS software receiver for the data set under analysis.

3.6   Front End and Other Practical Information

3.6.1   Front End

The MAXIM 2112 front end of the NSL Stereo described in Table 3.7 was selected due to the higher bandwidth required for processing GLONASS signals, as compared to GPS L1 C/A ones. The GLONASS L1OF time-domain plot, signal spectrum and bin distribution of the digitized IF samples are shown in Figure 3.6. We can observe a non-Gaussian shape in the histogram of the I and Q values from the ADC. We see this pattern when using the MAXIM 2112 front end, whose AGC configuration for complex I/Q data seems to saturate the digitized signal. This may lead to an additional quantization loss of around 1-2 dBs, with respect to the full 3-bit resolution of the ADC, depending on how severe this effect is. This loss does not lead to an appreciable degradation in the measurements and in the position computation. The reader can observe this effect in the signal samples recorded with the MAXIM 2112 accompanying this book, but it is not observed with samples captured with the MAXIM 2729B RFFE, whose I/Q histogram generally follows a Gaussian shape.

3.6.2   Data Collection

The GNSS data were collected on February 19, 2018, at around 11:23 AM UTC time at a static position with a roof antenna in Finnish Geospatial Research Institute, Kirkkonummi, Finland. The skyplot for the GLONASS constellation at the time of data collection was shown in Figure 3.1. The data were collected for about 100 s. There were eight visible GLONASS satellites at the moment of data collection.

Table 3.7 MAXIM 2112 configuration (NSL Stereo v2 front end) for collecting GLONASS data samples.

Parameter	Value	Unit
Chipset	MAX 2112	–
Bandwidth	10	MHz
Sampling frequency	26	MHz
Intermediate frequency	0	MHz
Number of quantization bits	3	–
Down-converted signal type	Complex	–
IQ signal orientation	Q, I	–
IQ sample size (I+Q)	8 + 8	bits

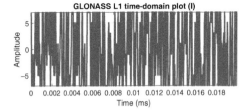

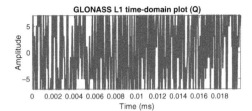

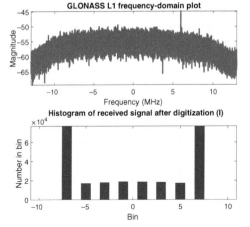

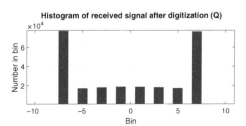

Figure 3.6 GLONASS L1 time-domain plot (up), signal spectrum (middle) and bin distribution of the digitized IF samples (down).

The FGI-GSRx receiver acquired, tracked and computed a navigation solution with six GLONASS satellites with frequency channel numbers 4, 6, 7, 8, 12 and 13.

3.6.3   MATLAB Configuration and Functions

The parameter file name for GLONASS L1 receiver is "default_param_GLONASSL1 _ Chapter3.txt." The user needs to pass this parameter file with the main function in the MATLAB command prompt as follows:

> > gsrx("default_param_GLONASSL1_Chapter3.txt")

The enabled GNSS signal was chosen as "glol1" in this case. The front-end configuration used to collect the example GLONASS raw IQ data set is given in Table 3.7. The MAXIM 2112 front end saved complex IQ data as Q, I (Quadphase data appearing before In-phase data), and the IQ sample size is 16 bits (i.e. 8 bits for Q, 8 bits for I). The intermediate frequency of the complex received GLONASS signal is 0 Hz.

The "plotAcquisition.m" function generates the acquisition metric for all the searched PRNs as can be seen in Figure 3.2. It also generates a two-dimensional signal correlation plot for all the successful PRNs that are above the acquisition threshold. Figure 3.3 can be obtained by calling the function "plotTracking.m." Figure 3.5 was generated within the function "calcStatistics.m" with a 1 Hz output rate for a standalone static GNSS user. Statistical calculations in terms of accuracy and availability can be found in this function.

References

[1] Y. Urlichich, V. Subbotin, G. Stupak, V. Dvorkin, A. Povalyaev, and S. Karutin, "GLONASS Modernization," in *Proceedings of ION GNSS, Portland, OR*, pp. 3125–3128, September 2011.

[2] Russian Institute of Space Device Engineering, "Global Navigation Satellite System GLONASS, Interface Control Document, Navigational Radiosignal in Bands L1, L2," 2008. Edition 5.1.

[3] Russian Space Systems, JSC, "Global Navigation Satellite System GLONASS, Interface Control Document, General Description of Code Division Multiple Access Signal System," 2016. Edition 1.0.

[4] Y. Urlichich, "Directions 2019: High-Orbit GLONASS and CDMA Signal." www.gpsworld.com/directions-2019-high-orbit-glonass-and-cdma-signal/. GPS World, December 12, 2018. Accessed: May 25, 2019.

[5] J. W. Betz, "On the Power Spectral Density of GNSS Signals, with Applications," in *Proceedings of ION International Technical Meeting (ITM), San Diego, CA*, pp. 859–871, January 2010.

[6] E. D. Kaplan and C. J. Hegarty, *Understanding GPS - Principles and Applications*. Artech House, Inc., 2006.

[7] A. Zueva, E. Novikov, D. Pleshakov, and I. Gusev, "System of Geodetic Parameters 'Parametry Zemli 1990' (PZ-90.11)." in *Proceedings of 9th Meeting of the International Committee on GNSS (ICG-9), Prague, Czech Republic* December 2014.

[8] K. Yan, H. Zhang, T. Zhang, and L. Xu and X. Niu, "Analysis and Verification to the Effects of NH Code for Beidou Signal Acquisition and Tracking," in *Proceedings of ION GNSS+, Nashville, TN*, pp. 107–113, September 2013.

[9] A. V. Dierendonck, P. Fenton, and T. Ford, "Theory and Performance of Narrow Correlator Spacing in a GPS Receiver," *Navigation*, vol. 39, no. 3, pp. 265–283, 1992.

[10] C. Ma, G. Lachapelle, and M. E. Cannon, "Implementation of a Software GPS Receiver," in *Proceedings of ION GNSS, Long Beach, CA*, pp. 21–24, September 2004.

4   Galileo E1 Receiver Processing

M. Zahidul H. Bhuiyan, Stefan Söderholm, Giorgia Ferrara,
Martti Kirkko-Jaakkola, Heidi Kuusniemi, José A. López-Salcedo
and Ignacio Fernández-Hernández

4.1   Introduction

Galileo is the European GNSS. The fully deployed Galileo system has 30 satellites
(24 operational + 6 active spares), positioned in three circular medium Earth orbit
(MEO) planes at 23,222 km altitude above the Earth, slightly higher than GLONASS
and GPS, and at an inclination of the orbital planes of 56 degrees with reference to the
equatorial plane.[1] Once full capability is achieved, the Galileo navigation signals will
provide good coverage even at latitudes up to 75 degrees north, which corresponds
to the North Cape and beyond. The real-time constellation status can be checked in
[2]. The Galileo system started by 1999/2000, by the European Space Agency and
the European Commission [3]. Over these more than 20 years, it has been designed,
developed, validated and deployed. At the time of writing this chapter, Galileo is in its
initial service phase and has 28 satellites in orbit, out of which 21 are operational. It is
planned to be fully operational sometime in the near future. It is designed to be compatible with all existing and planned GNSS and interoperable with GPS, GLONASS
and BeiDou [4]. For a more detailed description of the Galileo system, see [5]
or [6].

    Galileo is already providing an open service (OS), which results from the use of
open signals E1, E5a and E5b, with position and timing performances competitive
with other GNSSs [7]. In addition, it also includes the following services:

– A search and rescue service (SAR), which broadcasts globally the alert messages
   received from distress emitting beacons and contributes to enhance the
   performance of the international COSPAS-SARSAT (COsmicheskaya Sistyema
   Poiska Avariynich Sudov - Search And Rescue Satellite-Aided Tracking) SAR
   system. Galileo in particular provides a "Return Link" message to distressed users
   in its open signal message.

---

[1] The Galileo constellation also includes two satellites with an eccentricity of 0.162, a lower inclination
of 49.85 degrees, and a lower semimajor axis of 27,977.6 km versus the nominal one of 29,599.8 km,
which were put in this orbit due to a launch anomaly. Interestingly, the high clock stability combined
with the orbit eccentricity allowed testing with the highest accuracy until now the gravitational redshift
of Einstein's general theory of relativity [1].

- A high accuracy service (HAS), which consists of corrections to allow decimeter-level accuracy. It will be transmitted in the E6-B signal (1278.75 MHz) [8].
- A commercial authentication service (CAS) based on the E6-C signal code encryption allowing for increased robustness of professional applications. It will be based on the E6-C signal, pilot component (1278.75 MHz) [9].

At the time of writing this chapter, HAS is developed and under testing for some years [10]. It is currently available in the signal-in-space and already providing corrections worldwide and for free, and it will be declared operational very soon. CAS is under development and will be ready a few years later, and it was initially conceived to be offered at a fee. In addition to these, Galileo offers a Public Regulated Service (PRS), accessible to governmental users in the E1-A and E6-A signals, in quadrature to the E1-B/C and E6-B/C.

This chapter presents Galileo navigation satellite system receiver processing results obtained with the multi-GNSS software receiver accompanying this book. The chapter focuses on the description of the Galileo E1 signal, and the challenges related to Galileo-specific receiver implementation aspects.

4.2 Galileo E1 Signal Characteristics

4.2.1 Signal Structure

The Galileo E1 spreading codes are memory codes [12]. Galileo signals made a distinction between channels containing navigation data, or data channels, and channels carrying no data, or pilot channels. The data (E1-B) and pilot (E1-C) channels are transmitted in the same component, but with different codes, so the receiver can correlate them separately. The PRS E1-A component is shifted by 90 degrees and therefore transmitted in quadrature, which allows for its separated processing in the receivers. On the Galileo E1 OS data and pilot channels, the spreading code length for each satellite is 4,092 chips with a period of 4 milliseconds (ms). Both channels are processed in the accompanying software of this book.

For the pilot channel, the primary code is combined with a secondary code of 25 chips. The final code in the pilot channel is called the *tiered* code and is 100 ms long (i.e. 4 ms × 25 secondary code chips). Galileo E1 signal uses the composite BOC (CBOC) modulation, which is a combination of BOC(1,1) and BOC(6,1) modulations [4]. The Galileo E1-B and E1-C signal structure can be seen in Figure 4.1. Note that the nomenclature in this figure is that of Galileo's Signal-in-Space Interface Control Document (SIS ICD) [11]. It shows the spreading codes of the two components C_{E1-B} and C_{E1-C}, the latter already including the secondary code; the data stream in the data component D_{E1-B}; and the CBOC subcarrier modulations $sc_{E1-C,a}$ and $sc_{E1-C,b}$ for the pilot, and $sc_{E1-B,a}$ and $sc_{E1-B,b}$ for the data, each with two coefficients,

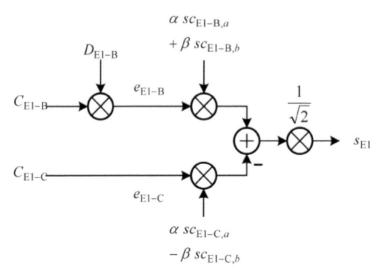

Figure 4.1 Galileo E1-B and E1-C modulation scheme, from [11].

α and β, for the BOC(1,1) and the BOC(6,1) components, where $\alpha = \sqrt{10/11}$ and $\beta = \sqrt{1/11}$.

4.2.2 Comparison with GPS L1 C/A

The spreading chipping rate is the same as for GPS L1 C/A (1.023 MHz), but Galileo E1-B/C has a code length which is 4 ms, four times that of GPS L1 C/A signal. In the E1-B component, a symbol is modulated every 4-ms code, for a total of 250 symbols per second. These symbols, excluding the 10-symbol synchronization pattern, encode a 120-bit-per-second message. This higher rate was foreseen to transmit safety-of-life integrity information, but this has been replaced by some service improvements described later in Section 4.6. As a result of this design, the Galileo E1-B coherent integration time is reduced to one code (4 ms), but this is partly compensated with the E1-C pilot tone.

Galileo E1-B symbols encode every second the 120 data bits at a coding rate of 1/2 into the 240 symbols though a FEC convolutional encoding scheme, depicted earlier in Figure 1.22. The resulting symbols are interleaved and modulated in the E1-B. In addition to error correction, every 2-second navigation page includes a 24-bit CRC for error detection, whereas the legacy GPS L1 C/A signal has no error detection and only employs a 6-bit parity check every 30 bits. FEC on the navigation message is an important improvement in overall signal robustness. It is also used in SBAS in a similar fashion. The Galileo E1 OS navigation message is called I/NAV. It is transmitted also, with a different sequence but the same data, in the E5b signal.

A summarized comparison of some key properties of GPS L1 C/A and Galileo E1-B is presented in Table 4.1, which focuses on the BOC(1,1) signal.

Table 4.1 Parameters of the general signal model in Eq. (1.13) particularized for the Galileo E1-B signal, considering the BOC(1,1) signal only and compared to the parameters of the GPS L1 C/A signal.

		GPS L1 C/A	GALILEO E1-B
Multiplexing technique		CDMA	CDMA
Center frequency		1575.42 MHz	1575.42 MHz
Primary (spreading) code type	$\{v[m]\}_{m=0}^{N_c-1}$	Gold code	Memory code
Primary code length	N_c	1,023	4,092
Primary code duration	T_{code}	1 ms	4 ms
Primary code chip rate	R_c	1.023 MHz	1.023 MHz
Primary code chip duration	$T_c = 1/R_c$	0.977 μs	0.977 μs
Primary code chip pulse shape	$p_{T_c}(t)$	$\Pi\left(\frac{t-T_c/2}{T_c}\right)$	$\Pi\left(\frac{t-T_c/4}{T_c/2}\right) - \Pi\left(\frac{t-3T_c/4}{T_c/2}\right)$
Secondary code type	$\{u[k]\}_{k=0}^{N_r-1}$	1	–
Secondary code length	N_r	20	–
Data modulation		BPSK	BOC(1,1)
Symbol (bit) rate	R_d	50 sps (50 bps)	250 sps (125 bps)
Symbol (bit) duration	T_d	20 ms (20 ms)	4 ms (8 ms)
Minimum receiver bandwidth requirement		~2 MHz	~4 MHz
Pilot channel available		No	Yes, the Galileo E1-C component
Error detection method		Extended (32,26) Hamming code	CRC (24 bits)
Error correction method		None	FEC, convolutional code with rate 1/2, constraint length 7
Coordinate system		WGS84	GTRF

4.3   Acquisition

As for other signals, the acquisition engine provided with this book as part of the companion FGI-GSRx software receiver performs the full parallel search method [13] for the acquisition of Galileo signals. This method was already described in the search step of Section 1.6, where it was shown to make extensive use of the FFT as a way to perform the correlation between the received signal and the local replica in the frequency domain. The search window, coherent integration time, noncoherent integration and detection thresholds are all configurable parameters. For the Galileo E1-B/C data-modulated components, the bit duration is $T_d = 4$ ms long and this limits the coherent integration time up to a maximum of $T = 4$ ms. For the E1-C pilot component, longer coherent integration beyond $T = 4$ ms can be performed, but they need to account for the presence of the secondary code, which needs to be acquired as well. This is a difference with respect to GPS L1 C/A signal, where each code lasts 1 ms, and each bit lasts 20 ms, so longer coherent integration can be performed, as mentioned in Chapter 2.

Table 4.2 Galileo signal acquisition parameter settings used to obtain the results presented herein.

Parameter	Value	Unit
Coherent integration time	4	ms
Noncoherent block number	2	–
Frequency search range	[± 6000]	Hz
Acquisition threshold	9	–

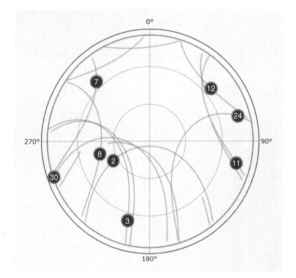

Figure 4.2 Skyplot of Galileo satellite navigation system on February 21, 2018, 13:12 UTC time at Finnish Geospatial Research Institute with an elevation cutoff angle of 10°. This figure is obtained with Trimble's online GNSS planning tool.

Similarly to previous chapters, experimental results are shown here using the multi-GNSS software receiver accompanying this book. In this case for the acquisition of the Galileo E1-B data channel, the skyplot for this acquisition data set is shown in Figure 4.2. The acquisition result is shown in Figure 4.3, and the acquisition values used to obtain these results are mentioned in Table 4.2. The acquisition function considers $T = 4$ ms coherent integration and $N_I = 2$ blocks of noncoherent integration to ensure that most of the visible Galileo satellites are acquired successfully. The integration parameters and the acquisition threshold were set such that low elevation satellites can also be acquired for tracking. The skyplot for this acquisition data set is shown in Figure 4.2. The companion software automatically sets the resolution of the frequency search to the recommended value $\Delta v_t' = 1/2T$ already discussed in Eq. (1.86), which for this example becomes $\Delta v_t' = 125$ Hz. The frequency search space then spans along a set of $N_{v'} = 97$ frequency bins according to Eq. (1.89), using the maximum frequency shift of 6,000 Hz considered in Table 4.2.

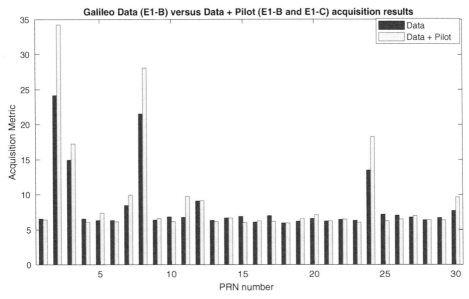

Figure 4.3 Results for the Galileo E1 signal acquisition with "Data"-only versus combined "Data+Pilot" channels using the companion multi-GNSS software receiver.

It can be seen from Figure 4.3 that eight Galileo satellites are acquired with PRN numbers 2, 3, 7, 8, 11, 12, 24 and 30. It is important to note here that a combined data and pilot acquisition are used in order to exploit the benefits of Galileo pilot channel. Data and pilot channels are noncoherently combined in order to increase the acquisition sensitivity. Figure 4.3 illustrates the fact that the signal energy acquired increased in case of combined "Data + Pilot" acquisition than that of "Data"-only acquisition.

4.4   Tracking

In case of Galileo E1-B signal tracking, the FGI-GSRx tracking module is only called after each code epoch, which is 4 ms for Galileo E1-B signal. As symbol transitions occur often in the Galileo E1-B signal, the FLL based on a four-quadrant arc-tangent discriminator [i.e. as in Eq. (1.102), using the atan2] may not be an ideal choice. A two-quadrant arc-tangent discriminator (i.e. using the atan) is considered instead for Galileo E1-B signal tracking [14]. The implemented two-quadrant arc-tangent discriminator is insensitive to data bit transition, but it has reduced tolerance of frequency uncertainty coming from the acquisition stage. It was shown in [14] that the frequency uncertainty tolerance is reduced by half, as compared to the conventional four-quadrant arc-tangent discriminator. This restriction in frequency uncertainty can be overcome by proper selection of coherent integration time (i.e. frequency bin size) at acquisition. The signal tracking is switched from FLL to a Costas PLL, once the FLL is locked.

Table 4.3 Galileo signal tracking parameter settings used to obtain the results presented herein.

Parameter	Value	Unit
Correlator spacing	0.1	chips
Number of tracking states	3	–
Tracking states	[PULL_IN; COARSE_TRACKING; FINE_TRACKING]	–
DLL loop filter bandwidth	[1; 1; 1]	Hz
FLL loop filter bandwidth	[75; 35; 5]	Hz
PLL loop filter bandwidth	[15; 10; 5]	Hz

The Galileo main E1 signal tracking parameters used are presented in Table 4.3. Figure 4.4 shows the tracking status of Galileo PRN 3 for 100 seconds long data obtained using a narrow correlator discriminator [15] with early and prompt correlator spacing of $\delta = 0.1$ chips. The tracking is carried out only on the Galileo E1-B data channel. If the user wishes to run Galileo E1-C, it is entirely possible in FGI-GSRx to run the pilot channel and to decode the 25-symbol secondary codes repeating after every 100 ms. The figure shows that it takes a few seconds for the FLL to lock the frequency. This can be observed in the FLL and PLL lock indicator figures, as well as in the estimated Doppler figure. This also leads to a higher amount of rotated I_P and Q_P values, compared to the equivalent figure in Chapter 2 for GPS.[2]

4.5   Computation of Position and Time

Once the bits are demodulated, the Galileo signal is synchronized by the 10-bit synchronization pattern (0101100000) transmitted in every 2-second I/NAV page. Once synchronized, the receiver can decode the I/NAV word transmitted in each page. In particular, words 1–4 transmit the clock and ephemeris data necessary to locate the satellite. Word 5 is also necessary as it transmits the Galileo System Time (GST), needed to time stamp the signal, the ionospheric corrections for the application of the NeQuick model, the BGDs, and the satellite status flags. As the clock corrections are transmitted for a dual-frequency E1-E5b user, an E1-only user needs to apply the BGD in word 5 to navigate with the E1 frequency only, as described in the Galileo ICD [11] and in Chapter 1. The same occurs for GPS L1 C/A, as the clock correction is referred to the L1-L2 observable.

As presented in Table 4.4, the navigation solution was computed every 100-ms interval, that is with a 10-Hz navigation update rate. It was further smoothed with 10 navigation solution samples per epoch. The position error variations in ENU frame

---

[2] Note that the BOC(1,1) signal has an autocorrelation function with several peaks, and there is a risk of locking to a peak that is not the main one, with the consequent accuracy error. In order to avoid this, unambiguous tracking techniques of BOC signals are dealt with in detail in Chapter 10.

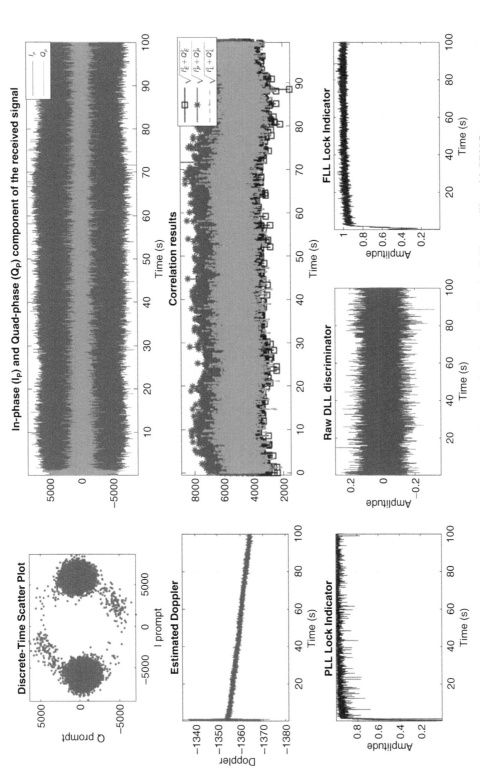

Figure 4.4 Tracking results provided by the companion multi-GNSS software receiver when tracking the Galileo satellite with PRN 7.

Table 4.4 Galileo navigation parameter settings used to obtain the results presented herein.

Parameter	Value	Unit
Navigation solution period	100	ms
Number of smoothing samples	10	–
Elevation mask	5	degree
C/N_0 mask	30	dB-Hz

Table 4.5 Position error statistics with respect to true position.

	Horizontal				Vertical				3D
GNSS Signal	RMS (m)	Mean (m)	Max (m)	Mean HDOP	RMS (m)	Mean (m)	Max (m)	Mean VDOP	RMS (m)
Galileo	2.43	2.74	4.82	1.40	2.67	1.77	5.09	1.26	3.61

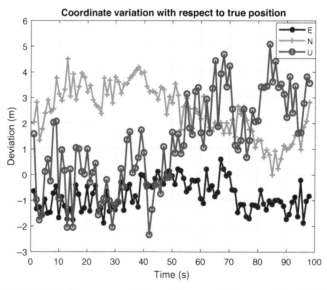

Figure 4.5 Position error variations with respect to true position in ENU frame, obtained by the multi-GNSS software receiver for the data set under analysis.

are shown in Figure 4.5, respectively, for Galileo-only position fix. The position error statistics for Galileo E1 are presented in Table 4.5. The error statistics were computed for a standalone code-based position solution. The NeQuick model is implemented to calculate the ionospheric corrections. The position error statistics were computed with respect to the true known position. The 3D RMS errors for Galileo E1 is 3.61 m.

4.6   Galileo E1 Additional Features

The current chapter and related receiver cover the standard implementation of Galileo E1 I/NAV. However, the bandwidth of the Galileo E1-B has been only partially used for standard navigation. This is because, initially, Galileo foresaw to provide a global safety-of-life service, and the message bandwidth was dimensioned for it, but the service was eventually not implemented. We describe here the two main features that are incorporated in the E1-B signal in this spare bandwidth. They are not yet processed in the current SDR, but we expect it will be the case in further updates.

4.6.1   Galileo Open Service Navigation Message Authentication (OSNMA)

The purpose of OSNMA is to authenticate the Galileo message, in order to ensure that the data origin is the Galileo system and not another source, for example, a spoofer or a carelessly operated testbed or signal simulator. OSNMA incorporates a lightweight authentication protocol called Timed-Efficient Stream Loss-Tolerant Authentication (TESLA) [16] and has adapted it to GNSS, including two main features: key chain sharing from all transmitting satellites and cross-authentication of some satellites from others [17]. The drawback of TESLA is that it needs an external loose time synchronization source. The latest available specification at the time of writing is [18]. The protocol is transmitted in the so-called "Reserved 1" field of the I/NAV in E1-B, occupying 40 bits every other page. In addition to data authentication, the bit unpredictability of OSNMA can be used to protect signals against replay attacks.

4.6.2   Galileo I/NAV Improvements

Some of the Galileo free pages will be used to improve the I/NAV performance. These functions are detailed in the latest Galileo ICD [11] and consist of:

- Reduced clock and ephemeris data: 112-bit compressed set of ephemeris and clock transmitted in one I/NAV word (Word 16). This allows computing a first position fix before the full ephemeris is received, with some loss in accuracy [19].
- FEC2 Reed-Solomon for CED: An outer forward error correction based on Reed-Solomon for the clock and ephemeris, in order to increase data demodulation threshold through the additional parity. It will be provided in I/NAV spare words, to be replaced by words 17–20.
- Secondary synchronization pattern (SSP): Currently, if a receiver is loosely synchronized, for example, with an uncertainty of few seconds and processes the I/NAV, it still needs to wait for the GST to synchronize with the signal, which is transmitted only once every 30 seconds. This delays the acquisition and synchronization process. To mitigate that, a SSP has been introduced. It consists of 8 bits added before the FEC encoding, providing a time measurement modulo 6, so a receiver with a time uncertainty of ± 3 seconds can determine its time without waiting for the GST.

4.7   Front End and Other Practical Information

4.7.1   Front End

The Stereo dual-frequency front end from NSL with the MAXIM 2769B front end (see Table 4.6) was used to capture the real Galileo E1 signal at 1575.42 MHz. The time-domain, signal spectrum and bin distribution plots of the digitized IF samples are shown in Figure 4.6 for the Galileo E1 signal.

4.7.2   Data Collection

The GNSS data were collected on February 21, 2018, at around 13:12 UTC time at a static position with a roof antenna in Finnish Geospatial Research Institute, Kirkkonummi, Finland. The data were collected for about 100 seconds. The skyplot for Galileo constellation at the time of data collection is shown in Figure 4.2. There were eight operational Galileo satellites (i.e. PRNs 2, 3, 7, 8, 11, 12, 24 and 30) visible at the moment of data collection. The FGI-GSRx receiver acquired, tracked and computed a navigation solution with all the visible Galileo satellites.

4.7.3   MATLAB Configuration and Functions

The parameter file name for Galileo E1-B receiver is "default_param_GalileoE1B_Chapter4.txt." The user needs to pass this parameter file with the main function in the MATLAB command prompt as follows:

> > gsrx("default_param_GalileoE1B_Chapter4.txt")

The enabled GNSS signal was chosen as "gale1b" in this particular case. The front-end configuration used to collect the example Galileo raw data set is given in Table 4.6. The MAXIM 2769B front-end saves real binary data with sample size of 8 bits, although quantized to 2 bits.

Figure 4.4 can be obtained by calling the function "plotTracking.m." This function also generates the C/N_0 plot for all the tracked Galileo E1-B satellites. Figure 4.5 was generated within the function "calcStatistics.m" with a 1-Hz output rate for a standalone static GNSS user.

Table 4.6 Configuration of MAXIM 2769B chipset within NSL Stereo v2 for collecting the Galileo raw samples.

Parameter	Value	Unit
Chipset	MAXIM 2769B	–
Bandwidth	4.2	MHz
Sampling frequency	26	MHz
Intermediate frequency	0	MHz
Number of quantization bits	2	–
Down-converted signal type	Real	–
Real sample size	8	bits

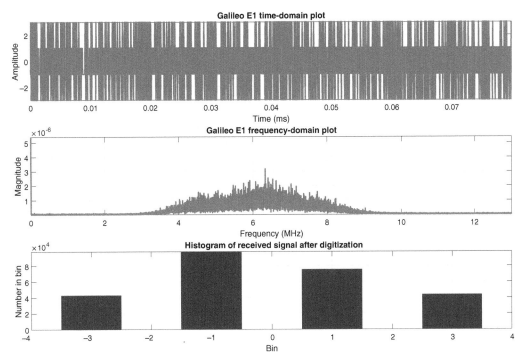

Figure 4.6 Galileo E1 time-domain plot (up), signal spectrum (middle) and bin distribution of the digitized IF samples (down).

References

[1] P. Delva, N. Puchades, E. Schönemann, F. Dilssner, C. Courde, S. Bertone, F. Gonzalez, A. Hees, C. Le Poncin-Lafitte, F. Meynadier, *et al.*, "Gravitational Redshift Test Using Eccentric Galileo Satellites," *Physical Review Letters*, vol. 121, no. 23, p. 231101, 2018.

[2] European GNSS Service Center. www.gsc-europa.eu. Accessed: July 13, 2022.

[3] J. Benedicto, S. Dinwiddy, G. Gatti, R. Lucas, and M. Lugert, "Galileo: Satellite System Design," in *European Space Agency*, Citeseer, 2000.

[4] J. Ávila Rodríguez, G. W. Hein, S. Wallner, J.-L. Issler, L. Ries, L. Lestarquit, A. de Latour, J. Godet, F. Bastide, T. Pratt, *et al.*, "The MBOC Modulation: The Final Touch to The Galileo Frequency and Signal Plan," *Navigation*, vol. 55, no. 1, pp. 15–28, 2008.

[5] M. Falcone, J. Hahn, and T. Burger, "Galileo," in P. J. G. Teunissen and O. Montenbruck (eds.), *Part B: Satellite Navigation Systems*, Springer Handbook of Global Navigation Satellite Systems, Ch. 9, Springer. 2017.

[6] J. P. Bartolomé, X. Maufroid, I. F. Hernández, J. A. López-Salcedo, and G. Seco-Granados, "Overview of Galileo System," in *GALILEO Positioning Technology*, Springer, pp. 9–33, 2015.

[7] European Commission, "Galileo Open Service - Service Definition Document", Issue 1.2, November 2021.

[8] I. Fernandez-Hernandez, A. Chamorro-Moreno, S. Cancela-Díaz, J. David Calle-Calle, P. Zoccarato, D. Blonski, T. Senni, F. Javier de Blas, C. Hernández, J. Simón, A. Mozo, "Galileo high accuracy service: initial definition and performance", *GPS Solutions*, vol. 26, no. 65, 2022.

[9] I. Fernandez-Hernandez, G. Vecchione, and F. Díaz-Pulido, "Galileo Authentication: A Programme and Policy Perspective," in *Proceedings of the International Astronautical Congress IAC, Bremen, Germany*, 2018.

[10] I. Fernández-Hernández, I. Rodriguez, G. Tobías, J. Calle, E. Carbonell, G. Seco-Granados, J. Simón, and R. Blasi, "Galileo's Commercial Service. Testing GNSS High Accuracy and Authentication," *Inside GNSS*, vol. 10, no. 1, pp. 38–48, 2015.

[11] European Union, "European GNSS (Galileo) Open Service, Signal-in-Space Interface Control Document", Issue 2.0, January 2021.

[12] J. O. Winkel, "Spreading Codes for a Satellite Navigation System." US8035555B2, February 2004.

[13] C. Ma, G. Lachapelle, and M. E. Cannon, "Implementation of a Software GPS Receiver," in *Proceedings of ION GNSS, Long Beach, CA*, pp. 21–24, September 2004.

[14] K. Yan, H. Zhang, T. Zhang, and L. Xu and X. Niu, "Analysis and Verification to the Effects of NH Code for Beidou Signal Acquisition and Tracking," in *Proceedings ION GNSS+, Nashville, TN*, pp. 107–113, September 2013.

[15] A. V. Dierendonck, P. Fenton, and T. Ford, "Theory and Performance of Narrow Correlator Spacing in a GPS Receiver," *Navigation*, vol. 39, no. 3, pp. 265–283, 1992.

[16] A. Perrig, R. Canetti, J. D. Tygar, and D. Song, "The TESLA Broadcast Authentication Protocol," *Rsa Cryptobytes*, vol. 5, no. 2, pp. 2–13, 2002.

[17] I. Fernández-Hernández, V. Rijmen, G. Seco-Granados, J. Simon, I. Rodríguez, and J. D. Calle, "A Navigation Message Authentication Proposal for the Galileo Open Service," *Navigation: Journal of the Institute of Navigation*, vol. 63, no. 1, pp. 85–102, 2016.

[18] European Commission, "Galileo Open Service Navigation Message Authentication (OSNMA) User ICD for the Test Phase", Issue 1.0, November 2021.

[19] M. Paonni, M. Anghileri, T. Burger, L. Ries, S. Schlötzer, B. Schotsch, M. Ouedraogo, S. Damy, E. Chatre, M. Jeannot, J. Godet, and D. Hayes, "Improving the Performance of Galileo E1-OS by Optimizing the I/NAV Navigation Message," in *Proceedings of the 32nd International Technical Meeting of the Satellite Division of the Institute of Navigation (ION GNSS+ 2019), Miami, FL, USA*, pp. 1134–1146, 2019.

5 BeiDou B1I Receiver Processing

M. Zahidul H. Bhuiyan, Stefan Söderholm, Sarang Thombre
and Heidi Kuusniemi

5.1 Introduction

BeiDou Navigation Satellite System (BDS) is the Chinese GNSS. It has been independently constructed, operated and maintained by China in order to fulfill the requirements of its own national security as well as economic and social development. Being space-come-land infrastructures of national significance, the BDS aims to provide all-time, all-weather high-accuracy positioning, navigation and timing services to global users.

The Chinese BDS has a mixed space constellation that will have, when fully deployed, five geostationary Earth orbit (GEO) satellites, twenty-seven medium Earth orbit (MEO) satellites and three inclined geosynchronous satellite orbit (IGSO) satellites. The GEO satellites are already operating in orbit at an altitude of 35,786 kilometers (km) and positioned at 58.75°E, 80°E, 110.5°E, 140°E and 160°E. The MEO satellites are operating in orbit at an altitude of 21,528 km and an inclination of 55° to the equatorial plane. The IGSO satellites are operating in orbit at an altitude of 35,786 km and an inclination of 55° to the equatorial plane. These satellites broadcast navigation signals and messages within three frequency bands. The BDS has been in development for more than a decade, and it is estimated to be operational with global coverage at the latest in 2020 [1]. In 2018, the BeiDou program performed 9 launches for 18 new satellites, the highest launch cadence in the history of GNSS at the time of writing this chapter. The BeiDou satellites transmit ranging signals based on CDMA principle, like GPS and Galileo. The mixed constellation structure of BeiDou is expected to result in a better observation geometry for positioning and orbit determination compared to current GPS, GLONASS and Galileo, especially in China and neighboring regions. The BeiDou system has already started contributing to the multi-GNSS benefits where increased accuracy, availability and integrity are possible when utilizing interoperable GNSS [2].

The next sections focus on the B1I signal characteristics, a description of the receiver processing algorithms and the results obtained with the FGI-GSRx. More details on the current status of BeiDou are introduced in [3].

5.2 BeiDou B1I Signal Characteristics

This book and associated software focus on the BeiDou B1I (B1 In-phase) signal. The characteristics of BeiDou B1I signal are compared in this section with the legacy GPS L1 C/A signal in order to show the similarities and differences between the two systems.

5.2.1   Signal Structure

The characteristics of BeiDou B1I signal (B1 In-phase) are compared in this section with the legacy GPS L1 C/A signal in order to realize the similarities and differences between the two signals. The BDS B1I and GPS L1 C/A signals have similar characteristics in general. B1I carrier frequency is 1561.098 MHz, not far from GPS's 1575.42 MHz. The periods of their spreading codes are both 1 ms long, and the coordinate systems (CGCS2000 in case of BDS) and the navigation message structures are also similar.

BeiDou B1I modulates a navigation message "D1" for its MEO and IGSO satellites at 50 bps like GPS L1 C/A and another one named "D2" for its GEO satellites at 500 bps. This eventually means that algorithms that are implemented for the GPS receiver can be readily available to the BeiDou receiver without any major modification [4]. But to improve the signal processing performance, modern GNSS signals, including BeiDou B1I, introduce a second layer of modulation between the navigation data and the PRN code chips, known as Neumann-Hoffman (NH) code modulation. NH modulation is also present in GPS L5 signals, as explained later in Chapter 8.

The legacy GPS L1 C/A signal has a symbol rate of 50 sps, which means that 1 symbol lasts for 20 ms (i.e. the PRN code repeats 20 times for each symbol). The symbol rate of the BeiDou D2 signal transmitted from GEO satellites is 500 sps, which means that 1 symbol lasts for only 2 ms (i.e. two spreading code periods). The symbol rate of the BeiDou D1 signals transmitted from MEO and IGSO satellites is originally 50 sps. However, each symbol is modulated with a NH code of length 20 bits (00000100110101001110), of 1 ms duration each, giving a total rate of 1 ksps. For this reason, the NH code-modulated symbol rate of the BeiDou signal increases significantly as compared to the GPS L1 signal. Due to the presence of NH code modulation, there are several symbol transitions within the symbol boundary. The use of the NH code and the resultant increase in the symbol rate has pros and cons. On the positive side, the NH code can boost the ability of antinarrowband interference and the cross-correlation property of satellite signals and the symbol synchronization [5], whereas on the negative side, the existence of NH code makes the acquisition and tracking more challenging [6, 7, 8].

5.2.2   Comparison with GPS L1 C/A

A comparison of the most relevant signal parameters for both BeiDou B1I and GPS L1 C/A is listed in Table 5.1. More details on the B1I signal can be found in [9].

Table 5.1 Parameters of the general signal model in Eq. (1.13) particularized for the BeiDou B1 signal and compared to those of the GPS L1 C/A signal.

		GPS L1 C/A	BeiDou B1-D1
Multiplexing technique		CDMA	CDMA
Center frequency		1575.24 MHz	1561.098 MHz
Primary (spreading) code type	$\{v(m)\}_{m=0}^{N_c-1}$	Gold code	Balanced, truncated Gold code
Primary code length	N_c	1,023	2,046
Primary code duration	T_{code}	1 ms	1 ms
Primary code chip rate	R_c	1.023 MHz	2.046 MHz
Primary code chip duration	$T_c = 1/R_c$	0.977 μs	0.488 μs
Primary code chip pulse shape	$p_{T_c}(t)$	$\Pi\left(\frac{t-T_c/2}{T_c}\right)$	$\Pi\left(\frac{t-T_c/2}{T_c}\right)$
Secondary code type	$\{u[k]\}_{k=0}^{N_r-1}$	1	Neumann-Hoffman
Secondary code length	N_r	20	20
Data modulation		BPSK	BPSK
Symbol (bit) rate	R_d	50 sps (50 bps)	50 sps ($\approx$ 37 bps)
Symbol (bit) duration	T_d	20 ms (20 ms)	20 ms ($\approx$ 15 ms)
Minimum receiver bandwidth requirement		~2 MHz	~4 MHz
Pilot channel available		No	No
Error detection method		Extended (32,26) Hamming code	BCH(15,11,1)
Error correction method		None	BCH(15,11,1)
Coordinate system		WGS84	CGCS2000

5.3 Acquisition

As for the other GNSS, the same principles described in Section 1.6 can be applied for the acquisition of the BeiDou B1I signal. In particular, the FFT-based signal acquisition technique described therein is implemented in the companion software receiver. The acquisition result is shown in Figure 5.2. BeiDou B1I signal was collected at the Finnish Geospatial Research Institute, Finland, on January 31, 2014, at around 9:30 AM UTC time at a static position with a roof antenna. A multifrequency active antenna from GPS Source was used for data collection. The skyplot for BeiDou constellation at the time of data collection is shown in Figure 5.1. The acquisition parameter values used to acquire the BeiDou satellites are presented in Table 5.2. The acquisition function considers $T = 2$ ms of coherent integration and six consecutive blocks of noncoherent integration, that is, $N_I = 6$, thus leading to a total integration time $T_{tot} = 12$ ms. Similarly to what already described for GPS L1 C/A, the companion software automatically sets the resolution of the frequency search to the recommended value $\Delta v_t' = 1/2T$ already discussed in Eq. (1.86), which for this example becomes $\Delta v_t' = 250$ Hz. The frequency search space then spans along a set of $N_{v'} = 49$ frequency bins according to Eq. (1.89), using the maximum frequency shift of 6,000 Hz considered in Table 2.2.

Table 5.2 BeiDou signal acquisition parameter settings used to obtain the results presented herein.

Parameter	Value	Unit
Coherent integration time	2	ms
Noncoherent block number	6	–
Frequency search range	[± 6000]	Hz
Acquisition threshold	10	–

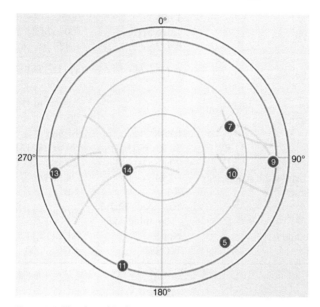

Figure 5.1 Skyplot of BeiDou on January 31, 2014, 9:30 AM UTC time at Finnish Geospatial Research Institute with elevation cutoff angle of 10°. Figure obtained with Trimble's online GNSS planning tool.

Seven BeiDou satellites with PRN numbers 5, 7, 9, 10, 11, 13 and 14 were acquired successfully. There were one GEO satellite (PRN 5), three IGSO satellites (PRNs 7, 9 and 10), and three MEO (PRNs 11, 13 and 14) satellites available at the moment of data collection.[1]

5.4   Tracking

The tracking stage of the companion multi-GNSS software receiver starts with an FLL processing the incoming signal to compensate for the carrier mismatches. One of the

---

[1] Due to their different orbits, the GEO, IGSO and MEO satellites have different Doppler. In particular, the contribution to the Doppler frequency effect due to the satellite movement in GEO satellites is very low, as the satellite appears as fixed in the sky. This may influence signal acquisition search space and tracking parameters.

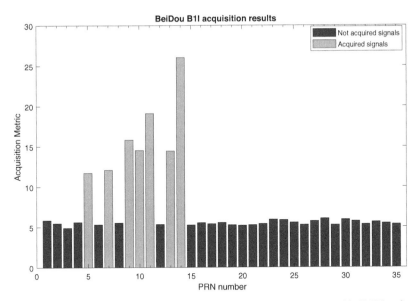

Figure 5.2 Results for BeiDou B1I acquisition using the companion multi-GNSS software receiver.

most commonly used frequency discriminators is the four-quadrant arctangent discriminator [i.e. as in Eq. (1.102), using the atan2]. But similarly to GLONASS, in case of BeiDou B1 or any modern GNSS signals with an extra tier of modulation on top of the transmitted symbols, sign transitioins usually occur rather quickly. Therefore, an FLL based on a four-quadrant arctangent discriminator is not an appropriate choice for BeiDou D1 signal tracking.

The same observation is true for BeiDou D2, where the symbol lasts for only 2 ms, with a maximum probability of symbol transition of 50%. Therefore, the BeiDou receiver should choose an FLL discriminator that is insensitive to symbol transitions. In view of this particular situation, a two-quadrant arctangent discriminator is implemented in the software receiver. The implemented two-quadrant arctangent discriminator is insensitive to symbol transition, but it has reduced tolerance of frequency uncertainty coming from the acquisition stage. This restriction in frequency uncertainty can be overcome by a proper selection of coherent integration time (i.e. frequency bin size) at the acquisition stage. The signal tracking is switched from FLL to a Costas PLL once the FLL is locked.

The main BeiDou B1 signal tracking parameters are presented in Table 5.3. In particular, the multi-GNSS software receiver used a narrow correlator discriminator [10] with early and prompt correlator spacing of $\delta = 0.25$ chips. Figure 5.3 shows the tracking status of BeiDou PRN 14 for 60 seconds long data.

Once the BeiDou receiver keeps tracking the carrier phase and the code offset of the incoming signal, the next phase is to detect the symbol boundary and then to wipe off the NH code. The purpose of symbol boundary detection is to avoid integration across a symbol edge. Algorithms for symbol boundary detection can be found in [11, 12]. The histogram method, for instance, senses the symbol sign changes and

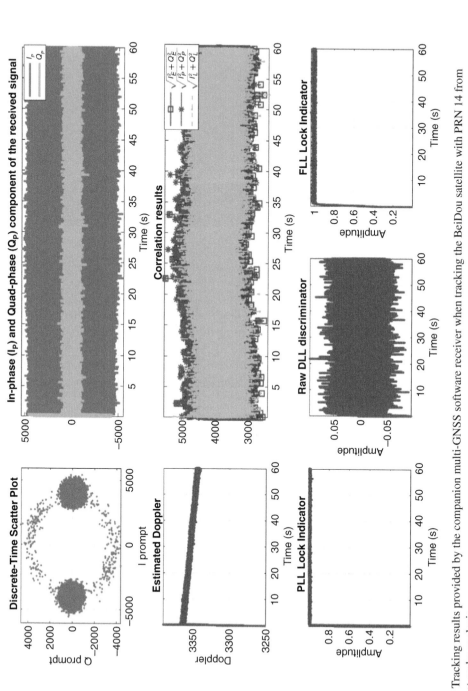

Figure 5.3 Tracking results provided by the companion multi-GNSS software receiver when tracking the BeiDou satellite with PRN 14 from the data set under analysis.

Table 5.3 BeiDou signal tracking parameter settings used to obtain the results presented herein.

Parameter	Value	Unit
Correlator spacing	0.1	chips
Number of tracking states	3	–
Tracking states	[PULL_IN; COARSE_TRACKING; FINE_TRACKING]	–
DLL loop filter bandwidth	[1; 1; 1]	Hz
FLL loop filter bandwidth	[300; 100; 5]	Hz
PLL loop filter bandwidth	[15; 15; 10]	Hz

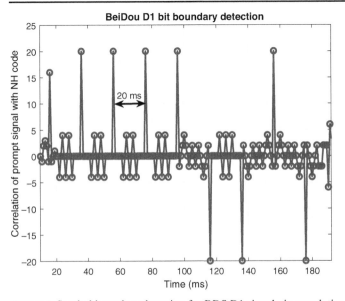

Figure 5.4 Symbol boundary detection for BDS D1 signal via correlation with NH code.

keeps a statistic of their position, but this approach will not work with the BeiDou D1 signal due to the presence of NH code. As the symbols are now modulated with the NH code, a simple correlation of the incoming NH code-modulated data with the locally generated NH code can be used to estimate the symbol edge, as shown in Figure 5.4. The index with a maximum correlation peak of 20 will be perfectly aligned with the NH code (i.e. the bit edge), and it can then be used as the symbol boundary index. In case of BeiDou D2 signal, a histogram-based approach is implemented for D2 symbol edge detection.

5.5   Computation of Position and Time

At the navigation message decoding phase, the first step is to detect the subframe preambles on the demodulated data. The BDS navigation message has both error correction coding and data interleaving. The error correction is performed by the Bose,

Table 5.4 BeiDou navigation parameter settings used to obtain the results presented herein.

Parameter	Value	Unit
Navigation solution period	100	ms
Number of smoothing samples	10	–
Elevation mask	5	degree
C/N_0 mask	30	dB-Hz

Chaudhuri and Hocquenghem (BCH 15, 11, 1) codes, which are capable of correcting one-bit error within every block of 15 bits. It is important to mention here that the BCH decoding and de-interleaving mechanism is the same for both D1 and D2 navigation messages, and in the first word of every subframe the first 15 bits are not BCH encoded. Only the next 11 bits are encoded, and hence, this word consists of 26 information bits and four parity bits. Also, data interleaving is not performed in this word. Therefore, the first word of every subframe has to be processed differently at the decoding stage within the receiver.

After successfully decoding the navigation message, a position can be computed with at least four visible satellites with the decoded ephemerides and estimated via least squares if more satellites are visible.

GPS and BDS both utilize Klobuchar model as the broadcast ionosphere correction model. The navigation message broadcast by the satellites contains a predicted ionospheric model (eight parameters, $\alpha_{0...3}$ and $\beta_{0...3}$) that can be used with the predefined ionospheric model to correct single frequency observations [13]. In case of GPS, the Klobuchar correction parameters are broadcast in subframe 4 for which the GPS receiver might need to wait for a maximum of 12.5 minutes to obtain the required coefficients. But, in case of BDS, the Klobuchar correction parameters are included already in the first subframe requiring only a maximum of 30 seconds in case of IGSO and MEO satellites and a maximum of 3 seconds in case of GEO satellites. A nice comparative study on the performance of BDS Klobuchar correction parameters is presented in [14].

The least squares-based position error variations in ENU frame are shown in Figure 5.5 for a BeiDou-only position fix. As shown in Table 5.4, a navigation solution was computed after every 100-ms interval with 10 Hz navigation update rate. The navigation solution was further smoothed with 10 navigation solution samples per epoch. The broadcasted Klobuchar ionospheric model parameters are utilized to calculate the ionospheric corrections. The position error statistics for BeiDou are finally presented in Table 5.5. They were computed with respect to the true known position, and for a standalone code-based position solution. The 3D RMS error for BDS is 7.46 m. The position error is partly due to the measurement errors and partly because the BDS satellite distribution is very asymmetric for the receiver position, since there are no visible BDS satellites in north-western sky as seen in Figure 5.1. The problem with satellite visibility is solved as BeiDou reaches full constellation in 2020. The worldwide performance for BeiDou-only receiver at present is much better than it was in 2014 due to larger number of available satellites.

Table 5.5 Position error statistics with respect to true position.

GNSS Signal	Horizontal				Vertical				3D
	RMS (m)	Mean (m)	Max (m)	Mean HDOP	RMS (m)	Mean (m)	Max (m)	Mean VDOP	RMS (m)
BDS	4.25	2.00	5.10	1.26	6.14	6.91	11.57	1.61	7.46

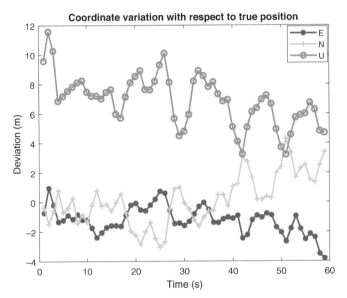

Figure 5.5 Position error variations with respect to true position in ENU frame, obtained by the multi-GNSS software receiver for the data set under analysis.

5.6   Front End and Other Practical Information

5.6.1   Front End

The Stereo dual-frequency front end from NSL with the MAXIM 2769B front end (see Table 5.6) was used to capture the real BeiDou B1I signal for the experimentation. The time-domain, signal spectrum and bin distribution plots of the digitized IF samples are shown in Figure 5.6.

5.6.2   Data Collection

The GNSS data were collected on January 31, 2014, at around 9:30 AM UTC time at a static position with a roof antenna in Finnish Geospatial Research Institute, Kirkkonummi, Finland. The data duration is 60 s.

Table 5.6 Configuration of MAXIM 2769B chipset of NSL Stereo v2 for collecting the BeiDou B1 samples.

Parameter	Value	Unit
Chipset	MAXIM 2769B	–
Bandwidth	8	MHz
Sampling frequency	26	MHz
Intermediate frequency	6.5	MHz
Number of quantization bits	2	–
Down-converted signal type	Real	–
Real data sample size	8	bits

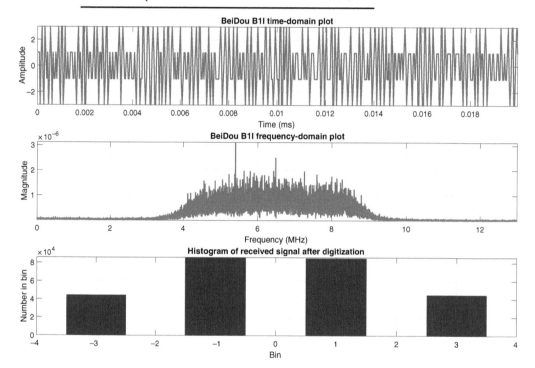

Figure 5.6 BeiDou B1I time-domain plot (up), signal spectrum plot (middle) and histogram plot of the digitized IF samples (down).

5.6.3   MATLAB Configuration and Functions

The parameter file name for BeiDou L1 receiver is "default_param_BeiDouB1_Chapter5.txt." The user needs to pass this parameter file with the main function in the MATLAB command prompt as follows:

> > gsrx("default_param_BeiDouB1_Chapter5.txt")

The enabled GNSS signal was "beib1" in this case. The front-end configuration used to collect the current samples is given in Table 5.6. The MAXIM 2769B front end saves real samples with sample size of 8 bits. Figure 5.3 can be obtained by calling the function "plotTracking.m." Figure 5.5 was generated within the function "calcStatistics.m" with a 1-Hz output rate for a standalone static GNSS user.

References

[1] A. Xu, Z. Xu, M. Ge, X. Xu, H. Zhu, and X. Sui, "Estimating Zenith Tropospheric Delays from BeiDou Navigation Satellite System Observations," *Sensors*, vol. 13, no. 4, pp. 4514–4526, 2013.

[2] N. Nadarajah, P. J. G. Teunissen, and N. Raziq, "BeiDou Inter-satellite-Type Bias Evaluation and Calibration for Mixed Receiver Attitude Determination," *Sensors*, vol. 13, no. 7, pp. 9435–9463, 2013.

[3] J. W. Betz, M. Lu, Y. J. Morton, and Y. Yang, "Introduction to the Special Issue on the BeiDou Navigation System," *Navigation*, vol. 66, no. 1, pp. 3–5.

[4] M. Z. Bhuiyan, S. Söderholm, S. Thombre, L. Ruotsalainen, and H. Kuusniemi, "Overcoming the Challenges of BeiDou Receiver Implementation," *Sensors*, vol. 14, no. 11, pp. 22082–22098, 2014.

[5] D. Zou, Z. Deng, J. Huang, H. Liu, and L. Yang, "A Study of Neuman Hoffman Codes for GNSS Application," in *Proceedings of Wireless Communications, Networking and Mobile Computing, Nanjing, China*, September 2009.

[6] C. Hegarty, M. Tran, and A. J. V. Dierendonck, "Acquisition Algorithms for the GPS L5 Signal," in *Proceedings of ION GNSS, Portland, OR*, pp. 165–177, September 2003.

[7] B. Zheng and G. Lachapelle, "Acquisition Schemes for a GPS L5 Software Receiver," in *Proceedings ION GNSS, Long Beach, CA*, September 2004.

[8] K. Yan, H. Zhang, T. Zhang, and L. X. and X. Niu, "Analysis and Verification to the Effects of NH Code for Beidou Signal Acquisition and Tracking," in *Proceedings ION GNSS+, Nashville, TN*, pp. 107–113, September 2013.

[9] China Satellite Navigation Office, "BeiDou Navigation Satellite System, Signal in Space Interface Control Document, Open Service Signal B1I," February 2019. Version 3.0.

[10] A. V. Dierendonck, P. Fenton, and T. Ford, "Theory and Performance of Narrow Correlator Spacing in a GPS Receiver," *Navigation*, vol. 39, no. 3, pp. 265–283, 1992.

[11] E. D. Kaplan and C. J. Hegarty, *Understanding GPS - Principles and Applications*. Artech House, Inc., 2006.

[12] C. Ma, G. Lachapelle, and M. E. Cannon, "Implementation of a Software GPS Receiver," in *Proceedings of ION GNSS, Long Beach, CA*, pp. 21–24, September 2004.

[13] J. Klobuchar, "Ionospheric Effects on GPS," in Spilker J.J. and Parkinson B.W. (eds.), Global Positioning System: Theory and Applications. 1996.

[14] M. Bhuiyan, S. Söderholm, S. Thombre, L. Ruotsalainen, and H. Kuusniemi, "Performance Analysis of a Dual-Frequency Software-Defined BeiDou Receiver with B1 and B2 Signals," in *Proceedings. of the China Satellite Navigation Conference (CSNC), Xian, China*, May 2015.

6   NavIC L5 Receiver Processing

Sarang Thombre, M. Zahidul H. Bhuiyan, Stefan Söderholm
and Heidi Kuusniemi

6.1   Introduction

The Indian Regional Navigation Satellite System (IRNSS), henceforth referred to by its official name Navigation with Indian Constellation (NavIC), is an independent navigation system operated by the Indian Space Research Organization. The latest version of the signal in space interface control document (SIS ICD version 1.1) was released in August 2017 [1] and describes its system architecture, satellite constellation, frequency spectrum, signal structure, modulation scheme and contents of the navigation payload. The NavIC space segment consists of seven satellites, three in geostationary (GEO) and four in inclined geosynchronous (IGSO) orbits. The three GEO satellites are positioned over 32.5° E, 83° E and 131.5° E, while the four IGSO satellites over longitudes of 55° E and 111.75° E. All seven satellites are already in space with the first being launched on x July 1, 2013, and the final one on April 12, 2018. Being a regional navigation system, NavIC primary coverage area extends to a circle approximately 1,500 km around India, wherein a positioning accuracy of better than 20 m is expected. The secondary coverage area extends approximately between latitudes 30° S and 50° N, and between longitudes 30° E and 130° E [2].

The ground segment consists of the NavIC Navigation Center which monitors and controls the overall system, along with NavIC Network Timing Center for accurate time reference. The satellite positions and orbits in space are monitored using the CDMA ranging and laser ranging stations, while their navigation data are updated using the Telemetry Tracking and Control and Navigation Uplink Centers which are a part of the Spacecraft Control Facility. The performance and integrity of the satellites are monitored by the NavIC Range and Integrity Monitoring Stations. Intermodule communication is facilitated through the Data Communication Network.

NavIC offers two services to users: an open service also called Standard Positioning Service (SPS), which is free of cost and utilizes unencrypted signals, and a restricted service (RS), which utilizes signal encryption and is available only for authorized users. Signals are offered on two frequency bands: L5-band (centered at 1176.45 MHz with bandwidth of 24 MHz) and S-band (centered at 2492.028 MHz with bandwidth of 16.5 MHz). NavIC is therefore, at the time of writing, the only satellite-based navigation system that is transmitting open signals in the S-band, which extends from 2 GHz to 4 GHz.

6.2   NavIC Signal Characteristics

The SPS signal transmitted in both the L5- and S-bands uses a CDMA scheme with BPSK(1) modulation, similarly to GPS L1 C/A. The RS signal is also transmitted in both the L5- and S-bands using a CDMA scheme, but it adopts a sinBOC(5,2) modulation for its data (in-phase) and pilot (quadrature) signal components [1]. SPS PRN codes are generated using maximum length shift registers, similar to GPS L1 C/A. The only difference is that for NavIC, a unique code for each satellite is generated using different initial bit sequences for the shift registers, rather than through different feedback paths as it is the case in GPS. The navigation data rate of the SPS signal is 25 bits per second with 1/2 FEC encoding resulting in 50 symbols (or encoded bits) per second. This means that each BPSK symbol is also 20 ms long as in GPS L1 C/A.

NavIC signals are right-hand circularly polarized (RHCP). The received power on ground using an ideally matched RHCP 0 dBi receiver antenna is between -154.0 and -159.0 dBW for the L5-band and between -157.3 and -162.3 dBW for the S-band. The use of the S-band is certainly one of the major innovations of NavIC, allowing it to provide better robustness against scintillation (because ionospheric effects are inversely proportional to the square of the center frequency). For the S-band, this means 2.5 times less ionospheric impairments compared to the conventional L-band. However, this advantage comes at the expense of higher propagation loss. The provision of positioning signals on two widely separated frequency bands also improves resilience to single-band radio frequency interference.

In this chapter, we focus exclusively on the NavIC SPS signals provided on the L5-band. A summarized comparison of these signals with GPS L1 C/A is presented in Table 6.1.

6.3   Acquisition

The acquisition of NavIC SPS signals follows the same general principles already presented in Section 1.6, with the companion multi-GNSS software receiver implementing the parallel code search acquisition based on the FFT described at the end of Section 1.6. The rest of the process is very similar to that followed by a GPS L1 C/A receiver, with the exception that the NavIC SPS spreading codes are generated slightly differently. They are actually obtained by the modulo-2 sum of two time-synchronized 10-bit maximum length feedback shift registers, called G1 and G2, each with its own generator polynomial [1]. The generator polynomial for register G1 is $X^{10} + X^3 + 1$ and for G2 is $X^{10} + X^9 + X^8 + X^6 + X^3 + X^2 + 1$. Every satellite has a unique initial condition for the G2 register [1], resulting in a unique PRN code sequence.

The NavIC spreading codes thus generated can be used in the parallel code delay search of the FFT-based acquisition algorithm [3]. For each acquired satellite, the coarse estimates of its Doppler frequency and code delay value are recorded by the multi-GNSS software receiver in order to initialize the tracking loop which will be allocated to that satellite. The acquisition search window, number of coherent and

Table 6.1 Parameters of the general signal model in Eq. (1.13) particularized for the NavIC L5-SPS signal and compared to those of the GPS L1 C/A signal.

		GPS L1 C/A	NavIC L5-SPS
Multiplexing technique		CDMA	CDMA
Center frequency		1575.24 MHz	1176.45 MHz
Primary (spreading) code type	$\{v[m]\}_{m=0}^{N_c-1}$	Gold code	Gold code
Primary code length	N_c	1,023	1,023
Primary code duration	T_{code}	1 ms	1 ms
Primary code chip rate	R_c	1.023 MHz	1.023 MHz
Primary code chip duration	$T_c = 1/R_c$	0.977 μs	0.977 μs
Primary code chip pulse shape	$p_{T_c}(t)$	$\Pi\left(\frac{t-T_c/2}{T_c}\right)$	$\Pi\left(\frac{t-T_c/2}{T_c}\right)$
Secondary code type	$\{u[k]\}_{k=0}^{N_r-1}$	1	1
Secondary code length	N_r	20	20
Data modulation		BPSK	BPSK
Symbol (bit) rate	R_d	50 sps (50 bps)	50 sps (25 bps)
Symbol (bit) duration	T_d	20 ms (20 ms)	20 ms (40 ms)
Minimum receiver bandwidth requirement		~2 MHz	~2 MHz
Pilot channel available		No	No
Error detection method		Extended (32,26) Hamming code	CRC (24 bits)
Error correction method		–	FEC (1/2, constraint length 7)
Coordinate system		WGS84	WGS84

noncoherent integration rounds and signal thresholds are all configurable parameters by the user. For the NavIC signal, one code epoch is T_{code} = 1 ms long and the data symbol is N_r = 20 epochs long, which limits the coherent integration time to T = 20 ms.

Figure 6.1 shows the skyplot for the GPS and NavIC constellations as observed over Kirkkonummi, Finland, at the time of data collection. There were 10 visible GPS satellites (PRNs 1, 3, 4, 8, 11, 12, 14, 17, 23, 25, 31 and 32) and 3 visible NavIC satellites (PRN 1 (IGSO), PRN 2 (IGSO) and PRN 3 (GEO)), mostly located in the south-east corner of the sky at quite low elevation angles (note that this is as is expected over Kirkkonummi, Finland, which falls outside even the secondary coverage area of NavIC). The acquisition parameters used to acquire the NavIC satellites are presented in Table 6.2. Relatively higher values were used for coherent and noncoherent integration time in order to acquire low elevated NavIC satellites. The acquisition function considers T = 4 ms of coherent integration and five consecutive blocks of noncoherent integration, that is, N_I = 5, thus leading to a total integration time T_{tot} = 100 ms. Similarly to what already described for GPS L1 C/A, the companion software automatically sets the resolution of the frequency search to the recommended value $\Delta v'_t$ = 1/2T already discussed in Eq. (1.86), which for this example becomes $\Delta v'_t$ = 100 Hz.

Table 6.2 NavIC signal acquisition parameter settings used to obtain the results presented herein.

Parameter	Value	Unit
Coherent integration time	5	ms
Noncoherent block number	20	–
Frequency search range	[± 6000]	Hz
Acquisition threshold	9	–

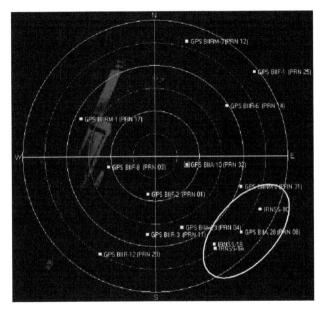

Figure 6.1 Skyplot of the recorded GPS and NavIC satellites at UTC time 07:30 AM on March 31, 2015, overhead FGI (elevation cutoff angle of 10°). Figure obtained using the Orbitron Satellite Tracking tool.

The frequency search space then spans along a set of $N_{v'} = 121$ frequency bins according to Eq. (1.89), using the maximum frequency shift of 6,000 Hz considered in Table 2.2. In case of GPS signal acquisition, same acquisition parameter values were used as mentioned in Table 2.2.

Figure 6.2 shows the comparison between the acquisition results for GPS and NavIC. Note that both signals will be combined when computing the user's position because our data do not contain sufficient NavIC satellites for a NavIC-only position solution. As mentioned earlier, the data were recorded over Kirkkonummi, Finland, which accounts for the acquisition of just three out of seven satellites. Notice also that the magnitude of the acquisition metric in Figure 6.2 is lower for NavIC compared to GPS. The results are expected to be much improved if the data are recorded from within the primary coverage area of NavIC.

Table 6.3 NavIC track signal tracking parameter settings used to obtain the results presented herein.

Parameter	Value	Unit
Correlator spacing	0.25	chips
Number of tracking states	3	–
Tracking states	[PULL_IN; COARSE_TRACKING; FINE_TRACKING]	–
DLL loop filter bandwidth	[1; 1; 1]	Hz
FLL loop filter bandwidth	[200; 100; 5]	Hz
PLL loop filter bandwidth	[15; 15; 10]	Hz

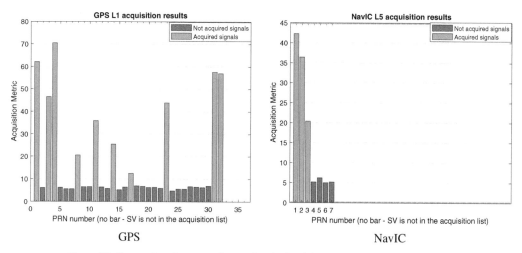

Figure 6.2 Comparison between the results obtained for the GPS L1 C/A and NavIC signal acquisition using the companion multi-GNSS software receiver. These data were recorded over Kirkkonummi, Finland, which is outside the primary and secondary coverage areas of NavIC.

6.4   Tracking

Once acquisition is completed, the results are handed over to initialize the individual tracking loops, and additional variables are defined and initialized. For carrier tracking, an FLL-assisted PLL is used to combine the robustness of the FLL with the phase accuracy of a PLL. Due to the presence of navigation bits, a two-quadrant arctangent discriminator is adopted in the multi-GNSS software receiver. For NavIC L5 code tracking, a DLL is used with a narrow correlator discriminator and early-prompt correlation spacing of $\delta = 0.25$ chips.

Some key NavIC L5 signal tracking parameters are mentioned in Table 6.3. Figure 6.3 shows the tracking status of NavIC PRN 2 over a duration of 120 seconds. Notice that since the NavIC data were recorded from outside the primary coverage area, the tracking results show increased noise compared to the other signals presented in earlier chapters.

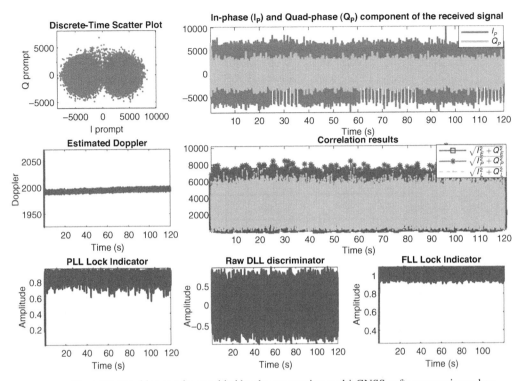

Figure 6.3 Tracking results provided by the companion multi-GNSS software receiver when tracking the NavIC PRN 2 from the data set under analysis.

6.5   Computation of Position and Time

The NavIC navigation data structure and contents are described in the SIS ICD [1]. One master frame of navigation data consists of 2,400 symbols and is divided into four subframes of 600 symbols each. Total length of a master frame is 48 seconds, and length of each subframe is 12 seconds. Sub frames 1 and 2 contain sufficient primary navigation parameters for obtaining a standard position solution. Subframes 3 and 4 contain secondary navigation parameters such as ionosphere correction coefficients, earth observation parameters, differential corrections and text messages.

The tracking loop outputs contain the navigation symbol stream from each tracked satellite. The first task in decoding the symbols is to locate the start of a subframe by searching for the first instance of the 16-bit sync word. The sync word pattern is EB90 Hex. Once a sync word has been located, the navigation symbol stream from this point onward can be divided into 600 symbol subframes and decoded sequentially subframe-by-subframe.

Each subframe of 600 symbols is decoded to extract 292 navigation data bits. First, the sync word is truncated leaving 584 symbols. Next, the symbols are passed through a 8 rows x 73 columns deinterleaver. The symbols are written along rows and read along columns. The resulting 584 deinterleaved symbols are then passed to a 171 (register G1, octal) x 133 (register G2, octal) Viterbi (1/2 convolution) decoder to

Table 6.4 Positioning accuracy of NavIC and GPS.

GNSS Signal	Horizontal				Vertical				3D
	RMS (m)	Mean (m)	Max (m)	Mean HDOP	RMS (m)	Mean (m)	Max (m)	Mean VDOP	RMS (m)
GPS	3.36	2.22	6.33	1.53	3.10	3.04	10.10	1.48	4.57
NavIC + GPS	2.50	2.00	7.51	1.53	2.39	2.02	6.23	1.44	3.46

Table 6.5 GPS + NavIC navigation parameter settings used to obtain the results presented herein.

Parameter	Value	Unit
Navigation solution period	100	ms
Number of smoothing samples	10	–
Elevation mask	5	degree
C/N_0 mask	30	dB-Hz

decode the FEC codes. The block schematic of the convolutional coding scheme is provided in [1]. It can be observed that this encoder is very similar to the convolutional encoder used for Galileo E1 navigation data. The only difference to note is that the G2 branch for the NavIC encoder does not contain an inverter. After the convolutional decoding, we are left with 292 navigation data bits. To ensure the correctness of the decoding steps performed until now, the 6 tail bits of the extracted data, needed for the FEC encoding, can be verified to be zeros. Furthermore, a 24-bit CRC can be performed based on the generator polynomial as shown in the SIS ICD [1].

After decoding the navigation message, the resulting parameters enter the main loop of the multi-GNSS software receiver for the navigation solution computation, performed using the traditional weighted least squares technique (see Section 1.10.2). The products of this final stage are structured data sets containing observations, satellite information and a navigation solution for all measurement epochs. The primary inputs to this stage are the results from track engine, the list of channels with ephemeris data available, the receiver configuration parameters and ephemeris data from NavIC.

Figure 6.4 shows the position accuracy of the GPS + NavIC solution using the software receiver platform. Some of the navigation parameters are listed in Table 6.5. The navigation solution was computed after every 100-ms interval with a 10-Hz navigation update rate. The navigation solution was further smoothed with 10 navigation solution samples per epoch.

Table 7.6 lists the different performance parameters – RMSE, mean and maximum position deviation in horizontal and vertical plane with respect to the true position. As mentioned earlier, the receiver is capable of a standalone NavIC-only position solution. However, because at most three satellites are visible from Finland, it is necessary here to demonstrate NavIC receiver operation in a multiconstellation setting. There are eight GPS satellites and three NavIC satellites used in the position computation. The 3D RMS error for this data set is 3.46 m.

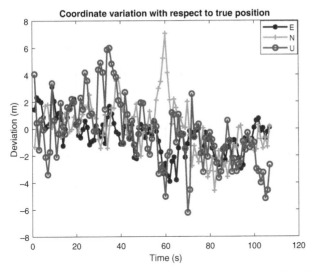

Figure 6.4 Position error variations with respect to true position in ENU frame, obtained by the multi-GNSS software receiver for the data set under analysis.

In conclusion, this chapter demonstrated the use of NavIC signals in position computation using a software receiver platform such as the one accompanying this book. In spite of falling outside the primary and secondary coverage areas, NavIC satellites are visible and usable in North Europe, and potentially in East and South-East Europe as well. The results with this particular data set showed that including NavIC in a multiconstellation solution with GPS can help to improve the GPS-only 3D RMS positioning accuracy by up to one meter.

6.6   Front End and Other Practical Information

6.6.1   Front End

The Stereo dual-frequency front-end MAXIM 2112 from NSL was used to simultaneously capture real NavIC L5 and GPS L1 signals for about 120 seconds from a static antenna installed on the roof of the Finnish Geospatial Research Institute in Kirkkonummi, Finland, on March 31, 2015, at UTC time 07:30 AM. To do so, front end 1 in Table 6.6 was configured to 4.2-MHz bandwidth to gather the GPS L1 C/A signals, while front end 2 was configured to 10 MHz to gather the NavIC L5 signals.

The NavIC L5 time-domain plot, signal spectrum and bin distribution of the digitized IF samples are shown in Figure 6.5. As already explained in Chapter 3, we can observe a non-Gaussian shape in the histogram of the I and Q values from the ADC coming from the MAXIM 2112 front end due to the AGC configuration, which may lead to a 1–2 dB additional quantization loss, without significant consequences on the positioning performance.

Table 6.6 Configuration of the two front ends included in the NSL Stereo v2 for collecting the GPS L1 and the NavIC L5 raw data samples.

Parameter	front end 1	front end 2
Chipset	MAXIM 2769B	MAXIM 2112
GNSS signal	GPS L1	NavIC L5
Bandwidth (MHz)	4.2	10
Sampling frequency (MHz)	26	26
Intermediate frequency (MHz)	6.39	0
Number of quantization bits	2	3
Down-converted signal type	Real	Complex
IQ signal orientation	–	Q,I
Raw sample size (bits)	8	16 (I+Q)

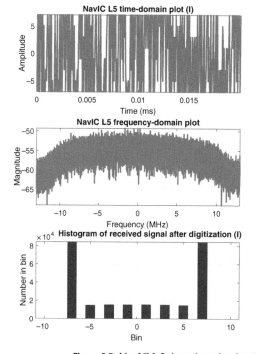
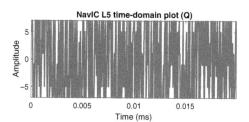
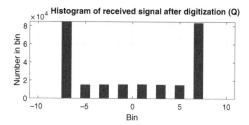

Figure 6.5 NavIC L5 time-domain plot (up), signal spectrum plot (middle), and histogram plot of the digitized IF samples (down).

6.6.2   Data Collection

Raw GNSS data were recorded at UTC time 07:30 AM on March 31, 2015, in Finnish Geospatial Research Institute, Finland. The data were collected for about 120 seconds with a static roof top antenna. The antenna was a multifrequency active antenna from GPS Source. The skyplot for GPS and NavIC constellation at the time of data collection is shown in Figure 6.1.

6.6.3 MATLAB Configuration and Functions

The parameter file name for NavIC L5 receiver is "default_param_NavICL5_ Chapter6.txt." The user needs to pass this parameter file with the main function in the MATLAB command prompt as follows:

> > gsrx("default_param_NavICL5_Chapter6.txt")

The enabled GNSS signal is chosen as "navicl5" in this particular case. The frontend configuration used to collect the example GPS L1, and the NavIC L5 raw data samples is given in Table 6.6. Out of the two front ends, the MAXIM 2112 was used to collect the GPS L1 signal and the MAXIM 2769B front end was used to save NavIC L5 signal. The GPS L1 signal is saved as real data with sample size of 8 bits, whereas NavIC L5 signal is saved as complex data with sample size of 16 bits (i.e. 8 bits for Q and 8 bits for I).

Figure 6.3 can be obtained by calling the function "plotTracking.m." Figure 6.4 was generated within the function "calcStatistics.m" with 1-Hz output rate for a standalone static GNSS user.

References

[1] Indian Space Research Organization, "IRNSS Signal in Space ICD for Standard Positioning Service." Published Online, August 2017.

[2] S. Thombre, M. Bhuiyan, S. Söderholm, M. Kirkko-Jaakkola, L. Ruotsalainen, and H. Kuusniemi, "A Software Multi-GNSS Receiver Implementation for the Indian Regional Navigation Satellite System," *IETE Journal of Research*, vol. 62, no. 2, pp. 246–256, 2015.

[3] C. Ma, G. Lachapelle, and M. E. Cannon, "Implementation of a Software GPS Receiver," in *Proceedings of ION GNSS, Long Beach, CA*, pp. 21–24, September 2004.

7   A Multi-GNSS Software Receiver

Stefan Söderholm, M. Zahidul H. Bhuiyan, Giorgia Ferrara,
Sarang Thombre and Heidi Kuusniemi

7.1   Introduction

In a software defined implementation, the main operations of a GNSS receiver, which have been described in detail in Chapter 1, are delivered by software, leading to a much more flexible design than in hardware architectures. This chapter presents the overall architecture of the multi-GNSS software receiver accompanying this book, the FGI-GSRx, used also in Chapters 2–6. In the sequel, we will describe how acquisition, tracking and PVT solution computation are implemented and then show some multi-GNSS processing experimental results obtained with a multiconstellation sample data set, which is also available to the reader as explained in Chapter 12.

The present software receiver is designed for postprocessing operations, and it does not support real-time processing. In other words, the FGI-GSRx processes stored IF samples that have been previously obtained through the use of a radio front end. The full GNSS receiver chain using the FGI-GSRx is shown in Figure 7.1 [1, 2]. The RF front end converts, in real-time, the received analog RF signals to digital IF data. These samples are stored in memory for later postprocessing with the FGI-GSRx, which performs acquisition, code and carrier tracking, navigation bit extraction, navigation data decoding, pseudorange estimation and position computation.

Within the receiver, the baseband processing block is responsible for processing the stored IF samples in order to extract navigation data and provides pseudorange estimates that are then utilized by the position computation block. The receiver is capable of performing acquisition and tracking of signals from GPS, GLONASS, Galileo, BeiDou and NavIC satellites. After the baseband processing block, the multiconstellation navigation solution block computes the multiconstellation position, velocity and time.

In the following sections, the acquisition, tracking, data decoding and multi-GNSS position computation are explained. In particular, we refer to the different configurable parameters in the receiver for each stage. In order to relate the theory of this chapter with the practical implementation in FGI-GSRx, the reader can refer in parallel to

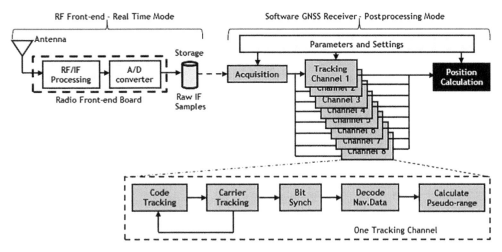

Figure 7.1 GNSS receiver chain using FGI-GSRx as postprocessing module.

Chapter 12, where the MATLAB configuration files and variables are presented, and the way to modify them to obtain different outcomes.

7.2   Multi-GNSS Signal Acquisition and Tracking

The receiver acquisition engine performs FFT-based parallel code search to identify visible satellites and obtain coarse estimates of their respective Doppler frequency and code delay values to initialize the individual tracking loops. The FFT-based parallel code delay search process was already described step by step at the end of Section 1.6.

For each navigation system, the search window, number of coherent and noncoherent rounds and detection threshold are all configurable parameters. Moreover, the user can also specify the PRN numbers of the satellites to search for. The acquisition is done for one navigation system at the time.

After acquisition has been completed for all systems, the results are handed over to initialize the individual tracking loops for all the acquired satellites. An FLL-assisted PLL for carrier signal tracking and a DLL for PRN code tracking are assigned to each visible satellite identified by the acquisition engine. Tracking is performed with a correlation interval of one code length duration, T_{code}, that can be different for each individual system (i.e. 1 ms for GPS, GLONASS, BeiDou and NavIC, and 4 ms for Galileo). The actual amount of data read from the file is adjusted for each epoch based on the true code frequency so that we always aim to process exactly one code epoch. Essentially, the code phase error is kept as close to zero as possible. In the software receiver, a two-quadrant arctangent discriminator-based FLL is implemented, which is insensitive to bit transition, but it has reduced tolerance of frequency uncertainty coming from the acquisition stage. For the phase error, the traditional Costas loop two-quadrant arctangent discriminator is used. Finally, in the code tracking loop,

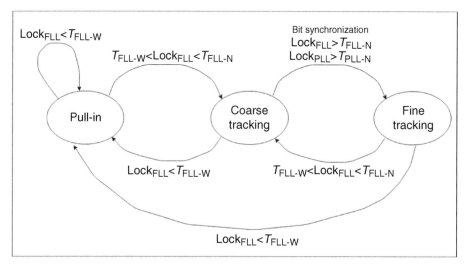

Figure 7.2 Tracking state transitions implemented in the FGI-GSRx multi-GNSS software receiver.

a Normalized Early Minus Late Noncoherent Envelope discriminator is implemented. They are discussed in Section 1.7.

FGI-GSRx supports three tracking states: pull-in, coarse tracking and fine tracking, and the tracking architecture has been designed to be highly configurable with support for the different tracking states. Within one state, the tracking loops use parameters that can be different from other states. Different channels of the receiver can be in different tracking states at any instant. Also, each channel may transit freely from one tracking state to another depending upon the prevalent signal and environmental conditions.

The transitions between states are regulated by the frequency and phase lock indicators ($Lock_{FLL}$ and $Lock_{PLL}$), which are implemented, respectively, based on [3] and [4]. For each indicator, two thresholds are used (T_{FLL-W} and T_{FLL-N} for the frequency lock indicator and T_{PLL-W} and T_{PLL-N} for the phase lock indicator). The value of such thresholds can be configured by the user in the parameter file, with $T_{FLL-W} < T_{FLL-N}$ and $T_{PLL-W} < T_{PLL-N}$. It is good to note that the subscripts "W" and "N" stand for "Wide" and "Narrow," respectively.

If the frequency lock indicator is smaller than T_{FLL-W}, the tracking is in "pull-in" state, where a wide loop filter bandwidth is used for the carrier tracking, in order to enable faster convergence to the actual Doppler frequency. When the frequency lock indicator value is between T_{FLL-W} and T_{FLL-N}, the tracking transitions into the more stable "coarse tracking" mode, where a smaller loop filter bandwidth is used. If the frequency errors increase beyond the predefined threshold, the loop reenters the "pull-in" state. Furthermore, upon successful bit synchronization, which is possible only when the receiver is tracking the carrier with sufficient accuracy, and if both frequency and phase lock indicators are greater than their respective thresholds T_{FLL-N} and T_{PLL-N}, the

Table 7.1 Example of "pull-in" state tracking table
for GPS L1 C/A signal.

Functions	Update interval (ms)
CN0fromSNR	1
freqDiscrimAtan2	1
freqLoopFilterWide	1
phaseDiscrim	1
phaseLoopFilterWide	1
bitSync	20
phaseFreqFilter	1
CN0fromNarrowWide	1
codeDiscrim	1
codeLoopFilter	1
gpsl1BitHandling	1
lockDetect	1
gpsl1UpdateChannelState	1

tracking state transitions to "fine tracking" where a very narrow loop filter bandwidth is used for the carrier tracking. If at any instant the frequency and phase errors reach unsustainable levels, the probability of data decoding starts to degrade, which in turn causes a loss in bit synchronization. The tracking reenters the more robust "pull-in" mode, and the process repeats. The possible tracking state transitions are shown in Figure 7.2.

In order to accommodate the diversity of tracking states and also to ensure that new states can be introduced without significant modification to existing software code, a table-based tracking architecture is adopted which improves its modularity and flexibility. An example of "pull-in" state tracking table for GPS L1 C/A signal is shown in Table 7.1. For a given tracking state, a set of functions is defined. Each function (column 1) defines the operation of a particular processing unit within the tracking engine, for example, C/N_0 estimator, correlator, integrator, discriminator and loop filter. The functions at different tracking states may differ in the loop filter bandwidth used for the carrier tracking. The second column of the tracking table denotes the update interval of the function (in ms). Utilizing this kind of an approach, the tracking engine can easily switch between, for example, different kinds of loop filters and accommodate for pull-in, high sensitivity or high dynamic states of the tracking without the risk of unmanageable code.

In case of multi-GNSS receiver tracking, the parameter values may be different from one constellation to the other depending on how the user would like to configure each particular signal tracking. This offers extra flexibility to the user in case he or she would like to configure receiver tracking on perconstellation basis. Some key GPS L1 C/A, Galileo E1-B and BeiDou B1I signal tracking parameters are presented in Tables 7.2, 7.3 and 7.4, respectively.

Table 7.2 GPS L1 signal tracking parameter settings in multi-GNSS receiver.

Parameter	Value	Unit
Correlator spacing	0.1	chips
Number of tracking states	3	–
Tracking states	[PULL_IN; COARSE_TRACKING; FINE_TRACKING]	–
DLL loop filter bandwidth	[1; 1; 1]	Hz
FLL loop filter bandwidth	[200; 100; 5]	Hz
PLL loop filter bandwidth	[15; 15; 10]	Hz

Table 7.3 Galileo E1-B signal tracking parameter settings in multi-GNSS receiver.

Parameter	Value	Unit
Correlator spacing	0.1	chips
Number of tracking states	3	–
Tracking states	[PULL_IN; COARSE_TRACKING; FINE_TRACKING]	–
DLL loop filter bandwidth	[1; 1; 1]	Hz
FLL loop filter bandwidth	[75; 35; 5]	Hz
PLL loop filter bandwidth	[15; 10; 10]	Hz

Table 7.4 BeiDou B1I signal tracking parameter settings in multi-GNSS receiver.

Parameter	Value	Unit
Correlator spacing	0.25	chips
Number of tracking states	3	–
Tracking states	[PULL_IN; COARSE_TRACKING; FINE_TRACKING]	–
DLL loop filter bandwidth	[1; 1; 1]	Hz
FLL loop filter bandwidth	[300; 100; 5]	Hz
PLL loop filter bandwidth	[15; 15; 10]	Hz

7.2.1 C/N_0 Computation

C/N_0 is used as a measure of strength for the received GNSS signal as already discussed in Section 1.7.2. The traditional C/N_0 estimation technique based on the Narrowband and the Wideband Power Ratio (NWPR), which is described in Section 1.7.3, is implemented in the multi-GNSS software receiver. Such technique works just perfectly for GPS L1 C/A signal. However, GNSS signals from other systems have different characteristics because of which they might no longer enjoy similar C/N_0 estimation performance that the NWPR-based technique offers for GPS L1 C/A. In particular, the performance degradation is caused by the frequent symbol transitions due to a higher symbol rate. For this reason, an additional C/N_0 estimation technique based on the SNPR is implemented in FGI-GSRx.

An estimate of the noise power μ_N is obtained by correlating the incoming signal with a +2 chips early locally generated replica. The properties of the PRN codes suggest that the autocorrelation values with the same PRN codes outside ±1 chip delay from the prompt correlator should either be zero or very close to zero due to the limiting length of the PRN codes themselves. A fair choice of +2 chips early correlation index is preferred, as this correlation index, +2 chips with respect to the prompt correlator, will have no impact from the multipath which usually comes as a delay in the late side of the correlation.

The signal plus noise power $\mu_{S,N}$ is estimated from the prompt correlation output. After each coherent integration period T, the estimate for the SNPR is computed as:

$$\widehat{\text{SNPR}}_i = \frac{\hat{\mu}_{S_i}}{\hat{\mu}_{N_i}} = \frac{\hat{\mu}_{S,N_i} - \hat{\mu}_{N_i}}{\hat{\mu}_{N_i}} \qquad (7.1)$$

and C/N_0 is estimated as:

$$\left(\frac{\widehat{C}}{N_0}\right)_i\Big|_{\text{dBHz}} = \left|10\log_{10}\left(\frac{\widehat{\text{SNPR}}_i}{T}\right)\right|. \qquad (7.2)$$

As presented in [5], the SNPR-based technique outperforms the NWPR-based estimator in case of higher data symbol rate, which is the case of the BeiDou B1I GEO signal and the Galileo E1 signal.

7.2.2   Multi-correlator Tracking

In the default configuration, only three correlator *fingers* are used for tracking and one finger for monitoring the noise level. The correlator spacing between the adjacent correlators is a configurable parameter. Additionally, the receiver has a feature called multicorrelator tracking, where the user can specify the number of correlators, correlator spacing between the fingers and correlator output rate (e.g. at 1 or 50 or a maximum of 1,000 Hz rate). This feature is intended only for analyzing the cross-correlation function of the received signal in more detail, and it does not interfere with the actual tracking. A typical multicorrelator function for a code delay window of ±2 chips is shown in Figure 7.3. In this figure, there are altogether 17 correlators with a correlator spacing of $\delta = 0.25$ chips. The correlation function resembles very well with a BOC-modulated Galileo signal.

7.2.3   Data Decoding

The sign of the prompt finger is copied into the data decoding buffer and when the buffer is full the receiver correlates with the data frame preamble for the respective system. After successful correlation, the start of a data frame is found, and the raw data bits can be extracted from the signal. The final decoded data contain the transmission time for the beginning of the data frame for each channel. Since we know the sample at which the data frame started, we can precisely link the transmission time for each signal to a specific sample count.

7.3   Computation of Multi constellation Position and Time

When tracking is completed, the position, velocity and time can be computed. The input is the decoded data and the observations for each channel from the tracking engine. Since the observations from each channel are only aligned with the bit edge of that channel at this stage, but these observations are not taken at the same time instant. The decoded data frame in each channel n will provide the transmission time T_n for the sample S_n that the receiver acquired at the beginning of the frame. In order to obtain synchronization, it is necessary to extrapolate the transmission time for all channels to one common sample S_0. This is shown in Figure 7.4.

The obtained transmission time for each channel T'_n refers then to the same sample S_0 in the incoming data. It is worth noting that since the transmission times are extrapolated from the time stamp in the data frames for each channel, they are in different time domains, namely GPS Time (GPST), GLONASS Time (GLONASST), Galileo System Time (GST, BeiDou Navigation Satellite System Time (BDT) and NavIC system time, depending on the signal which occupies that channel.

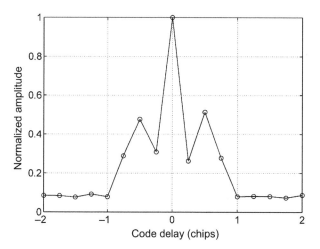

Figure 7.3 Example of Galileo E1-B correlation function on the BOC(1,1) contribution.

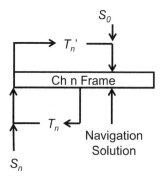

Figure 7.4 Transmission time T_n for channel n extrapolated to the sample at which we will compute our position. Note that T'_n is channel specific and system-dependent.

To obtain the initial receiver time estimate T_{rx} at sample S_0, it is assumed that the signal with the shortest travelling distance for each system has travelled for 80 ms. The accuracy of this receiver time estimate is not critical for the position solution, and our time solution will give the final accurate receiver time. The estimated receiver time is a vector with as many components as the satellite systems. For example, when processing signals from GPS, Galileo and BeiDou, such vector is:

$$
T_{rx} = \begin{bmatrix} T_{rx}^{(G)} \\ T_{rx}^{(E)} \\ T_{rx}^{(B)} \end{bmatrix}
\tag{7.3}
$$

which are the estimated receiver times in GPST, GST and BDT for the same sample S_0. The superindices in Eq. (7.3) indicate the satellite systems (G for GPS, E for Galileo and B for BeiDou).

The pseudoranges $\rho_n^{(k)}$ can then be computed as:

$$
\rho_n^{(k)} = (T_{rx}^{(k)} - T_n')c,
\tag{7.4}
$$

where n is the channel index, k the system index (GPS, Galileo, or BeiDou) and c is the speed of light. Note that, as we are discriminating between satellite systems, we use a slightly different notation than in Chapter 1.

7.3.1 Position Solution

Using an a priori estimate for the user's position $\mathbf{p}_{t-1} = [x_{t-1}, y_{t-1}, z_{t-1}]^T$ such that $\mathbf{p}_t = [x_{t-1} + \Delta x_t, y_{t-1} + \Delta y_t, z_{t-1} + \Delta z_t]^T$ for some coordinates increments $[\Delta x_t, \Delta y_t, \Delta z_t]^T$ and using the decoded ephemeris, it is possible to iteratively solve for the user position as already described in Section 1.10. In the present case, however, we need to extend the formulation to the multi-GNSS position solution. To this end the satellite positions and the estimated pseudorange between the user and each satellite, $\hat{\rho}_i^{(k)}$, are needed in order to form the observation vector (observed minus computed values),

$$
\Delta\rho = \begin{bmatrix} \rho_1^{(G)} - \hat{\rho}_1^{(G)} \\ \vdots \\ \rho_n^{(G)} - \hat{\rho}_n^{(G)} \\ \rho_1^{(E)} - \hat{\rho}_1^{(E)} \\ \vdots \\ \rho_m^{(E)} - \hat{\rho}_m^{(E)} \\ \rho_1^{(B)} - \hat{\rho}_1^{(B)} \\ \vdots \\ \rho_k^{(B)} - \hat{\rho}_k^{(B)} \end{bmatrix}
\tag{7.5}
$$

In $\hat{\rho}_i^{(k)}$, the subscript i indicates the satellite number and superscript k indicates the satellite system: G for GPS, E for Galileo and B for BeiDou. The observation vector is identical regardless of the number of systems we have. In the geometry matrix $\mathbf{A}$ containing the directional cosines, we need to add one clock term for each enabled system. A typical row in $\mathbf{A}$ can therefore be written as:

$$[\mathbf{A}]_{i,:}^{j} = [\; \frac{\delta x_i^j}{\rho_i^j} \quad \frac{\delta y_i^j}{\rho_i^j} \quad \frac{\delta z_i^j}{\rho_i^j} \quad g \quad e \quad b \;], \tag{7.6}$$

with δX, δY and δZ are the differences between the satellite coordinates and the a priori user coordinates. The parameter $g = 1$ if $j = $ (G) else zero $e = 1$ if $j = $ (E) else zero, and $b = 1$ if $j = $ (B) else zero. To obtain the updates to the a priori user position, it is necessary to solve the observation equations similarly to the procedure already described in Section 1.10. First, the observation vector is modeled to be linearly related to the vector $\mathbf{x}$ containing the updates from the previous to the current time instant,

$$\Delta \rho = \mathbf{A} \Delta \mathbf{x} \tag{7.7}$$

with

$$\Delta \mathbf{x} = [\; \Delta x \quad \Delta y \quad \Delta z \quad c\Delta b^G \quad c\Delta b^E \quad c\Delta b^B \;]^T, \tag{7.8}$$

containing the user's position increments from the a priori estimate to the current one, Δt^j are the different clock offsets for the three GNSS systems and c is the speed of light. Finally, the error vector ϵ is assumed to be zero-mean, and the solution to Eq. (7.7) can be found through the Least Squares (LS) principle to be given by

$$\hat{\Delta \mathbf{x}} = (\mathbf{A}^T \mathbf{A})^{-1} \mathbf{A}^T \Delta \rho. \tag{7.9}$$

The a-priori user position estimate $\mathbf{p}_{t-1}$ is now updated by the estimated coordinate entries in Eq. (7.9):

$$\hat{\mathbf{p}}_t = [x_{t-1} + \Delta x_t, y_{t-1} + \Delta y_t, z_{t-1} + \Delta z_t]^T, \tag{7.10}$$

The iterations (7.3)–(7.10) are repeated until the norm of the change in the estimated $\hat{\Delta \mathbf{x}}$ is sufficiently small. Note that we have one additional unknown per additional GNSS due to the time offset, and therefore, we need a measurement from an additional satellite per additional GNSS. In order to avoid that, a receiver can also use the inter-GNSS time offset broadcast in the navigation message, as for example the GGTO (Galileo-GPS Time Offset) in the case of Galileo. Notice also that, as shown in Chapter 1, Table 1.1, different GNSS give their orbits with respect to different reference frames, for example, WGS84 for GPS and NAVIC, and GTRF for Galileo. The differences in reference frames may introduce an error in a multiconstellation position if the frame offset is not taken into account. However, this error is very small and generally it is neglected for code-based positioning.

7.3.2   Time Solution

The observations from all systems are aligned to the same sample count, but the receiver time is a vector such as the one in Eq. (7.3) with the time in each system separately. The navigation solution will provide us with the clock offset, Δt for each system as shown in Eq. (7.11), and we can accurately determine the true time for that particular sample count in each system's time domain by correcting the initial estimate with Δt.

$$T_{rx}^{true} = \begin{bmatrix} T_{rx}^{(G)} - \Delta t^{(G)} \\ T_{rx}^{(E)} - \Delta t^{(E)} \\ T_{rx}^{(B)} - \Delta t^{(B)} \end{bmatrix}. \tag{7.11}$$

The GPST is semisynchronized to UTC time is such a way that the time difference can be defined as:

$$UTC - GPST = -\text{leap}_G + C_0, \tag{7.12}$$

where $\text{leap}_{(G)}$ is the number of leap seconds specified for a particular time and date. At the time of writing, the number of leap seconds for GPS was 18. The value of the constant C_0 is continuously monitored by the GPS ground segment, and parameters for a UTC model are broadcast as part of the GPS almanacs. The value of C_0 is targeted to be less than $1\ \mu s$, but it is typically smaller than $100\ ns$. GST is defined in a similar fashion:

$$UTC - GST = -\text{leap}_{(E)} + C_1. \tag{7.13}$$

The number of leap seconds is the same for Galileo as for GPS and the difference between C_0 and C_1 is typically less than $50\ ns$. Similarly, BeiDou time is defined as:

$$UTC - BDT = -\text{leap}_{(B)} + C_2. \tag{7.14}$$

The value of the constant C_2 is kept less than $100\ ns$, and for BeiDou, the number of leap seconds is 4. An example of time-domain differences are shown in Figures 7.5 and 7.6. As shown in Figure 7.6, the difference between the two constants C_0 and C_2 for this test was about $190\ ns$.

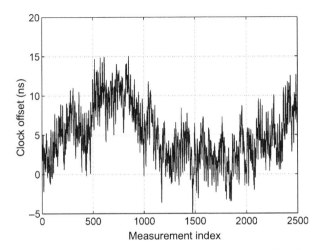

Figure 7.5 Difference between GPS and Galileo clock offsets in ns.

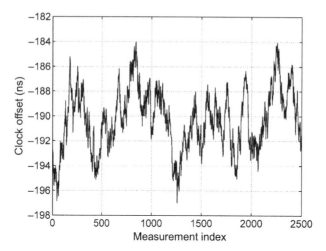

Figure 7.6 Difference between GPS and BeiDou clock offsets in ns. The leap second difference between GPS and BeiDou has been removed.

7.3.3   Velocity Solution

The velocity solution is computed in an analogue fashion as the position solution. The observation matrix in this case is the difference between the observed Doppler frequency obtained directly from the PLL and the theoretical Doppler computed from the a priori user velocity and the satellite ephemeris. The geometry matrix $\mathbf{A}$ in Eq. (7.6) is the same as for the position computation, and the solution is obtained using Eqs. (7.5)–(7.7). Instead of obtaining a position solution, we obtain a velocity solution:

$$\mathbf{\Delta v} = \begin{bmatrix} \Delta v_x & \Delta v_y & \Delta v_z & \frac{c}{L_G}\Delta f^{(G)} & \frac{c}{L_E}\Delta f^{(E)} & \frac{c}{L_B}\Delta f^{(B)} \end{bmatrix}^T, \qquad (7.15)$$

where c is the velocity of light, L is the center frequency for each system and Δf is the frequency offset from the nominal intermediate frequency. After every iteration, we obtain the true user velocity and the frequency offsets for each system.

7.3.4   Experimental Results

This subsection shows multi-GNSS processing experimental results obtained with a multiconstellation sample data set which is available with the book as described in Chapter 12. The navigation solution is computed after every 100-ms interval, that is, with a 10-Hz navigation update rate, as shown in Table 7.5. The navigation solution is further smoothed with 10 navigation solution samples per epoch. Statistics in Table 7.6 are generated with different combinations of GNSS constellations for a standalone static GNSS user with a 1-Hz navigation update rate. At the end of the processing, coordinate transformations, time corrections, and satellite elevation and azimuth angles are computed. The satellite elevation is used after the initial position estimate in order to omit satellites below a user defined cutoff angle. The update rate

Table 7.5 Multi-GNSS navigation parameter settings.

Parameter	Value	Unit
Navigation solution period	100	ms
Number of smoothing samples	10	–
Elevation mask	5	degree
C/N_0 mask	30	dB-Hz

Table 7.6 Positioning accuracy of NavIC and GPS.

	Horizontal				Vertical				3D
GNSS	RMS	Mean	Max	Mean	RMS	Mean	Max	Mean	RMS
Signal	(m)	(m)	(m)	HDOP	(m)	(m)	(m)	VDOP	(m)
GPS	1.84	1.84	4.39	1.38	1.89	1.26	4.43	1.25	2.64
Galileo	6.59	4.69	10.56	2.59	3.78	4.43	11.22	2.55	7.60
BeiDou	4.25	2.00	5.10	1.26	6.14	6.91	11.57	1.61	7.46
GPS & Galileo	1.43	1.39	3.25	1.20	2.02	1.47	5.07	1.12	2.48
GPS & BeiDou	1.78	1.48	2.93	0.91	3.61	3.55	5.44	0.96	4.02
Galileo & BeiDou	4.79	1.82	3.46	1.13	5.59	6.88	11.86	1.34	7.36
GPS & Galileo & BeiDou	1.95	1.05	2.03	0.86	3.50	3.73	5.75	0.90	4.01

for the position computation is defined by the user. The default rate is 10 Hz, that is, one position every 100 ms.

Analyses were performed with various combinations of satellite systems, and the performance metrics are presented in Table 7.6. The number of satellites in view was 8 for GPS, 4 for Galileo and 7 for BeiDou. This is reflected in the HDOP and VDOP values, which are higher for the cases with less satellites, leading also to higher errors. For the ionospheric corrections, the receiver used the default values, based on GPS's Klobuchar model [6].

From Table 7.6, we see the vertical component is offset relatively much for the Galileo-only solution and for the BeiDou-only solution which is most likely due to the geometry. This affects the vertical offset of the overall multi-GNSS solution. However, when using the three-GNSS solution, an improvement in the horizontal component is achieved, as it can also be seen from Figure 7.7. The addition of more GNSS generally increases accuracy, but this may not be the case if the added GNSS has higher measurement errors, which seems to be the case in this data capture particularly for the BeiDou case. Another interesting effect is that, with good visibility conditions, moving from two constellations with enough visible satellites (GPS and BeiDou in our case) to three (GPS, BeiDou and Galileo) does not greatly improve geometry, as seen in VDOP/HDOP values.

Notice that the purpose of the multi-GNSS experiment is to show the reader what can be performed with the receiver, but not to optimize the position solution, which the readers are encouraged to do. These results also illustrate that, while an all-GNSS position can improve the position performance, it may not always be the case.

Table 7.7 Configuration of the two front ends included in the NSL Stereo v2 for collecting GPS L1, Galileo E1-B and BeiDou B1I raw data samples.

Parameter	Front end 1	Front end 2
Chipset	MAXIM 2769B	MAXIM 2112
GNSS signal	BeiDou B1I	GPS L1, Galileo E1-B
Bandwidth (MHz)	8	10
Sampling frequency (MHz)	26	26
Intermediate frequency (Hz)	6500000	353
Number of quantization bits	2	3
Down-converted signal type	Real	Complex
IQ signal orientation	–	Q.I
Raw sample size (I+Q)	8	8 bits

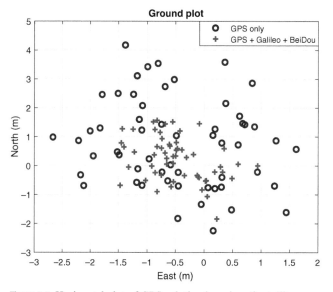

Figure 7.7 Horizontal plot of GPS solution based on 6 satellites versus a combined solution based on GPS & Galileo & BeiDou on 14 satellites

7.4   Front End and Other Practical Information

7.4.1   Front End

As explained in the previous Chapters 2–6, the dual-frequency Stereo front end from NSL was used to capture the real GNSS data. Among the two front ends in the Stereo, the MAXIM 2112 front end was configured to receive GPS L1 and Galileo E1 at 1575.42 MHz and the MAXIM 2769B was configured to receive BeiDou B1I signal at 1561.098 MHz, as mentioned in Table 7.7. The front-end configurations used to collect GPS L1, Galileo E1-B and BeiDou B1I raw data samples are also given in this table. The GPS L1 and Galileo E1-B signals are saved as real data with sample size of 8 bits, whereas BeiDou B1I signal is saved as complex data with sample size of 16 bits (i.e. 8 bits for Q and 8 bits for I).

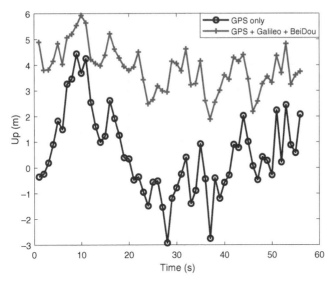

Figure 7.8 Altitude error of GPS solution based on 6 satellites versus a combined GPS & Galileo & BeiDou on 14 satellites.

7.4.2 Data Collection

The signals were captured using a G5Ant-3AT1 active antenna located on the roof of the Finnish Geospatial Research Institute in Kirkkonummi, Finland.

GPS, Galileo and BeiDou processing were enabled in FGI-GSRx. A suitable time for the test was selected so that a minimum of four satellites were visible from all three systems. Sixty seconds of data were logged, and the positions were computed at a rate of 10 Hz. The ground plot and the altitude variations are shown in Figure 7.7 and Figure 7.8, respectively. Both figures show the GPS-only and multi-GNSS solution results.

7.4.3 MATLAB Configuration and Functions

The parameter file name for multi-GNSS receiver is "default_param_MultiGNSS_Chapter7.txt." The user needs to pass this parameter file with the main function in the MATLAB command prompt as follows:

> > gsrx("default_param_MultiGNSS_Chapter7.txt")

In case of multi-GNSS signal processing, the enabled GNSS signals are "gpsl1", "gale1b" and "beib1."

References

[1] S. Söderholm, M. Bhuiyan, S. Thombre, L. Ruotsalainen, and H. Kuusniemi, "A Multi-GNSS Software-Defined Receiver: Design, Implementation, and Performance Benefits," *Annals of Telecommunications*, vol. 71, pp. 399–410, 2016.

[2] S. Thombre, M. Bhuiyan, S. Söderholm, M. Kirkko-Jaakkola, L. Ruotsalainen, and H. Kuusniemi, "A Software Multi-GNSS Receiver Implementation for the Indian Regional Navigation Satellite System," *IETE Journal of Research*, vol. 62, no. 2, pp. 246–256, 2015.

[3] C. Ma, G. Lachapelle, and M. E. Cannon, "Implementation of a Software GPS Receiver," in *Proceedings of ION GNSS, Long Beach, CA*, pp. 21–24, September 2004.

[4] B. W. Parkinson and J. J. Spilker, *Global Positioning System: Theory and Applications Volume I*. American Institute of Astronautics. 1996.

[5] M. Bhuiyan, S. Söderholm, S. Thombre, L. Ruotsalainen, M. Kirkko-Jaakkola, and H. Kuusniemi, "Performance Evaluation of Carrier-to-Noise Density Ratio Estimation Techniques for BeiDou B1 Signal," in *Proceedings of Ubiquitous Positioning, Indoor Navigation and Location-Based Services Conference, TX, USA*, November 2014.

[6] J. A. Klobuchar, Global Positioning System: Theory and Applications, Ch. *Ionospheric Effects on GPS*. American Institute of Aeronautics and Astronautics (AIAA). 1996.

8 A Dual-Frequency Software Receiver

Padma Bolla and Kai Borre

8.1 Introduction

The accuracy of GNSS-based positioning is decided by two factors: the geometry between the user and the satellites, and the observation quality. The geometry of the satellite constellation can be improved by the availability of more satellites. The observation accuracy of a standard single-frequency GNSS receiver is limited to a few meters due to some significant error sources, that is, satellite orbital parameter accuracy, satellite clock estimation accuracy, ionospheric and tropospheric delays, multipath and receiver noise. A major source of pseudorange deviation is ionospheric delay, but interestingly, this deviation can be mostly corrected by dual-frequency receivers. Another advantage of dual-frequency receivers is that they can continue operating if one of the frequencies suffers from interference. Due to these reasons, the original GPS receivers were actually conceived as L1–L2 dual-frequency receivers. In fact, the GPS receiver clock corrections in L1 C/A message are given for dual-frequency receivers, and a group delay has to be applied to navigate with L1 C/A only. The same occurs for Galileo E1-E5b I/NAV message. SBAS has functioned in one frequency (L1), but the current standard under definition at the time of writing this chapter, dual-frequency multi constellation, is also L1–L5 [1].

The parameter responsible for ionospheric delay is the total number of electrons encountered by the radio wave on its path from satellite to GNSS receiver or TEC. The difference between the arrival times of two carrier frequencies allows an explicit algebraic solution for the delay [2]. In single-frequency receivers, ionospheric delay can be corrected by several approaches. One of the simplest and most popular methods is to utilize the Klobuchar model, which includes eight ionospheric coefficients broadcasted as part of navigation message. During normal operation, the parameters of the model are updated at least once every 6 days. This algorithm can be used in real-time, and it was designed to provide a correction for approximately 50 % root mean square of the ionospheric range delay, see Section 1.9.1 or [3, Chapter 12]. The alternative is applying corrections from SBAS, but this requires being in an SBAS coverage area, and the ionospheric message may take some minutes to be received, and still the error will not be fully corrected. In spite of these limitations, most of the early civilian receivers were based on a single frequency because of inaccessibility of code on a second carrier frequency (unknown P/Y code over L2 signals) to civilians.

However, from the past decade, GNSS infrastructure has been modernized with multiple-frequency signal transmissions to overcome the inherent performance limitations of single-frequency navigation. The GPS modernization program has added new civil signals L2C and L5 to the suite of signals already transmitted by the satellites. This has enabled civilian users to use dual-frequency (L1 and L2C or L1 and L5) GPS observations in a wide spectrum of applications, including smartphones [4, 5]. The spectral spacing between the L1 and L2 frequencies is 347.82 MHz, and between the L1 and L5 frequencies is 398.97 MHz. They are large enough to facilitate estimation of the ionospheric group delay. Therefore, by tracking at least two carriers, a multiple-frequency receiver can model and remove a significant portion of the ionospheric bias. Out of the possible frequency combinations, L1 and L5 have added advantage, because of the broad bandwidth of L5 signals compared to L1 and new L2C signals [6], and the fact that all other systems are also transmitting in the L5-band. Another advantage of L5 for some users compared to L2 is that it is in the aeronautical radionavigation services (ARNS) bands of the ITU, so it can be used in civil aviation.

This chapter provides an overview of a dual-frequency GPS L1 C/A and GPS L5 software receiver implemented in MATLAB, which is also provided with this book. It also includes a brief description of the signal structure, RF front end, acquisition, tracking and accuracy improvement using dual frequency.

8.2   GPS L5 Signal Characteristics

Since GPS L1C/A signal structure was already described in Chapter 2, the present chapter will focus on the GPS L5 civil navigation signal as the counterpart for enabling dual-frequency positioning.

8.2.1   Introduction

GPS satellites transmit carrier signals at three different frequencies in the L-band, L1: 1575.42 MHz L2: 1227.60 MHz and L5: 1176.45 MHz. Until 2010, the satellites which include Blocks I, II, IIA, IIR and IIR-M, broadcasted signals at two frequencies L1 and L2. Block IIF satellites, launched first in May 2010, introduced a new civil signal at L5 to broadcast in the protected ARNS band centered at 1176.45 MHz. The civil navigation signal L5, claimed to be a "safety-of-life" signal, was designed with features of higher power, greater bandwidth and advanced signal structure to enhance civil navigation performance and utility. The use of L5 in combination with L1 improves accuracy, via ionospheric correction, and robustness via signal redundancy. At the time of writing this chapter, a total of 13 satellites are transmitting L5 signals along with L1 C/A and L2 signals and 8 satellites are transmitting preoperational civil navigation (CNAV) data over L5 carrier with status unhealthy. By 2024, the US DoD expects the L5 constellation to be complete with 24 satellites for global coverage. The L5 signal is already described in the last GPS Standard Positioning Service, Performance Standard (April 2020), at the time of writing this chapter [7].

8.2.2   Signal Structure

The L5 signal carrier has two components in quadrature, the in-phase I and the quadrature Q, both with equal signal power. The minimum received power in each component is -157.9 dBW that is 0.6 dB more than the L1 C/A code signal.

Both signal components are BPSK modulated with different PRN codes, each of length $N_c = 10{,}230$ chips generated at a rate $R_c = 10.23$ Mchip/s, resulting in a PRN repetition period of $T_{code} = 1$ ms. The in-phase signal component is modulated by a data message at $R_d = 50$ bps with FEC to create 100 symbols per second (sps) and the quadrature signal component is dataless.

In addition, the data and pilot signal components are modulated with a Neumann-Hoffman (NH) code with length $N_r = 10$ and $N_r = 20$, respectively. The data component NH transition period is thus 10 ms, and the pilot component NH transition period is 20 ms. The secondary NH codes help to reduce the code cross-correlation minor peaks, to resolve the bit timing clock and to reduce the code spectral line spacing from 1 kHz to 100 Hz in the data component and 50 Hz in dataless component.

The L5 signal is transmitted at a higher power than the current civil GPS signals and has a wider bandwidth. Its higher received power and lower carrier frequency enhance the signal reception in indoor environments. The high code rate in the L5 signal also provides improved multipath performance and protection against narrow band interference. The presence of a dataless component allows significant improvement in the signal phase tracking sensitivity, accuracy and robustness to RF interference. The GPS L5 signal is designed to be compatible and interoperable with Galileo, GLONASS and QZSS [8].

8.3   Architecture of a Dual-Frequency Receiver

A dual-frequency receiver processes two different frequency signals from the same satellite. This enables to take observations at two different frequencies and to generate a linear combination of observations that can be used for different purposes. For example, the *iono-free* combination, explained later, eliminates the ionospheric delay error, at the expense of amplification of the other uncorrelated errors, principally noise and multipath. The *wide-lane* combination has a long wavelength, and it is used for carrier phase ambiguity resolution in high-accuracy applications. The *narrow-lane* combination generates low-noise observations, and it is also used to fix the integer ambiguities in carrier phase observations. Therefore, L1–L5 receivers generally provide more accurate positioning than L1-only receivers. In open sky, iono-free observations provide very accurate positioning, as shown later. In high multipath environments, other linear combinations of L1 and L5 can be used, such as a Weighted Least Squares combination [9].

In any case, a wideband RF front-end and multichannel digital signal processing unit is required to process signals from two RF frequency bands. The RF front end receives RF signals from the antenna and, in our case, downconverts them to an intermediate frequency (IF), and finally digitizes the analog data at IF, as discussed in

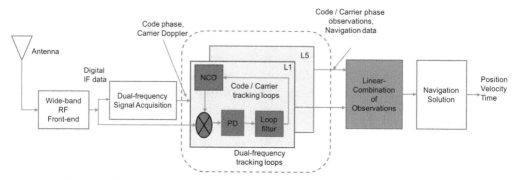

Figure 8.1 Dual-frequency receiver system architecture.

Section 1.4.3. A digital signal processing unit carries out the entire signal and data processing including acquisition, tracking and position solution. Figure 8.1 shows the functional block diagram of dual-frequency receiver system architecture.

Two digital signals are produced, one for each frequency L1 and L5. Digitized complex IF data, I and Q from the RF front end will be processed in three stages in the receiver: acquisition, tracking and position solution, similarly to what already discussed in Section 1.5 for single-frequency receivers. The signal processing module, involving acquisition and tracking, is the core of the dual-frequency software receiver, and it is computationally intensive due to the continuous processing of multiple parallel channels tracking signals at two different frequencies. This part includes multiple digital receiver channels – usually 8 to 12 – each of them tracking one of the visible satellites and collecting navigation data transmitted by them. Finally, the retrieved data along with timing information are passed to the navigation block. It extracts ephemerides from navigation data and performs a position computation based on an ionosphere-free linear combination of pseudorange observations from the L1 and L5 signals.

8.4   Acquisition

In our dual-frequency receiver, GPS L1 C/A and L5 signals can be acquired in different ways by considering the fact that the code and carrier frequencies are coherently related on these two signals for a given GPS satellite. One way is to simultaneously perform the acquisition process on both frequency channels, so as to obtain the code delay and Doppler frequency parameters for the visible satellites. Another way is to acquire signals at one frequency, so the estimated acquisition parameters $(\hat{\tau}, \hat{v})$ can be scaled and used interchangeably to acquire signals at the other frequency.

The latter approach is preferable to save computation time. In this approach, either L1 or L5 signal can be acquired and then acquisition of the second signal can be aided from the earlier acquired signal. In general, it is always advisable to acquire L1 signals and then derive L5 signal observations. The reason is that the L5 primary code chipping rate is 10 times higher than L1 code chip rate, thus making the acquisition processing time of L5 considerably longer than that of L1.

Before proceeding to detail the combined L1 and L5 signal acquisition process, we will introduce the individual L1 and L5 acquisition. Since the GPS L1 acquisition process is well discussed in Chapter 2, the following subsection focuses on a brief introduction to L5 acquisition.

8.4.1   Single- versus Dual-Channel GPS L5 Acquisition

Acquisition strategies in dual-component or dual-channel GPS L5 signals (i.e. I&Q) are slightly different from the conventional single-component or single-channel GPS L1 C/A signals.[1] This is because GPS L5 signals can be acquired either by using *data or pilot* components, or both together. The same occurs with the Galileo E1 signal, as explained in Chapter 4, with the difference that Galileo open pilot and data signals are not in phase quadrature. Some research analyzing the benefits of acquiring single-channel versus dual-channel acquisition at different scenarios was for instance carried out in [10]. Many strategies can be used to implement signal detection and acquisition. They fall in two main categories:

1. Single-channel acquisition, typically, the pilot channel.
2. Dual-channel acquisition, when both the data and the pilot channels are processed, and the energy summed.

Some of the parameters that influence the selection of a suitable acquisition method are predetection integration time, signal power and acquisition time:

– Predetection integration time, which is actually the coherent integration time T for which the received signal is correlated with the local replica. It is a compromise design parameter to process weak and high dynamic signals as discussed in Section 1.6. The predetection integration time can be as low as one code period to process strong and high-dynamics signals to as long as possible (as allowed by the quality of the receiver clock and the user dynamics) to process weak signals. We must take into account that long integration times lead to smaller frequency bins, as detailed in Section 1.6. The coherent integration time in the data or data + pilot channel acquisition is limited by the navigation data symbol period, T_d, while the dataless pilot channel acquisition in GPS L5 signal has the benefit of a longer coherent integration time. In the latter case, if the receiver is not a priori synchronized with the start of the 20 ms NH code, it should synchronize with it prior to the longer coherent integration of signal, for example, by trying different NH code starts and looking for the one providing the maximum correlation.

---

[1] Note that in this chapter for dual-frequency processing, the terms *component* and *channel* are often used indistinctly.

– Signal power in the composite GPS L5 signal is divided equally between data and pilot channels. So, processing of the combined data and pilot channels has overall 3-dB power gain compared to a single channel.
– Acquisition time is the time to detect and obtain the coarse parameters of the detected satellites. This parameter is directly proportional to computation complexity of the acquisition process. In dual (data and pilot) channel processing, computation complexity is exactly twice to that of single-channel processing because it needs to correlate the received signal with both data and pilot code replicas.

The schemes for different scenarios of GPS L5 signal acquisition are discussed below.

Single-Channel Acquisition

In general, single-channel acquisition (i.e. acquisition on I or Q components) can be done either by data or pilot channels, but the data channel processing suffers from the unknown data symbol transitions. On the other hand, pilot channel processing benefits from the ease of implementation to coherently integrate the signal for a very long period of time, comprised of the tiered primary code of the L5Q component with the secondary 20 ms NH code. This, however, requires hybrid (Primary + secondary NH) code acquisition to apply 20 ms long coherent integration. Acquisition over longer coherent integration time enables to detect signals in very poor SNR conditions such as indoors, as discussed in Section 1.6.

Regarding the L5 signal acquisition, the same operational principles as those already described in Section 1.6 apply. That is, when the local code and carrier replica are aligned to the received signal with some tentative time-delay and carrier frequency (τ', ν'), the coherent correlation output is given by Eq. (1.69). As it was already mentioned at that time, longer coherent integration needs finer Doppler frequency resolution in order to avoid the SNR loss that is incurred when coherent integration is carried out in the presence of some residual frequency error. The effect was already shown in Figure 1.20 for illustration purposes. Thus, a longer coherent integration time incurs into higher computation complexity and longer acquisition time, which has an impact on the time to first fix. Thus, unless a coarse prior knowledge of the Doppler shift and NH code delay is available, increasing the integration time costs longer acquisition time, or it requires numerous correlators [10], since more frequency bins need to be tested as indicated in Eqs. (1.88)–(1.89). This computational complexity can be reduced by taking Doppler and code-phase aiding. So, this approach is more suitable for reacquisition where there is prior coarse position information.

Dual-Channel Acquisition

In this case, the input signal is correlated with both the data and pilot local code replica obtaining two complex correlations. Subsequently, data and pilot complex correlator outputs can be combined coherently or noncoherently as discussed in [10] to generate decision statistics. This combined data and pilot channel processing increases the overall SNR. However, this approach suffers from the presence of unknown data symbol

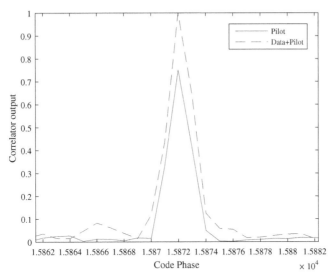

Figure 8.2 GPS L5 acquisition results: combined data and pilot channel versus pilot channel processing with $T = 1$ ms coherent integration.

transitions on the data channel and from the discrepancies between the data and pilot NH codes periods. To avoid ambiguities in symbol boundaries, one can process combined data and pilot channels with coherent integration time of $T = 1$ ms and further to use noncoherent integration. In case of $T = 1$ ms, NH code wipeoff can be avoided and the decision statistic can tolerate an error in Doppler estimate up to $v_{t,\epsilon,\max} = 500$ Hz, actually $\pm v_{t,\epsilon,\max}$, in case of using a frequency search bin width $\Delta v'_t = 1/T = 1$ kHz. This choice was already shown in Section 1.6 to incur in $\sim$4-dB loss in signal power, but it may not hurt in good visibility conditions as far as speeding up the acquisition time is the top priority, as it may be the case in the cold start mode of a receiver.

Combined data and pilot channel processing improves the acquisition gain compared to the single-channel case. As an example, correlation results from GPS L5 acquisition using single-and dual-channel approach with $T = 1$ ms coherent integration time are shown in Figure 8.2.

In summary, combined pilot and data channel acquisition with a short 1 ms coherent integration and further noncoherent integration is a suitable approach in cold start mode of L5 acquisition, with less computational complexity and time. On the other hand, pilot-only channel acquisition with longer coherent integration is suitable for acquiring signals with low SNR, if prior Doppler information is available.

Regarding the computational complexity, the following can be claimed for L1 and L5 signal acquisition. In a static case, with maximum Doppler uncertainty of ± 5 kHz and code uncertainty equal to the code length, that is 10,230 chips in L5, and 1,023 chips in L1, all frequency and code bins need to be explored. With the common assumptions that coherent integration time is $T = 1$ ms, frequency bin width $\Delta v'_t = 1/2T = 500$ Hz and code bin width half a code chip $\Delta \tau'_t/T_c = 0.5$, then the maximum Doppler uncertainty becomes $v_{t,\epsilon,\max} = \Delta v'_t/2 = 250$ Hz and the maximum code uncertainty $\tau_{t,\epsilon,\max}/T_c = (\Delta\tau_t'/2)/T_c = 0.25$. In this situation, the number of code delay cells to be searched for L5 is

$$N_{\tau|L5}' = 10{,}230/(\Delta\tau_t'/T_c) = 10{,}230 \times 2 = 20{,}460, \tag{8.1}$$

whereas for L1 is $N_{\tau|L1}' = 2{,}046$ due to the 10 times smaller chip rate. The number of frequency cells to be searched becomes

$$N_v = \lfloor 2 \times 5000/\Delta v_t' \rfloor + 1 = 21, \tag{8.2}$$

and therefore the total number of search cells for L5 is

$$N_{\tau|L5}' N_v' = 429660, \tag{8.3}$$

while for L1 it is 10 times smaller, $N_{\tau|L1}' N_v' = 42966$.

After considering the computational complexity of GPS L1 and L5 signal acquisition, a suitable approach for fast acquisition in dual-frequency receiver is to acquire L1 signal prior to L5. Subsequently, L5 acquisition process can be aided from the L1 results. This approach will bring down the computational complexity and time which is the most crucial parameter in the cold start mode of operation.

8.4.2   GPS L1-Aided GPS L5 Acquisition

The coherent generation of the GPS L1 and L5 codes from the common clock reference is described in [8]. This signal coherency feature can be explored to develop an acquisition aiding mechanism between two signals. The code delay and Doppler acquired during the acquisition of L1 signals can thus be used to find the code delay and Doppler of L5 signals. By this aiding mechanism, the L5 acquisition time reduces drastically compared to standard procedure. Figure 8.3 depicts the example of two-dimensional code delay and frequency search processes, which come as the result of the following steps.

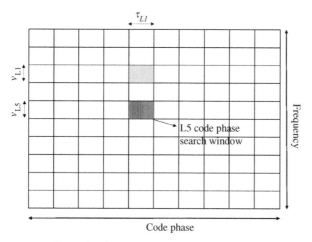

Figure 8.3 Example of GPS L1 and L5 acquisition process in code delay and frequency dimension.

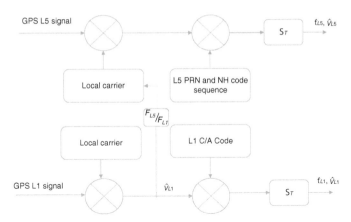

Figure 8.4 Example of GPS L1-aided L5 acquisition process.

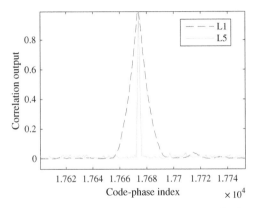

Figure 8.5 Correlation peak of the GPS L1 and L5 acquired signals for a given satellite.

First, the true GPS L1 code delay and Doppler values (τ_{L1}, ν_{L1}) need to be found using, for instance, the parallel code delay search algorithm described in Section 1.6. Then, the Doppler frequency of the L5 signal can be obtained by scaling the L1 Doppler estimated value as follows:

$$\hat{\nu}_{L5} = \hat{\nu}_{L1} \frac{F_{L5}}{F_{L1}},\qquad(8.4)$$

where $F_{L1} = 1575.42\,\text{MHz}$ and $F_{L5} = 1176.45\,\text{MHz}$.

Next, the L5 carrier is generated using the Doppler value in Eq. (8.4), and the carrier is wiped off from the composite received signal. Now the acquisition process has reduced to a one-dimension code delay search problem. At this point, the L5 primary code can be generated and wiped off from the received signal through the correlation process. Figure 8.4 shows the resulting functional diagram of combined GPS L1 and L5 acquisition process.

Figure 8.5 shows the L1 and L5 signal correlation output in the code delay dimension, as obtained during the acquisition stage. Both L1 and L5 signals are detected in

Table 8.1 Signal parameters for the acquisition of
GPS L1 and L5 signals used herein.

Frequency	GPS L1	GPS L5
Coherent integration time	1 ms	1 ms
Acquisition search range	10 kHz	10 kHz
Acquisition threshold	3	3

the same code delay cell, but with different precision based on the respective signal chip rate. The GPS L5 signal with 10 times higher chip rate than L1 has less influence of observation noise and multipath. Hence, the spread in the L5 signal correlator output is low as compared that of L1 signal. As a result, the L5 code delay observations are more precise.

In case of GPS L1 and L5 combined signal acquisition, the typical values used to generate Figure 8.5 are mentioned in Table 8.1. The user has the option to select the required coherent integration time for a particular signal acquisition. For GPS L1 C/A signal, the user could moderately extend the coherent integration time beyond 1 ms due to the nonexistence of secondary code and the duration of navigation data symbol lasting for 20 ms. In the GPS L5 signal, the secondary code lasts for 10 ms in the I-channel and 20 ms in the Q-channel. Hence, the coherent integration time beyond 1 ms requires hybrid code acquisition (i.e. primary + secondary code), which can extend up to 10 ms for I-channel and 20 ms for Q-channel. The received signal will be retrieved through correlation process depending on the selection of these acquisition parameters (i.e. choice of coherent integration, search bandwidth).

8.5   Tracking

The acquisition process obtains coarse code delay and Doppler values of all visible satellites. Based on these values, the tracking loops commence and keep track of the code and carrier phase changes of all the satellite signals continually. In dual-channel GPS L5 signal tracking, many strategies can be adapted based on the intended application.

As mentioned earlier in the acquisition process, the GPS L5 composite signal, pilot and data, can be tracked as single- or dual-channel. Both approaches have their merits and demerits. To benefit from the total incoming power, the receiver needs to process both pilot and data channel, with separate correlators and discriminators, and then to combine results appropriately. This approach is computationally expensive. Moreover, coherent integration time in data + pilot channel tracking is limited to navigation symbol period due to unknown symbol transition in data channel. In case of single-channel tracking, it is of course more advantageous to process the pilot channel and to use the code and carrier phase observations to demodulate the navigation message from the in-phase channel. Although each component has only half the total power, processing the

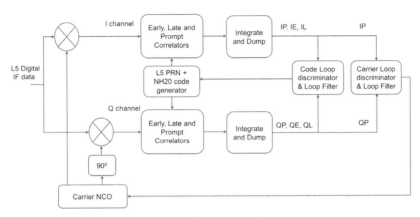

Figure 8.6 GPS L5 pilot (Q) channel tracking loop architecture.

pilot channel allows using a four-quadrant (i.e. atan2) instead of a symbol-insensitive Costas (i.e. atan) phase loop discriminator. This gives a 6-dB advantage in tracking threshold, offering an overall gain of 3 dB [11]. Details of the GPS L5 signal tracking loop is discussed in the following paragraph.

The GPS L5 data or pilot channel tracking loop follows a similar structure to that of the general tracking architecture of GPS L1C/A. But if the coherent integration time is above 1 ms, an additional step is required to synchronize the secondary (NH) code and wipe off the secondary codes as shown in Figure 8.6. In case of 1 ms coherent integration, secondary code wipe-off is not required at the tracking stage and that can be done in the navigation stage as part of the symbol synchronization process. Figure 8.7 shows the pilot channel tracking loop output after wiping off primary code and secondary codes at 20-ms signal integration time. The plot depicts the L5 pilot channel tracking loop results tapped at the output of code and carrier phase discriminators, loop filters and early/late/prompt correlators.

Some important tracking loop parameters for dual-frequency receiver are presented in Table 8.2. There are a good number of other relevant tracking parameters specific to each signal. The signal-specific parameters are usually known from the interface control document [8].

8.6   Generation of GPS Observations

The receiver navigation process uses the code and carrier tracking loop accumulator outputs to generate pseudorange, delta pseudorange, and carrier phase or integrated Doppler observations. Pseudorange observations are generated from code delay values as described already in Chapters 1 and 2. The dual-frequency software receiver accompanying this book uses carrier phase observations to smooth out the code error, so we will focus now on carrier phase observations.

Table 8.2 Signal parameters for the tracking of GPS
L1 and L5 signals used herein.

Frequency	GPS L1	GPS L5
Correlator spacing	0.1	0.1
Tracking integration time	20 ms	20 ms
DLL loop BW	1 Hz	1 Hz
PLL loop BW	15 Hz	15 Hz
FLL loop BW	10 Hz	10 Hz

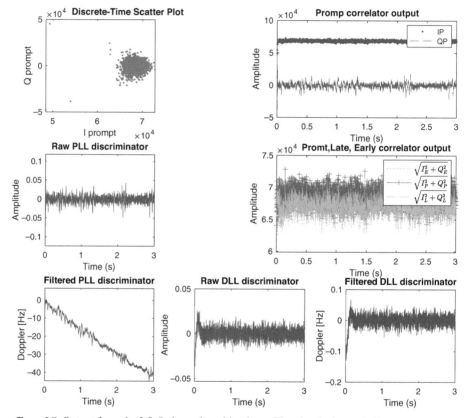

Figure 8.7 Output from the L5 Q channel tracking loop. The plot depicts L5 pilot channel
tracking loop results tapped at the output of code and carrier phase discriminators, loop filters
and early/late/prompt correlators.

8.6.1   Carrier-Phase Observation

It is also called carrier beat phase observation, and it measures the phase of the GNSS
signal frequency carrier. The carrier phase measurement estimation is based on the
difference between the phases of the received carrier signal and the carrier signal generated by the receiver in the PLL. It is about a hundred times more precise than the

code-based pseudorange measurement, but it is ambiguous, as the receiver does not know a priori the integer number of cycles between the satellite and the receiver. If the receiver can solve this ambiguity, it can obtain a position accurate to the centimeter level.

The carrier phase observation in cycles (or number of wavelengths) can be written as

$$\Phi(t) = \Phi_u(t) - \Phi_s(t - \tau) + N, \tag{8.5}$$

where $\Phi_u(t)$ is phase of the receiver generated signal, $\Phi_s(t - \tau)$ is the phase of the received signal from satellite at time $t - \tau$ and N is the integer ambiguity. Carrier observations are generated using a carrier accumulator in the carrier tracking loop, which tracks the integer number of carrier cycles and fractions of a cycle since the carrier lock is established. It is updated at the Numerically Controlled Oscillator update rate.

Both code and carrier phase observations are corrupted by the same errors, although the ionospheric error is added to the code, delaying it and subtracted to the carrier. A detailed explanation of this a priori not very intuitive effect is provided in [12, Chapter 5]. Another important difference is that the code delay observations are coarse but unambiguous and thus can be used for absolute positioning. On the other hand, carrier phase observations are precise, but ambiguous, as abovementioned, and they are generally used in relative (RTK) and absolute (PPP) precise positioning.

8.6.2 Doppler Observation

The relative motion of satellite and user receiver results as Doppler shift v_t in the frequency of the satellite signal. Given the satellite velocity, the Doppler shift can be used to compute the user velocity, as shown in Chapter 7. This Doppler shift is observed traditionally by the carrier tracking loops (FLL, PLL). Doppler frequency is equivalent to rate of change of pseudorange observation or rate of change of carrier phase observation over a time interval as given below,

$$v_t = \frac{\rho(t_2) - \rho(t_1)}{(t_2 - t_1)\lambda} = \frac{\Phi(t_2) - \Phi(t_1)}{t_2 - t_1}, \tag{8.6}$$

where t_1 and t_2 are two time instants.

8.7 Dual-Frequency Receiver Observations

By accounting for clock biases, initial phase offsets, atmospheric propagation delays and observational errors, the *measured* pseudorange ρ_i and the carrier phase range

Φ_i at a given carrier frequency F_i can be modeled as observation equations in units of [m] as[2]

$$\rho_i = r + c(b - dt_i) + I_i + T_i + \varepsilon_{\rho_i}$$
$$\Phi_i = r + c(b - dt_i) - I_i + T_i + \lambda_i N_i + \varepsilon_{\Phi_i}, \qquad (8.7)$$

where r is the true geometric distance between the satellite and receiver [m], c is the speed of light [m/s], b and dt are the receiver and satellite clock errors [s], I_i is the ionospheric delay [m] at the subscripted frequency, T_i is the tropospheric delay [m], N_i is the integer carrier ambiguity [cycles], λ_i is the wavelength of the subscripted frequency, and ε_{ρ_i} and ε_{Φ_i} are remaining observation errors in code and carrier phase for the subscripted frequency, including multipath and receiver noise.[3]

We now remove from Eq. (8.7) satellite clock errors with the broadcast clock parameters and estimate tropospheric delay using atmospheric parameters. Then our observational model Eq. (8.7) simplifies to:

$$\rho_i = r + cb + I_i + \varepsilon_{\rho_i}$$
$$\Phi_i = r + cb - I_i + \lambda_i N_i + \varepsilon_{\Phi_i}, \qquad (8.8)$$

Considering two pseudorange observations obtained by tracking two distinct signals of the same satellite at frequencies F_i and F_j, most of the terms in Eq. (8.7) cancel when differencing both observations. The code and carrier pseudorange difference of two signals is

$$\rho_i - \rho_j = I_i - I_j + \Delta\varepsilon_{\rho_{ij}}$$
$$\Phi_i - \Phi_j = I_j - I_i + \lambda_i N_i - \lambda_j N_j + \Delta\varepsilon_{\Phi_{ij}} \qquad (8.9)$$

where $\Delta\varepsilon_{\rho_{ij}}$, $\Delta\varepsilon_{\Phi_{ij}}$ are differential code and carrier phase observation bias due to receiver noise and multipath.

The ionospheric signal delay in Eq. (8.8) is significant, and thus, it is of major concern for satellite navigation. The frequency dependence of the ionospheric effect I_i is described in Eq. (8.10), see also Chapter 1 or [3, Chapter 12]:

$$I_i = \frac{40.3}{F_i^2}\text{TEC.} \qquad (8.10)$$

The ionospheric delay in Eq. (8.10), given in meters, is dependent on the carrier frequency F_i and the TEC. Hence, the frequency dispersive nature of the ionosphere can be exploited to estimate and eliminate ionosphere delay in code and carrier phase

---

[2] Following Chapter 1 notation, we make a distinction between the measured pseudorange ρ_m, before correcting propagation and satellite clock errors, and the corrected pseudorange ρ after correcting these errors. In this chapter, we add the underscript i to denote the frequency, and for simplicity, we omit the superscript (p) which indicates the satellite. Furthermore, we assume here that the term b, the receiver clock bias, is common to both frequencies. Any additional error from this assumption, for example, due to receiver group biases, will be estimated as part of the other errors, or remain in the residual errors.

[3] As in Chapter 1, the multipath error (M_{ρ_i} and M_{Φ_i} in code and carrier phase pseudoranges [m]) is incorporated into the noise term.

range observations. The ionosphere delay in signal path can be estimated by taking a linear combination of the code and carrier phase observations at two frequencies. From Eqs. (8.9) and (8.10), the linear combination of ionosphere-free code and carrier phase pseudorange observations for F_i and F_j frequencies can be expressed as:

$$\rho_{\text{iono-free}} = \rho_i - \frac{1}{1 - \gamma_{ij}}(\rho_i - \rho_j) \tag{8.11}$$

$$\Phi_{\text{iono-free}} = (\Phi_i - \lambda_i N_i) - \frac{1}{1 - \gamma_{ij}}((\Phi_i - \Phi_j) - (\lambda_i N_i - \lambda_j N_j)), \tag{8.12}$$

where $\gamma_{ij} = F_i^2/F_j^2$, while ρ_i and Φ_i are the code and carrier phase pseudoranges observed on the frequency indicated by the corresponding subscript, and finally c is speed of light.

From the mathematical model for ionospheric delay in Eqs. (8.10) and (8.9), ionosphere error in code and carrier phase pseudoranges can be estimated as:

$$I_\rho = \frac{F_i^2 F_j^2}{F_i^2 - F_j^2}(\rho_i - \rho_j) \tag{8.13}$$

$$I_\Phi = \frac{F_i^2 F_j^2}{F_i^2 - F_j^2}((\Phi_i - \Phi_j) - (\lambda_i N_i - \lambda_j N_j)). \tag{8.14}$$

8.7.1   Dual-Frequency Pseudorange Equations

The ionosphere-free pseudorange in Eq. (8.12) needs to account for the multiple carrier frequencies and for satellite-specific signal alignment errors in the new L2C and L5I and L5Q signals [13]. The broadcast GPS ephemeris data are based on an ionosphere-free pseudorange computed from dual L1P/L2P observations, but the new civil signals are not all perfectly aligned to it. It is for this reason that new broadcast parameters and a new modernized dual-frequency algorithm are needed in order to align new signals with the dual L1P/L2P signal. New intersignal correction (ISC) broadcast parameters and the modernized dual-frequency algorithm were published in [8]. Here, ISCs are defined as differential delays between all signals relative to L1P. Each satellite has its unique set of ISC. Newly added signals from IIR-M and IIF satellites are now broadcasting CNAV messages along with ISCs as MSG 30.

As per [8] the two signals, L1 C/A and L5 user shall correct for the group delay and ionospheric effects, as follows. Let c denote the speed of light, and set $\gamma_{1,5} = F_1^2/F_5^2$, the intersignal group delay be T_{GD} and ISC_i the ISC for the channel indicated by the subscript. Then, a user exploiting the combination of L1 C/A and L5-I shall correct for the group delay and ionospheric effects by applying the following correction formula:

$$\rho_{1,5} = \frac{(\rho_{\text{L5I}} - \gamma_{1,5}\rho_{\text{L1C/A}}) + c(\text{ISC}_{\text{L5I}} - \gamma_{1,5}\text{ISC}_{\text{L1C/A}})}{1 - \gamma_{1,5}} - cT_{\text{GD}}, \tag{8.15}$$

while a user exploiting the L1C/A and L5 Q5 signals shall use the formula,

$$\rho_{1,5} = \frac{(\rho_{\text{L5Q}} - \gamma_{1,5}\rho_{\text{L1C/A}}) + c(\text{ISC}_{\text{L5Q}} - \gamma_{1,5}\text{ISC}_{\text{L1C/A}})}{1 - \gamma_{1,5}} - cT_{\text{GD}}. \tag{8.16}$$

Now, $\rho_{1,5}$ is the pseudorange corrected for ionospheric effects, and ρ_i is specifically the pseudorange observed on the channel indicated by the subscript.

From Eq. (8.12), the noise variance in ionospheric-free pseudorange observations is:

$$\sigma_{\text{iono-free}}^2 = \sigma_1^2 + (\frac{1}{1 - \gamma_{1,5}})^2(\sigma_1^2 + \sigma_5^2).$$

(8.17)

From the noise variance expression Eq. (8.17), it is inferred that the linear combination of ionosphere-free pseudorange observations amplifies the observation noise. Hence, the linear combination of pseudorange observations has eliminated common-mode ionosphere delay at the expense of a raise in the other uncorrelated observation noise. Other uncorrelated receiver errors that are not eliminated in the linear combination of dual-frequency observations are due to receiver thermal noise, multipath and interference. This is the major limitation of dual-frequency ionosphere-free combination. In case of an L1 and L5 dual-frequency receiver, this error variance is increased by a factor of three, approximately. The high precision in the L5 signal code range observations attained through higher chip rate will be degraded while combined with a low precision L1 code range observations. This needs to be addressed through proper signal processing algorithms to get mutual benefit of two signals.

As mentioned earlier, GPS L1 and L5 signals are coherently generated from the same satellite, hence this feature can be exploited to develop collaboration between two frequency signal tracking loops. In a GPS L1 and L5 signal linear combination, the L5 signal with a higher chip rate has lower observation noise and multipath than the L1 signal. Hence, the Doppler aid from L5 signal tracking loop improves L1 signal tracking noise performance. As can be seen in the literature, the Doppler aid to a GPS receiver is not a new concept and was inspired from the use of external inertial devices to improve the receiver tracking loop noise performance under high dynamic conditions. A similar performance improvement can be achieved through Doppler-aided tracking in a dual-frequency receiver, as discussed in [6]. Another approach to reduce the noise in code pseudorange observations is by making use of carrier phase observations as explained in the next section.

8.7.2   Dual-Frequency Carrier-Phase Equations

Carrier phase observations can be used to reduce the code pseudorange observation noise. In the early days of GPS navigation, Ron Hatch at Magnavox designed a filter that combined delta carrier phase observations and code pseudoranges into a single noise reduced observation [14]. The Hatch filter, also known as carrier smoothing algorithm, is commonly used to filter code pseudorange observation noise using precise delta carrier phase observations. The change in pseudorange-related geometrical change in the satellite's position is represented by the change in the carrier phase observations from satellite to the receiver. The advantage of this procedure is that the delta carrier phase observations are more precise and have lower noise than code observations. Moreover, the integer ambiguity and cycle slip occurrence in carrier phase observations can be neglected in delta carrier phase observations. The Hatch filter algorithm uses a time-dependent weighting factor for a proper weighting of code and carrier phase observations. The carrier smoothed code pseudorange $\rho_{sm}(n)$ can be expressed as:

$$\rho_{sm}(n) = \frac{1}{\alpha}\rho(n) + (\frac{\alpha - 1}{\alpha})(\rho_{sm}(n-1) + (\Phi(n) - \Phi(n-1))), \tag{8.18}$$

where $\rho(n)$ and $\Phi(n)$ are the code and carrier phase range observations in meters for the current epoch and α represents a weight factor. For the first epoch, the weight factor set to 1 makes the full weight to the code pseudorange. For the subsequent epochs, the weight of the code range is continuously reduced by a value $\Delta\alpha$ and the relative weight of carrier phase observations is thus increased. The amount of reduction of the weight factor from epoch to epoch controls the behavior of the algorithm.

Combining code pseudorange and delta carrier phase observations can reduce the noise in the pseudorange observation significantly. However, the carrier smoothing procedure is practically limited to short time intervals in single-frequency receiver due to ionospheric error divergence in code and carrier phase observations. The ionospheric phase advance is equal and opposite to the ionospheric group delay, so over time the change in pseudorange deviates from the change in the carrier phase according to the ionosphere change. Hence, the effectiveness of the phase smoothing procedure is limited in a kinematic environment.

The ionosphere divergence in carrier smoothing procedure can be eliminated using the ionosphere error-free linear combination of code and carrier phase observations in a dual-frequency receiver. In particular, the ionosphere-free pseudorange observations can be smoothed using ionosphere-free delta carrier phase observations as follows:

$$\rho_{sm,\,iono\text{-}free}(n) = \frac{1}{\alpha}\rho_{iono\text{-}free}(n) + \left(\frac{\alpha - 1}{\alpha}\right)\left(\rho_{sm,\,iono\text{-}free}(n-1) + T\,\Delta\Phi_{iono\text{-}free}(n)\right),$$
$$\tag{8.19}$$

where $\Delta\Phi_{iono\text{-}free}(n)$ is the ionosphere-free delta carrier phase observation in meters/sec and T is the observation time interval. The carrier smoothing algorithm for GPS L1 and L5 dual-frequency receiver can thus be expressed as

$$\rho_{sm,\,1,5}(n) = \frac{1}{\alpha}\rho_{1,5}(n) + \left(\frac{\alpha - 1}{\alpha}\right)\left(\rho_{sm,\,1,5,}(n-1) + T\,\Delta\Phi_{1,5}(n)\right), \tag{8.20}$$

where $\rho_{1,5}(n)$ is the ionosphere-free pseudorange and $\Delta\Phi_{1,5}(n)$ is the ionosphere-free delta carrier phase observation in GPS L1/ L5 dual-frequency receiver.

It is to be noted that carrier phase observations at longer signal integration time have less fluctuations in challenging signal environment [15]. Hence, the carrier phase smoothing algorithm performance is better with longer signal integration time.

8.8   Computation of Position and Time

The role of the navigation processor can be divided into four tasks. The first one is to estimate the pseudorange and range rate (Doppler) observations from the tracking loop accumulators. The next task is to decode the navigation data from the data channel. After, the receiver can compute the satellite position. Finally, with at least four measurements and satellite positions, the receiver can compute the user position, as explained in Chapter 1. In particular, the dual-frequency software receiver in this book performs the following sequence of operations in the navigation processor:

1. Collect the code delay and carrier phase observations from L1 and L5 signal tracking loops.
2. Decode the navigation data from one of two-frequency channels, that is, either from L1 or L5 tracking loop in-phase component. Then, extract ephemeris data and satellite transmit time. During the time of writing this chapter, GPS L5 signal is not transmitting operational navigation data; hence, L1 signal is chosen to decode data and extract ephemeris in our software receiver.
3. Compute the pseudorange and range rate observations for L1 and L5 signals, based on the transmit time, code delay offset and carrier Doppler values.
4. Compute satellite positions at a reference epoch, here the start of subframe 1 of the GPS L1 signal.
5. Take L1 and L5 code and carrier phase observations and compute ionosphere-free pseudorange and range rate observations as per the algorithm given in [8] and explained before.
6. Apply carrier smoothing on ionosphere-free pseudorange observations using ionosphere-free delta carrier phase observations.
7. Compute position using a least-square solution or a Kalman filter based on satellite positions and ionosphere delay-free pseudorange observation.

Some parameters relevant to the navigation solution are presented in Table 8.3. The user can choose the satellite elevation mask. The navigation output rate is 2 Hz with 500 ms of navigation solution period. The navigation engine will then decide on the satellites which pass this criteria. The user can modify this value depending on the requirement of the underlying GNSS application.

Position errors in East, North and Up (ENU) directions for real-time observations at Samara University roof top on July 6, 2016, at 12 noon are shown in Figure 8.8(a). It is observed that the error variance in the ionosphere-free position is significantly

Table 8.3 Dual-frequency navigation solution parameters.

Parameter	Value
Navigation solution period	500 ms
Elevation mask	10 degree

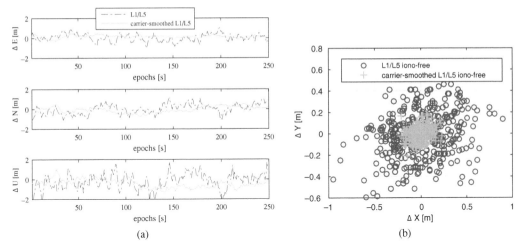

Figure 8.8 (a) Dual-frequency position error variation in ENU coordinates. (b) Position results of the ionosphere-free combination of L1 and L5 observations from six GPS satellites. Data obtained at Samara University, July 6, 2016, 12:30.

reduced using carrier phase smoothing. As a result, the precision in position solution improved as shown in Figure 8.8(b), which shows a scatter plot of the static positions. Position values are computed by taking pseudorange observations from L1 and L5 combination of signals at an epoch interval of 500 ms, over a period of 200 s. Position values were computed using six satellite signals with a PDOP = 2.4 and ionosphere-free linear combination of L1 and L5 pseudorange observations and spread with a standard deviation of 0.8 m. After code pseudorange observations are smoothed using ionosphere-free delta carrier phase observations, precision in position solution is improved to a standard deviation of 0.2 m. These results show the great advantages of using dual-frequency receivers, and in particular carrier-smoothed measurements, especially in good visibility and low multipath environments as this one.

8.9 Front End and Other Practical Information

8.9.1 Front End

A four-channel wideband RF front-end SdrNav40 [16] was used to record the signals processed in this chapter. An example of the front-end configuration is given in Table 8.4. In order for the receiver to work properly, the user should specify the following front-end parameters: sampling frequency in hertz (Hz), sample size in bits (i.e. number of bits in one sample), intermediate frequency, down-converted signal type (real or complex) and IQ data orientation (i.e.-in case of a complex signal type, the IQ data can be written as I, then Q, or swapped and written as Q, then I).

Table 8.4 GPS L1 and L5 dual-frequency receiver front-end configuration.

Frequency	GPS L1	GPS L5
Sampling frequency	27.456 MHz	27.456 MHz
Intermediate frequency	132 kHz	418 kHz
Bandwidth	20.46 MHz	20.46 MHz
Data format	I, Q	I, Q
Number of quantization bits	8	8

8.9.2   Data Collection

The GNSS data processed herein were collected using a dual-band antenna, Javad G5T, along with the SdrNav40 front end. The data were recorded at Samara University roof top on July 6, 2016, at 12 noon, leading to the results shown before in Figure 8.8(a).

8.9.3   MATLAB Configuration and Functions

For the ease of understanding, the users of DF-GSRx software receiver are provided with parameter files to process L1 and L5 frequency channels separately. For example, the parameter file name for GPS L1 receiver is "SettingsL1.m" and GPS L5 is "SettingsL5.m." These parameter files for two-frequency channels are read through a function call in the main function "DFSRx.m." The DFSRx processes the raw digital IF data provided in the specified file name, and the user must have to specify a minimum set of data configuration parameters in the parameter file, as per Table 8.4.

The acquisition results can be plotted by the function call "plotAcquisition.m"; the plotAcquisition function generates the acquisition metric for all the searched PRNs.

The users can find tracking parameters in the function called "settngsLx.m" for each signal. Figure. 8.7 can be obtained by calling the function "plotTracking.m." The plotTracking function produces one plot for each satellite tracking channel showing the performance of different key tracking parameters. The signal integration time in tracking loop can be extended beyond 1 ms after finding the bit boundary of the received signal through a function called "bitSynchronisation.m." The DLL/PLL/FLL tracking loop bandwidth changes as a function of signal coherent integration time and expected signal dynamics. Hence, the user has to consider all these design issues while setting the tracking loop parameters.

Figure 8.8 is generated within the function "plotNavigation.m" with 2-Hz output rate for a stand-alone static GNSS user.

References

[1] T. Walter, J. Blanch, and P. Enge, "L1-L5 SBAS MOPS to Support Multiple Constellations," in *Proceedings of the 25th International Technical Meeting of the Satellite Division of the Institute of Navigation (ION GNSS), Nashville, TN*, Citeseer, 2012.

[2] K. Borre and G. Strang, *Algorithms for Global Positioning*. Wellesley-Cambridge Press , 2012.

[3] B. W. Parkinson and J. J. Spilker, *Global Positioning System: Theory and Applications Volume I.* American Institute of Astronautics, 1996.

[4] P. Crosta, P. Zoccarato, R. Lucas, and G. De Pasquale, "Dual Frequency Mass-Market Chips: Test Results and Ways to Optimize PVT Performance," in *Proceedings ION GNSS+, Miami, FL*, pp. 323–333, September 2018.

[5] U. Robustelli, V. Baiocchi, and G. Pugliano, "Assessment of Dual Frequency GNSS Observations from a Xiaomi Mi 8 Android Smartphone and Positioning Performance Analysis," *Sensors*, vol. 91, no. 8, pp. 1–16, 2019.

[6] P. Bolla and K. Borre, "Performance Analysis of Dual-Frequency Receiver Using Combinations of GPS L1, L5, and L2 Civil Signals," *Journal of Geodesy*, vol. 93, pp. 437–447, March 2019.

[7] GPS Navstar Joint Program Office, "Global Positioning System Standard Positioning Service Performance Standard," 5th Edition, April 2020.

[8] GPS Navstar Joint Program Office, "Navstar GPS Space Segment/Navigation User Segment Interfaces, IS-GPS-200," Revision J, May 22, 2018.

[9] C. S. Jones, J. M. Finn, and N. Hengartner, "Regression with Strongly Correlated Data," *Journal of Multivariate Analysis*, vol. 99, no. 9, pp. 2136–2153, 2008.

[10] C. Yang, C. Hegarty, and M. Tran, "Acquisition of the GPS L5 Signal Using Coherent Combining of I5 and Q5," in *Proceedings of ION GNSS, Long Beach, CA*, pp. 2184–2195, 2004.

[11] L. Ries, C. Macabiau, O. Nouvel, Q. Jeandel, W. Vigneau, V. Calmettes, and J.-L. Issler, "A Software Receiver for GPS-IIF L5 Signal," in *Proceedings of ION GPS, Portland, OR*, pp. 1540–1553, September 2002.

[12] P. Misra and P. Enge, *Global Positioning System: Signals, Measurements and Performance, Revised 2nd Edition*. Ganga-Jamuna Press, 2011.

[13] G. Okerson, J. Ross, A. Tetewsky, A. Stoltz, J. Anszperger, and S. R. Smith, "Inter-Signal Correction Sensitivity Analysis," *Inside GNSS*, vol. 11, no. 3, pp. 44–53, May/June 2016.

[14] R. Hatch, "The Synergism of GPS Code and Carrier Measurements," in *Proceedings International Geodetic Symposium on Satellite Doppler Positioning, Las Cruces, NM*, pp. 1213–1231, 1983.

[15] L. Zhang, Y. Morton, and T. Beach, "Characterization of GNSS Signal Parameters under Ionosphere Scintillation Conditions Using Software-Based Tracking Algorithms," in *Proceedings IEEE/ION Position, Location and Navigation Symposium, Indian Wells, CA*, pp. 267–275, April 2010.

[16] "SdrNav40 – OneTalent GNSS." www.onetalent-gnss.com/ideas/software-defined-radio/sdrnav40. Accessed: May 25, 2019.

9 Snapshot Receivers

Ignacio Fernández-Hernández, José A. López-Salcedo
and Gonzalo Seco-Granados

9.1 Introduction

This book has dealt so far with conventional receivers, which sample the received signal impinging on the antenna, acquire the visible satellites and track them through the code, carrier phase and frequency tracking loops to continuously deliver position, velocity and time. This architecture serves well to many user groups, like aviation or transport. However, it may not serve well the needs of many radiolocalization users nowadays, which require an instantaneous, and sometimes sporadic location.

There are some reasons for that: first, standard receivers need to demodulate the data from the signal to compute the position. This may take several seconds, around 30 seconds in the case of GPS L1 C/A. Second, some users just need the position sporadically, and continuous positioning consumes too much power. Snapshot positioning solves this problem. It is based on saving digital samples from the GNSS receiver front end, getting the satellite data from another channel and computing a position and time solution.

Snapshot positioning often relies on assisted GNSS (A-GNSS) to get the satellite data. Often, getting the rough position and time improves acquisition by shortening acquisition time and increasing sensitivity. On the one hand, the search space can be reduced, and on the other hand, the dwell time per code-frequency bin can be increased. In addition to shortening acquisition, A-GNSS also avoids the need to demodulate the data from the signals. A thorough description of A-GNSS can be found in [1].

There is a wealth of inventions, some patented, in the domain of A-GNSS, many of them allowing location with mobile phones in cities, where users cannot demodulate the data from the signals, and have no time to wait for half a minute to get a first fix. However, there are not many references devoted to the underlying principles of snapshot positioning. For example, techniques and performance of direct, open-loop acquisition, including no tracking stage, are not much explained in the GNSS literature. The same occurs with positioning methods. This chapter intends to cover this gap. It is divided into two main blocks. The first block deals with signal processing, whereas the second block deals with position computation. We also refer to a MATLAB implementation of a GPS L1 C/A snapshot receiver that accompanies the book, as for the other chapters. More details about operating the snapshot receiver are provided in Chapter 12.

9.2 Snapshot Signal Processing

Conventional GNSS receivers are designed to operate continuously, and they provide position and time in a regular manner at a given output rate, say 1 Hz. This approach is aligned with the continuous transmission of signals from the GNSS satellites and leads to receiver architectures based on closed-loop (i.e. feedback) estimation techniques for the determination of the signal-level parameters such as the code delay, carrier phase and carrier frequency shift.

In contrast, snapshot GNSS receivers are designed to operate on demand, when requested by the user, and this approach involves a paradigm shift in the receiver design. At the time of request, the receiver must gather a piece of received signal, the so-called *snapshot*, and provide the best possible estimate of the parameters required to determine the user's position. This constraint leads to receiver architectures based on open-loop (i.e. feed-forward) techniques for the estimation of these parameters. This means that there is no tracking stage implemented in snapshot GNSS receivers, since snapshots are processed independently one from another. This is due to the fact that no continuity can often be guaranteed between consecutive snapshots, since one snapshot may have been requested at a given time, under some given working conditions, and the next one at a different time and different working conditions. Furthermore, open-loop architectures involve a block-wise implementation where samples are processed in batches of a given length, and efficient algorithms are used to speed up the processing time. This is the case of the extensive use of the FFT for implementing the code correlation, and the implementation of interpolation techniques to obtain sub-sample code delay accuracy. An example of a snapshot GNSS receiver architecture is depicted in Figure 9.1.

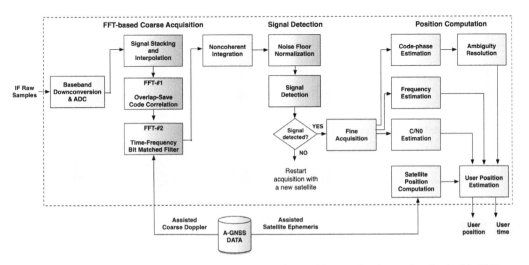

Figure 9.1 Example of a snapshot GNSS receiver architecture implementing the double-FFT acquisition algorithm.

This section will focus on how the code correlation is implemented in an efficient manner using FFT processors, and how the output samples are coherently and noncoherently accumulated in a GNSS receiver operating in snapshot mode. The combination of coherent and noncoherent integrations allows us to extend the total integration time beyond the limitations of the symbol period and thus to obtain an increased SNR gain that helps in improving the receiver sensitivity. This approach is the basis for the so-called high-sensitivity GNSS (HS-GNSS) receivers, which operate in snapshot mode to avoid the tracking limitations of conventional GNSS receivers in harsh working conditions. HS-GNSS receivers have been widely studied in the literature in the context of indoor positioning [2, 3, 4] and also proposed for more exotic applications such as lunar missions, due to the weak-signal similarities of both scenarios [5, 6]. Once coherent and noncoherent integrations have been presented, we will follow with some practical considerations such as the fact that a noninteger number of samples is usually contained in a received code period. However, efficient signal processing using the FFT makes use of integer-size blocks of data, so some adaptation is required. This problem will require a specific treatment in the form of subsample interpolation techniques. Finally, we will revisit the signal detection problem by providing the formulation for determining the signal threshold and detection performance when using parallel search acquisition, thus complementing the results in Section 1.6.

9.2.1   Coherent Integration

As explained in Chapter 1, GNSS signals follow a hierarchical structure whereby a primary spreading code is repeated several times with a sign determined by the chips of a so-called secondary code. The receiver must first align its primary code local replica to the received signal, a process that is carried out using the coherent correlation already introduced in Eq. (1.65), reproduced here for the sake of clarity, using a pair of tentative code delay and Doppler frequency shift (τ', v'). That is,

$$R_C(\tau', v') = \sum_{n=0}^{N_T-1} r[n]c^*[n - \tau']_{N_{scode}} e^{-j2\pi v'n}, \tag{9.1}$$

where a total of N_T samples are involved, corresponding to a coherent integration time of $T = N_T T_s$ seconds, with T_s the sampling time.

From a theoretical point of view, there is no remarkable difference between the primary code correlation in a snapshot receiver compared to that in a conventional receiver. Therefore, all the available acquisition strategies, namely combinations of serial, sequential or parallel search in the time and frequency domains, are possible. However, many snapshot receivers are often implemented in software making extensive use of the FFT for efficiently implementing the correlation in Eq. (9.1). Even some hardware implementations are also making use of the FFT, particularly when this operation is already implemented for other purposes. This is the case in chipsets for wireless communications where FFT processors are already used for multicarrier (OFDM) signal processing, since OFDM is widely adopted nowadays in many communication systems such as xDSL, WiFi, 4G-LTE and 5G.

Snapshot receivers often implement the correlation in the frequency domain, which means that they first take the FFT of the batch of input samples to be correlated and the complex-conjugate of the FFT of the code replica. Each FFT output is stored into a vector, and then both vectors are element-wise multiplied. Finally, the IFFT is applied to the result thus turning back to the time domain. This process is conceptually implementing the correlation in the frequency domain, with the exception that the frequency domain transformation is brought by the FFT instead of the discrete-time Fourier transform (DTFT), as it would correspond to discrete-time signals such as $r[n]$ and $c[n]$ involved in Eq. (9.1). The fact of using the FFT/IFFT instead of the DTFT/IDTFT involves that the correlation being obtained is actually the circular correlation, not the linear (i.e. conventional) one. This has some implications such as causing the resulting correlation to be periodic with period the FFT size, which may lead time-aliased replicas to eventually overlap one with each other. To avoid this effect and to efficiently implement the correlation for a long stream of input samples, two methods referred to as *overlap-add* and *overlap-save* are available [7]. Both of them are valid options for performing the correlation with the primary code. Nonetheless, for the typical dimensions of the vectors found in GNSS, the overlap-and-save method offers a slight computational advantage. This method is implemented in the step-by-step description of the double-FFT method discussed in Section 9.2.8, and it is also discussed in detail in Section 9.2.9.

The use of the FFT-based correlation also provides a very natural way to realize the frequency search of the acquisition stage. Different trial values of the frequency shift can easily be applied to the local replica of the primary code by simply circularly shifting its FFT. The circular shift of one position of the frequency-domain vector is equivalent to a frequency shift of

$$\Delta v_{t,\text{FFT}} = \frac{F_s}{N_{\text{FFT}}} = \frac{1}{T_{\text{FFT}}} \quad \text{(Hz)}, \tag{9.2}$$

where N_{FFT} is the number of bins of the FFT and T_{FFT} the time duration of such block of samples. If the overlap-save method is used, T_{FFT} should be at least twice the duration of the primary code, $T_{\text{FFT}} \geq 2T_{\text{code}}$. This must be borne in mind when constraints are imposed on the FFT size that can be implemented, as described in Section 9.2.9. The aforementioned choice for T_{FFT} typically provides sufficient resolution of the frequency grid to perform the acquisition with small worst-case losses.

9.2.2   Noncoherent Integration

In order to be able to detect weak signals, there is no choice but to accumulate the signal energy during long observation intervals. This is the so-called *high-sensitivity* principle in GNSS, and it is based on extending the correlation time by combining coherent and noncoherent integrations. By doing so, we can get rid of potential symbol transitions and accumulated phase errors. As already described in Chapter 1 and Eq.

(1.75), the noncoherent correlation is defined as

$$R_{NC}(\tau', \nu') = \sum_{k=0}^{N_I-1} |R_C(\tau', \nu'; k)|^2, \tag{9.3}$$

where N_I is the number of noncoherent integrations. The total integration time thus becomes

$$T_{tot} = N_I T \quad \text{(s)}, \tag{9.4}$$

with T the coherent integration time.

At the acquisition stage of a high-sensitivity GNSS receiver, we already saw in Chapter 1 that the satellite under analysis is declared to be present whenever $R_{NC}(\tau', \nu') > \gamma$, for a given threshold γ specified by the user-defined probability of false alarm P_{FA}. This detection process automatically provides coarse estimates for the unknown time-delay and frequency shift errors as:

$$(\hat{\tau}, \hat{\nu}) = \arg\max_{\tau', \nu'} R_{NC}(\tau', \nu') \quad \text{subject to} \quad \max_{\tau', \nu'} R_{NC}(\tau', \nu') > \gamma. \tag{9.5}$$

It is important to note that since we are correlating with a one-code-period signal replica, the time-delay estimates $\hat{\tau}$ actually correspond to the fractional part of the true time-delay τ within one code period. So strictly speaking, we should formally refer to $\hat{\tau}$ as $\hat{\tau}_{N_{scode}}$, with

$$\tau_{N_{scode}} = (\tau)_{N_{scode}}, \tag{9.6}$$

where $(\cdot)_N$ the represents the modulus N operation, and thus, $\tau_{N_{scode}} \in [0, N_{scode} - 1)$ is contained within a one code period length. The GNSS receiver has to determine the number of elapsed code periods since the signal was transmitted and then add this offset to the result from Eq. (9.5). This problem, which is similar to carrier phase ambiguity resolution, is solved at the navigation module of the receiver, and it will be discussed later on.

The total integration time T_{tot} implemented by Eq. (9.3) is the result of noncoherently accumulating N_I coherent correlations, each of them being the result of coherently integrating for T seconds. The total integration time is therefore a function of N_I and T. These are two parameters that the receiver can control to increase the total integration time and thus fulfill a given requirement in terms of acquisition sensitivity. Nevertheless, there are some fundamental limits that prevent us from increasing the noncoherent and coherent integration time without bound, as already discussed in Section 1.6 and briefly summarized next.

On the one hand, the presence of unknown symbol transitions limits the coherent integration time to a fraction of the symbol interval. A common practice in this case is to adopt the half-bit acquisition method, which boils down at using two consecutive pieces of signal, each of them with length equal to half the symbol period, for implementing two separate coherent correlations [8]. This guarantees that at least one of the two correlations will not be affected by the symbol transition and one coherent correlation of half the symbol period length can at least be assured. An alternative is to synchronize with the symbol transition and therefore be able to implement a coherent correlation of the full symbol period. The problem can be formulated as a multiple hypothesis testing, where each hypothesis corresponds to a time shift equal to the primary code length. Finally, a coherent correlation beyond one symbol period could also be possible if precise knowledge of the navigation symbols was available, so that the symbols could be wiped off from the received signal. This option would be possible in case of having access to external assistance data like A-GNSS, so that coherent correlations beyond one symbol period would actually be possible. A simpler approach is that of acquiring the pilot component available in most modernized GNSS signals, instead of the data component. This has the advantage that no unexpected symbol transitions must be circumvented, and thus, long coherent correlation times can easily be implemented. In that case, the only physical limitation is due to the clock instabilities of the receiver, since the accumulated phase noise for a very long period of time severely degrades the result of such coherent correlation. As a rule of thumb, the maximum coherent integration time is limited up to $100\,\text{ms}$ for conventional TCXO clocks, and up to $1000\,\text{ms}$ for medium quality OCXO clocks, as already pointed out in Section 1.6.

On the other hand, limitations are also encountered when trying to extend the noncoherent integration length. The main one is due to the code Doppler, which corresponds to the term v_t/F_c appearing in the received signal model in Eq. (1.52). Code Doppler is a negligible effect for short correlation intervals but becomes relevant for very long correlation intervals due to the mismatch between the received signal and the local code replica. When this mismatch is carried over a long integration interval, the result is a small time shift between consecutive correlation peaks to be noncoherently integrated. This blurs the resulting total correlation peak, thus introducing a bias in the final estimate of the time-delay. The total time-delay error caused by code Doppler was already mentioned in Chapter 1 to be given by Eq. (1.78). Unfortunately, even if the Doppler estimation was improved and a more precise estimate was available, the code Doppler error over a long integration interval would still be subject to variations caused by the clock instabilities. The way to overcome this problem would be by using methods to estimate the clock dynamics, but this is a difficult task to carry out in mass-market GNSS receivers and remains within the realm of professional ones.

9.2.3   Block-wise Sampling Correction

In the previous sections, we addressed the problem of correlating the input signal with the local code replica in a block-wise or snapshot-based manner. This approach is well-suited for efficient implementations, since the input signal is processed in batches of a fixed length and the correlation with the primary code can be carried out in the frequency domain using the FFT. The length or period of the primary code was already shown to be $N_{\text{scode}} = T_{\text{code}}/T_s$ samples, which for the sake of simplicity, we assumed was an integer number. This means that the sampling rate was assumed to be an integer multiple of the primary code chip rate. Unfortunately, this will hardly ever be the case in practice. One reason is the code Doppler effect, which depending on its sign makes the primary code period to be slightly compressed or expanded with respect to its nominal value. Another reason is the sampling frequency itself, which from a performance point of view should actually not be an integer multiple of the primary code period to avoid degrading the code delay estimation. The use of a noncommensurate sampling rate is a guideline among experienced GNSS receiver designers, see [9].

Based on the above reasons, it is clear that a snapshot receiver needs to cope with the presence of noninteger code periods, even though this may enter into conflict with the snapshot-based philosophy. Namely, processing batches of samples having *all the same fixed length*. A practical way to cope with this situation is to work with the rounded value of the actual code period,

$$N'_{\text{scode}} = \text{round}(N_{\text{scode}}),\qquad(9.7)$$

which provides the integer value we need. However, the price to be paid is that we need to keep a record of the samples that are missing or exceeding with respect to the previous batch, in order to compensate for them. Assuming that we are processing batches of samples every code period, the code Doppler effect from one batch to the next becomes

$$\delta\tau_{\text{d},t} = \frac{v_t}{F_{\text{c}}}T_{\text{code}} \;\; (\text{s}) \quad \rightarrow \quad \delta\tau_{\text{d}} = \frac{v_t}{F_{\text{c}}}N_{\text{scode}} \;\; (\text{samples}).\qquad(9.8)$$

Since we do not know the Doppler frequency yet, because we are still in the middle of the acquisition process, it is customary to replace v with its tentative value v'. Regarding the sampling frequency, the fractional difference between the rounded code period and the true one must also be compensated from one batch to the next. The goal is to make sure that the input signal in all batches starts at the correct (and same) position. This can be achieved by applying the following delay from one batch to the next,

$$\delta\tau_{\text{s}} = N'_{\text{scode}} - N_{\text{scode}}.\qquad(9.9)$$

In summary, the total delay to be compensated at the k-th batch of samples is given by

$$\tau_{\text{corr}}[k] = (\delta\tau_{\text{d}} + \delta\tau_{\text{s}})(k - 1),\qquad(9.10)$$

which is typically implemented using an interpolation filter in the frequency domain, thus benefiting from the use of the FFT processor that is already used for the code correlation at the acquisition stage.

9.2.4   Signal Detection

This section describes how to decide whether a satellite of interest is present or not in the samples being processed. This task is carried out using the samples at the correlation output, once the received signal is despread. As already pointed out in Eq. (9.5), the detection criterion is based upon the following condition:

$$\max_{\tau',v'} R_{\text{NC}}(\tau',v') > \gamma,\qquad(9.11)$$

for some detection threshold γ.

As with any decision to be taken, one may be right or wrong, and this is evaluated through the so-called probability of detection and probability of false alarm, already introduced in Section 1.6. They will be denoted herein by P_D and P_{FA}, respectively. Furthermore, the choice of γ has a direct impact onto the detection performance, since a very large γ will make it difficult to detect the presence of a satellite signal with finite received power, thus causing $P_D \rightarrow 0$. At the same time, setting $\gamma \rightarrow 0$ will make the detector extremely sensitive to the very tiniest correlation value, even if it is just due to background noise, thus causing $P_{FA} \rightarrow 1$. In practice, we would like to have errors under control. It is for this reason that P_{FA} is often provided as a design requirement, and then, the problem is finding the appropriate γ to achieve such P_{FA}. Once γ is fixed, we just need to make sure that the resulting P_D is acceptable for the application under consideration.

The procedure for finding γ for a given P_{FA} starts with the statistical characterization of the detection metric, $R_{NC}(\tau', v')$, in Eq. (9.11). This metric is the accumulation of N_I values of the coherent correlation $R_C(\tau', v'; k)$ that is modeled as:

$$R_C(\tau', v'; k) = \begin{cases} \mathcal{N}(\sqrt{\Psi(\tau_\epsilon, v_\epsilon)}e^{j\varphi(k)}, \sigma_w^2) & : \mathcal{H}_1 \\ \mathcal{N}(0, \sigma_w^2) & : \mathcal{H}_0, \end{cases} \tag{9.12}$$

where $\mathcal{H}_1$ stands for the satellite present hypothesis, $\mathcal{H}_0$ is the satellite absent hypothesis, and $(\tau_\epsilon, v_\epsilon)$ denote the time-delay and frequency shift errors between the true values (τ, v) and the tentative ones (τ', v'). That is, $\tau_\epsilon = \tau - \tau'$ and $v_\epsilon = v - v'$, as defined in Chapter 1. Under $\mathcal{H}_1$, the amplitude is affected by some phase rotation $\varphi(k)$, and it is embedded onto complex Gaussian noise with zero mean and variance σ_w^2. Under $\mathcal{H}_0$, the correlation output is dominated by noise, and therefore complex Gaussian distributed ($\mathcal{N}$) with zero mean and variance σ_w^2. Based on this model, the normalized noncoherent correlation becomes χ^2 (*Chi-squared*) distributed as follows:

$$\xi(\tau', v') = \frac{R_{NC}(\tau', v')}{\sigma_w^2/2} = \begin{cases} \chi_{2N_I}^2(N_I \text{SNR}_C(\tau_\epsilon, v_\epsilon)) & : \mathcal{H}_1 \\ \chi_{2N_I}^2 & : \mathcal{H}_0, \end{cases} \tag{9.13}$$

with $\chi_{2N_I}^2$ the central χ^2 distribution with $2N_I$ degrees of freedom, whereas $\chi_{2N_I}^2(\lambda)$ stands for the noncentral χ^2 distribution with $2N_I$ degrees of freedom and noncentrality parameter λ. In this case, $\lambda = N_I \text{SNR}_C(\tau_\epsilon, v_\epsilon)$, where SNR_C is the SNR at the output of the coherent correlation,

$$\text{SNR}_C(\tau_\epsilon, v_\epsilon) = \frac{\Psi(\tau_\epsilon, v_\epsilon)}{\sigma_w^2}, \tag{9.14}$$

with $\Psi(\tau_\epsilon, v_\epsilon)$ the ambiguity function already introduced in Eq. (1.69), which includes the received signal power.

It is important to note that the normalization in Eq. (9.13) by the one-dimensional noise power is required in order to be compliant with the χ^2 definition, which assumes the accumulation of squared unit-variance and real-valued Gaussian random variables. To do so, σ_w^2 can be easily estimated by from the cells outside of the main correlation peak, which do only contain noise and therefore $\mathcal{H}_0$ holds true. Note that such noise floor normalization is carried out individually for each of the frequency bins under analysis, in such a way that σ_w^2 actually becomes $\sigma_w^2(v')$. This is motivated by the variability of the cross-correlation between different satellites when evaluated at different frequency shifts. This variability translates into different noise floor levels at different frequency bins of the acquisition search, namely $\sigma_w^2(v_1') \neq \sigma_w^2(v_2')$. This observation indicates that a frequency-dependent noise floor normalization is actually required in Eq. (9.13) in order to compare all frequency bins in a fair and equal manner.

Regarding the detection performance, two performance metrics need to be determined, namely the probability of false alarm P_{FA} and the probability of detection P_D. Now, let us assume that there is only one search cell in Eq. (9.11) so that the maximum value is not required. In that case, the global P_{FA} would correspond to the probability of false alarm of a single cell which is denoted by P_{fa},

$$P_{fa}(\gamma) = \text{prob}(\xi(\tau', v') > \gamma | \mathcal{H}_0) = 1 - F_{\xi, \chi_c^2}(\gamma; 2N_I), \qquad (9.15)$$

with $F_{\xi, \chi_c^2}(\gamma; n)$ the cumulative density function of a central χ^2-random variable ξ with n degrees of freedom. The detection threshold can then be determined using the relationship in Eq. (9.15), resulting in:

$$\gamma = F_{\xi, \chi_c^2}^{-1}(1 - P_{fa}; 2N_I). \qquad (9.16)$$

In practice, though, there is not only one single cell to be evaluated in Eq. (9.11), but many. They may range from thousands in case that assistance information is not available during the acquisition stage to a few tenths when it is available, see [1]. Assuming that the search space for τ' is divided into $N_{\tau'}$ independent steps and the search space for v' into $N_{v'}$ independent steps, there is a total of $N_{\tau'}N_{v'}$ independent cells whose maximum value needs to be compared with some detection threshold. Therefore, the *global* probability of false alarm for the whole detection process can be determined as

$$P_{FA}(\gamma) = \text{prob}(\max_{\tau', v'} \xi(\tau', v') > \gamma | \mathcal{H}_0)$$

$$\equiv \text{prob}(\text{at least one } \xi(\tau', v') > \gamma | \mathcal{H}_0) \qquad (9.17)$$

$$= 1 - (1 - P_{fa}(\gamma))^{N_{\tau'}N_{v'}} \approx P_{fa}(\gamma)N_{\tau'}N_{v'}.$$

This probability is often provided as part of the user requirements that the receiver must be compliant with. Based on this value, we can then solve for $P_{fa}(\gamma)$ by using Eq. (9.17) in order to find the single-cell probability of false alarm we need and thus the threshold to be used. This leads us to the following relationship:

$$P_{fa}(\gamma) \approx \frac{P_{FA}(\gamma)}{N_{\tau'}N_{v'}}, \qquad (9.18)$$

whose value can be substituted in Eq. (9.16) to obtain the detection threshold as a function of the global P_{FA}, which is the requirement we will be given in practice. The result is:

$$\gamma \approx F_{\xi, \chi_c^2}^{-1}\left(1 - \frac{P_{FA}(\gamma)}{N_{\tau'}N_{v'}}; 2N_I\right). \qquad (9.19)$$

Once the detection threshold is set for a given P_{FA}, the probability of detection can be found using the following expression,

$$P_D(\gamma) = \text{prob}(\max_{\tau',\nu'} \xi(\tau',\nu') > \gamma|\mathcal{H}_1)$$

$$\approx \text{prob}(\xi(\hat{\tau},\hat{\nu}) > \gamma|\mathcal{H}_1) \tag{9.20}$$

$$= 1 - F_{\xi,\chi^2_{nc}}(\gamma; 2N_{\text{I}}; N_{\text{I}}\text{SNR}_C(\tau_{\epsilon|\tau'=\hat{\tau}}, \nu_{\epsilon|\nu'=\hat{\nu}})),$$

where $F_{\gamma,\chi^2_{nc}}(x; n; \lambda)$ is the cumulative distribution of a noncentral χ^2 random variable ξ with n degrees of freedom and noncentrality parameter λ, as a function of γ. The pair $(\tau_{\epsilon|\tau'=\hat{\tau}}, \nu_{\epsilon|\nu'=\hat{\nu}})$ is the time-delay and frequency error corresponding to the cell where the signal peak is located. It can be seen that the accuracy of these estimates has an impact on the probability of detection through the SNR of the coherent correlation output.

The result in Eq. (9.20) provides some hints on how to dimension the acquisition search grid such that the resulting time-frequency granularity incurs in a small enough degradation. That is, the wider the time-frequency bins of the acquisition stage, the larger the time-frequency errors that are incurred. This makes more difficult to pick exactly the maximum peak of the ambiguity function, thus introducing some loss in terms of SNR with respect to the ideal case where exactly the maximum peak was picked (i.e. if a finer enough grid was used). This topic has been further discussed in Chapter 1 when presenting the results of Figures 1.25(a) and 1.25(b).

9.2.5 Interpolation of the Correlation Peak

Coarse time-delay and frequency estimates can readily be obtained as a by-product of the signal detector in Eq. (9.11) as follows:

$$(\hat{\tau},\hat{\nu}) = \arg\max_{\tau',\nu'} R_{\text{NC}}(\tau',\nu') > \gamma, \tag{9.21}$$

where the accuracy of the coarse estimates $(\hat{\tau},\hat{\nu})$ is determined by the grid spacing of the time-frequency search. In the time domain, the grid spacing is done using a fraction of the chip period, which typically ranges from one half to one tenth of chip. This means that the resolution of the time-delay estimate in Eq. (9.21) when converted into distance units would be of the order of 150 to 30 meters for a GPS L1 C/A signal, respectively. In the frequency domain, the grid spacing is determined by the inverse of the coherent integration time, and at least one half of this value is typically needed. This would lead to a frequency resolution up to 50 Hz in case of using the conventional 10-ms integration time to process GPS L1 C/A signals. While these resolutions may seem too coarse, they are fairly good for initializing the tracking stage of a conventional GNSS receiver where the fine estimation of time-delay and frequency is ultimately carried out using closed-loop tracking techniques. Unfortunately, snapshot receivers have no tracking stage and thus the accuracy of their time-frequency estimates would remain the one already discussed for Eq. (9.21). In order to circumvent this limitation, some refinements are discussed below:

1. *Resampling.* The coarse frequency estimate at the acquisition output can be used to refine our initial guess about the frequency shift. This is important because a more precise knowledge of the frequency shift provides a more precise estimate of

the code chip rate of the received signal. This refined chip rate is in turn used to generate a new replica of the local code and to recompute a few correlation points around the maximum peak where a more precise time-delay estimate can now be obtained.

2. *Algebraic correction.* This is an alternative approach to resampling that considers the effect of Doppler as a time-delay onto the code. This effect is somehow compensated by generating the local code replica with the proper chip rate, including the code Doppler effect. However, the frequency shift information we had available when the local code replica was first generated did not take into account the refined estimate $\hat{v}$ in Eq. (9.21). It is for this reason that instead of resampling the local code, we can directly compensate the current time-delay estimate $\hat{\tau}$ with the term corresponding to the refined code Doppler correction,

$$\hat{\tau}_{ac} = \hat{\tau} - \frac{\hat{v}_t}{F_c} T_{tot},\qquad(9.22)$$

where $T_{tot} = N_I T$ is the total snapshot length we have processed.

3. *Peak interpolation.* The time-delay resolution was mentioned to be limited by the sampling period, so one may think of improving this resolution by increasing the sampling rate. In fact, this is actually unnecessary because the key information on the position of the correlation peak is contained in the shape of the correlation points. This shape can be inferred even if few correlation points are available, by means of interpolation techniques. In this way, we can interpolate the missing correlation points and infer the position where the true peak is located. Different interpolation methods can be implemented. The piecewise linear interpolator proposed in [3] is one of the simplest and most effective ones. The method can be implemented using just three correlation points and leads to the following refined time-delay estimate:

$$\hat{\tau}_{interp} = \hat{\tau} + \frac{1}{2} \frac{R_{NC}(\hat{\tau}+1,\hat{v}) - R_{NC}(\hat{\tau}-1,\hat{v})}{R_{NC}(\hat{\tau},\hat{v}) - \min\{R_{NC}(\hat{\tau}+1,\hat{v}), R_{NC}(\hat{\tau}-1,\hat{v})\}}.\qquad(9.23)$$

Another interpolation method is the quadratic interpolation. It fits the correlation results with a parabola, and the code delay estimate corresponds to the parabola maximum. Its performance is similar to the linear interpolation in the presence of noise. Figure 9.2 shows an example of quadratic peak interpolation, where the only available samples are in values "0," "1" and "2" [10].

9.2.6   Snapshot-Based C/N_0 Estimation

Observing the received signal strength is an essential part of any GNSS receiver. It is required, for instance, when computing the user's position based on the set of available time-delay estimates. In this case, a weighted least-squares approach is adopted where the time-delay estimate from each satellite is weighted by the square root of the corresponding signal strength. Signal strength monitoring is also required as a form of lock detection in GNSS tracking loops, in order to ascertain whether the GNSS signal is being tracked or not. For high-sensitivity GNSS, near-far mitigation techniques often

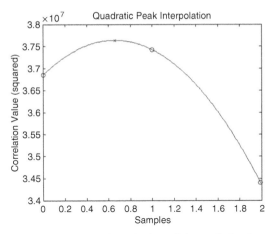

Figure 9.2 Example of quadratic peak interpolation from GPS PRN 2, L1 C/A and $F_s = 16.386$ MHz.

make use of signal strength estimates. They are used to distinguish between weak and strong satellites, thus being able to detect potentially harmful interference sources to the acquisition of weak signals.

Being related to the SNR, as discussed in Section 1.4.1, signal strength estimates in terms of C/N_0 suffer from the same limitations of SNR estimates. That is, they are generally biased and exhibit a large variance due to their implementation as ratios of quadratic forms. In practice, this limitation is overcome by increasing the processed signal length until bias and variance are much smaller than the actual values of SNR or C/N_0 to be observed. However, as it occurs at the acquisition stage, this approach may be not feasible at all because the length of the available signal may be too short. An important limitation of C/N_0 estimators is their poor performance when implemented in indoor GNSS receivers, where biases on the order of 20 dB are commonly found. The reason for such a degradation lies in the very fundamental structure of most C/N_0 estimators. They are based on a coherent postcorrelation approach in which the C/N_0 is estimated at the correlation output, where both code delay and Doppler shift are optimistically assumed to have been accurately compensated. This is the case of narrowband-wideband power ratio (NWPR) method described in Section 1.7.3.

Apart from the residual errors in the delay and Doppler estimates, which inevitably degrade the accuracy of the NBP and WBP metrics, the main problem with the NWPR method is that NBP and WBP are too noisy estimates of the corresponding powers due to a lack of sufficient averaging. Note also that the NBP depends on coherent integration, which cannot be further extended because of the symbol transitions and code Doppler limitations, as already discussed in Section 9.2.2. At this point, we already know that, for high-sensitivity techniques, noncoherent correlations are needed to obtain an acceptable level of signal growth above the noise. So, the question is to devise a C/N_0 estimator that makes use of the noncoherent integrations. The following noncoherent postcorrelation estimator has been proposed in [11] and is used in the MATLAB snapshot receiver accompanying this book,

$$\left(\frac{\hat{C}}{N_0}\right)_{\mathrm{NC}} = \frac{R_{\mathrm{NC}}(\hat{\tau}, \hat{v})B_{\mathrm{n}} - T\hat{P}/T_{\mathrm{s}}}{T^2\hat{P} - R_{\mathrm{NC}}(\hat{\tau}, \hat{v})}, \tag{9.24}$$

with B_{n} being the receiver noise bandwidth and $\hat{P}$ being an estimate of the input signal power. This estimator employs only quantities that are readily available in an snapshot receiver, and thus, it does not require any additional complexity. The bias is found to be smaller than 1 dB-Hz for all possible values of C/N_0 down to the receiver sensitivity (normally between 10 and 20 dB-Hz), and the variances become very close to the Cramér-Rao bound for large C/N_0 values [12].

Some other estimators have been proposed in the literature where the relationship between some intermediate values computed in the receiver and C/N_0 are exploited. The interested reader may find a valuable summary of these and other techniques in [13].

9.2.7   Low-Complexity Multipath Detection

We have introduced multipath in Chapter 1. It is one of the major sources of error in GNSS receivers, particularly those operating in urban areas. This problem has been the focus of significant research efforts in the past decades, and there exists a vast literature of different techniques for detecting, compensating and mitigating these errors. In particular, multipath detection turns out to be a simple and effective approach as compared to other families of techniques that try to mitigate it. In many practical situations, identifying the multipath-affected satellites and removing them from the computation of the navigation solution often suffices to get rid of a significant multipath degradation. This is the reason why multipath detection techniques are becoming widely adopted in many GNSS receivers.

A review of multipath detection and mitigation techniques is presented in Chapter 10. Here, we focus on a specific detection technique that is based on the analysis of the correlation samples and therefore can be of interest for snapshot receivers. It is actually one of the most popular approaches for multipath detection, which consists in taking advantage of the geometrical properties of the correlation peak. In particular, it is known that multipath components are always delayed with respect to the line-of-sight, which means that the slope of the right side of the correlation peak will be much more affected by multipath than the slope of the left side. This underlying principle is also exploited by signal quality monitoring techniques, which are intended to detect anomalies in the satellite signals by looking at the shape of the correlation peak.

Based on this principle, a low-complexity multipath detection technique can be implemented just by comparing the left and right side slopes of the correlation peak. The technique is referred to as *slope asymmetry metric* (SAM), and it can be computed as follows [14]:

$$\mathrm{SAM}[k] = \hat{a}_+[k] + \hat{a}_-[k], \tag{9.25}$$

where $\hat{a}_+[k]$ and $\hat{a}_-[k]$ are the least-squares estimates of the left and right side slopes of the correlation peak corresponding to the k-th snapshot. These estimates can be obtained by fitting each of the sides of the correlation peak to a straight line of the form $a\tau + b$ whose unknown parameters are the slope a and offset b. Let us stack the line parameters into a vector $\theta = [a,b]^T$, and the set of L samples of a given side of the correlation peak into the corresponding vector,

$$y = [R_{NC}(\tau_1, \hat{v}), R_{NC}(\tau_2, \hat{v}), \dots, R_{NC}(\tau_L, \hat{v})]^T. \tag{9.26}$$

Then, the least-squares estimate of the line parameters in one of the two sides of the correlation peak is obtained as follows:

$$\hat{\theta} = (\mathbf{M}^T\mathbf{M})^{-1}\mathbf{M}^T y, \tag{9.27}$$

where the matrix $\mathbf{M}$ is given by

$$\mathbf{M} = \begin{bmatrix} 1 & 1 & 1 & 1 \\ \tau_1 & \tau_2 & \dots & \tau_L \end{bmatrix}^T. \tag{9.28}$$

The solution in Eq. (9.27) must be evaluated twice using the correlation samples from each of the two sides of the correlation peak. That is, with $\{\tau_1^+, \tau_2^+, \dots, \tau_L^+\}$ for the left hand side and $\{\tau_1^-, \tau_2^-, \dots, \tau_L^-\}$ for the right hand side, as illustrated in Figure 9.3. Finally, the SAM metric in Eq. (9.25) can be evaluated at consecutive time instants resulting in the following stream of values as a function of time,

$$\text{SAM}[k] = [\hat{\theta}_+[k]]_1 + [\hat{\theta}_-[k]]_1, \tag{9.29}$$

where $[\cdot]_i$ denotes the i-th component of a vector.

The result in Eq. (9.29) should provide a zero-mean random process when no multipath is present, since both the left and right slopes should be nearly equal but with opposite signs. However, the presence of multipath can still be unveiled by computing the variance of the SAM metric, since the multipath time-variations will cause the slopes of the correlation peak to fluctuate as well. Different types of multipath can be identified by computing the autocorrelation of the SAM metric and measuring the

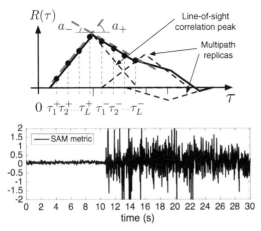

Figure 9.3 (Up) Correlation peak of a GNSS received signal affected by multipath. (Down) SAM metric as a function of time when a multipath ray appears at $t = 11$ s.

width of the main peak. A very narrow peak will indicate a fast time-varying multi-path, such as the one experienced indoors or in the presence of moving obstacles. A wide peak, in contrast, will indicate a slow time-varying multipath, such as the case of the specular reflection of the direct signal. In summary, Eq. (9.29) provides a simple means to detect the presence of multipath by monitoring the samples of the correlation peak. The interested reader will find more details in [14].

9.2.8　Double-FFT Implementation

The correlation between the input signal and the local replica is certainly the most computationally demanding task of a GNSS receiver, particularly when long integration times need to be implemented. It is for this reason that snapshot receivers typically resort to the use of FFT processors for implementing the code correlation (i.e. the time search) in the frequency domain, as previously discussed at different points of this book. This approach leads to the so-called parallel code phase search acquisition (PCS) [15], or just full parallel search as described in Section 1.6, where significant computational savings are achieved. For a snapshot of N signal samples, the complexity of implementing the correlation moves from $O(N^2)$, in the time domain, to the much lower $O(N \log N)$ in the frequency domain. So actually implementing the correlation and thus the time search in the frequency domain pays off.

Another computationally intensive task is that of fine Doppler estimation at the output of the code correlator. This problem involves finding the frequency of a sinusoid that may range from a maximum of ±5 kHz (or actually more depending on the local clock stability) with the requirement of incurring in just a few Hz of estimation error. The optimal maximum likelihood frequency estimator for this problem turns out to be based on a search along the frequency axis [16], which leads naturally to the FFT as well. So both the code correlation and the fine frequency search can actually be implemented in an efficient manner by making use of the FFT. This dual search is the underlying principle for the so-called double-FFT technique proposed in [17], and described next.

The double-FFT technique can be understood as the optimal implementation of the time-frequency matched filter to the input signal, with the aim of making an extensive use of FFT processors. The technique is based on using two FFT operations: the first one for the correlation between the input signal and the local replica and the second one for implementing the ML estimation of the frequency shift. This second FFT is actually implemented on a sliding window spanning over one symbol period, so the frequency estimation is carried out simultaneously with the symbol synchronization. This is an important feature because by doing so, the technique is able to extend the coherent correlation time up to the symbol period, thus providing a remarkable advantage in high-sensitivity scenarios. Note that this approach is rather different to conventional software receiver implementation, where coherent correlations at the acquisition stage are typically lacking any symbol synchronization. It is for this reason that they conservatively choose a short coherent integration time in order to contain as much as possible the losses that inevitably will be incurred when bit transitions appear, as indicated in Eq. (1.79). Furthermore, the symbol synchronization implemented by the double-FFT algorithm allows the receiver to be aligned with the symbol transitions of the navigation message. This makes possible to infer the navigation data symbols if the resulting E_b/N_0 is good enough and thus to have an acceptable symbol error rate given by Eq. (1.50). In case the symbols could reliably be inferred, they could eventually be used to remove the symbol contribution from the received signal samples with the aim of extending the coherent integration time far beyond the symbol period.

At high level, the implementation of the double-FFT method follows these steps:

1. Initialize a hypercube $\mathbf{H}$ with dimensions $(2N_r \times N'_{\text{scode}} \times N_r \times N_{v'})$ with N_r the number of primary codes within a symbol period (i.e. $N_r = 20$ for GPS L1 C/A) and $N_{v'} = \lfloor \frac{2v_{t,\max}}{\Delta v_t} \rfloor + 1$ the number of coarse frequency bins to be evaluated during the acquisition stage, with $v_{t,\max}$ the maximum Doppler span and Δv_t the frequency search bin spacing, according to Eq. (1.86).

2. `for n = 1:`N_I noncoherent accumulation

   (a) Load one block of received signal samples corresponding to twice the symbol period (i.e. 40 ms for GPS L1 C/A), as discussed in Section 9.2.9. This block length allows the observation of at least one whole symbol period irrespective of the position where the symbol transition may appear, thus facilitating the symbol synchronization.

   (b) Generate a matrix $\mathbf{M}_v$ with dimensions $(2N'_{\text{scode}} \times 2N_r - 1)$ whose columns are filled with the $2N'_{\text{scode}}$ points FFT of pieces of the loaded signal comprising two consecutive code periods. From column to column, there is a time shift of N'_{scode} samples.

   (c) `for i = 1:`N_v coarse Doppler search

      i. Set the current coarse Doppler frequency to $v_i = (i - 1)\Delta v - v_{\max}$.

      ii. Generate matrix $\mathbf{C}_{v_i}$ with dimensions $(2N'_{\text{scode}} \times 2N_r - 1)$ containing in each column the conjugated $2N'_{\text{scode}}$ points FFT of the local code replica including the tentative Doppler v_i.

      iii. Generate a matrix $\mathbf{T}_v$ with dimensions $(2N'_{\text{scode}} \times 2N_r - 1)$ containing the block-wise interpolation corrections, described in Section 9.2.3, in order to keep the correlation peaks from each column of $\mathbf{M}_v$ aligned.

      iv. Obtain matrix $\mathbf{Y}_i$ as the column-wise inverse FFT of the resulting matrix from the element-wise product of $\mathbf{M}_v$, $\mathbf{C}_{v_i}$ and $\mathbf{T}_v$. That is,

$$\mathbf{Y}_i = \mathbf{F}^H_{2N'_{\text{scode}}} [\mathbf{M}_v \odot \mathbf{C}_{v_i} \odot \mathbf{T}_v], \qquad (9.30)$$

      with $\mathbf{F}_N$ the N-points FFT matrix. The superscript ^{H} stands for complex conjugate transposition, and the $\odot$ denotes the Hadamard or element-wise product.

      v. Update the interpolation matrix $\mathbf{T}_v$ with the time-delay to be compensated in the next block of samples to be loaded in step 2.1.

      vi. `for p = 1:`N_r symbol synchronization

         A. Obtain matrix $\mathbf{R}_{v_i}$ by selecting columns p to $p + N_r - 1$ and rows 1 to N'_{scode} from matrix $\mathbf{Y}_i$, and applying the row-wise $2N_r$ point FFT.

That is,

$$\mathbf{R}_{v_i} = [\mathbf{Y}_i]_{(1:N'_{\text{scode}}, p:p+N_r-1)} \mathbf{F}^T_{2N_r}.$$ (9.31)

The size of this FFT is twice the row size of $\mathbf{Y}_i$, which is equivalent to zero padding each row with N_r zeros. Since the samples in each row correspond to downsampling at the primary code rate (i.e. 1 kHz in GPS L1 C/A), the frequency resolution of this row-wise FFT is $\Delta v_t = 1\,\text{kHz}/2N_r = 25\,\text{Hz}$ for GPS L1 C/A. In other words, the frequency resolution of this second FFT is the one corresponding to a coherent integration of $N_r T_{\text{code}}$, which leads to the general result

$$\Delta v_{t,\text{fine}} = \frac{1}{2N_r T_{\text{code}}}.$$ (9.32)

B. Take the element-wise square of $\mathbf{R}_{v_i}$ and add the result to the hypercube $\mathbf{H}$ as follows:

$$\mathbf{H}(:,:,p,i) = \mathbf{H}(:,:,p,i) + |\mathbf{R}_{v_i}|^2.$$ (9.33)

3. Search for the maximum correlation peak:

$$\{\hat{v}_{\text{fine}}, \hat{\tau}, \hat{p}, \hat{i}\} = \arg\max_{v'_{\text{fine}}, \tau', p, i} \mathbf{H}(v'_{\text{fine}}, \tau', p, i).$$ (9.34)

It is important to recall that the time-delay estimates we obtain are always circumscribed to the observation length of the input signal we are processing. Since the tentative τ' searches for the time-delay within the range $0 \leq \tau' \leq N'_{\text{scode}}$, which coincides with the second dimension of matrix $\mathbf{H}$, it would be better to refer to this value as $\tau'_{N_{\text{scode}}}$ in line with Eq. (9.6). We will henceforth use this notation for the last step to be described next.

4. Determine whether the satellite being searched is present or not:

(a) If $\mathbf{H}(\hat{v}_{\text{fine}}, \hat{\tau}_{N_{\text{scode}}}, \hat{p}, \hat{i}) > \gamma$, the satellite being searched is present with the following time-delay and frequency shift:

$$\hat{\tau}_{N_r N_{\text{scode}}} = \hat{\tau}_{N_{\text{scode}}} + \hat{p} N'_{\text{scode}}$$ (9.35)

$$\hat{v} = \hat{v}_{\text{fine}} \Delta v_{t,\text{fine}} + \hat{i}\Delta v - v_{\text{max}},$$ (9.36)

where $\hat{p} \in \{0, 1, \ldots, N_r - 1\}$ and $\hat{i} \in \{0, 1, \ldots, N_v - 1\}$. Note that the time-delay estimate in Eq. (9.35) is now contained within the symbol period once the symbol synchronization has been estimated as well through the search in p. The estimated time-delay $\hat{\tau}$ and frequency shift $\hat{v}$ can further be refined using the interpolation strategies discussed in Section 9.2.5, and then, we can proceed to compute the PVT. Please note that $\hat{\tau}$ is always a relative time-delay measurement, so we have ambiguous pseudoranges, that are limited to the distance corresponding to either the code or the symbol period, depending on whether we consider $\hat{\tau}_{N_{\text{scode}}}$ or $\hat{\tau}_{N_r N_{\text{scode}}}$, respectively. To solve for the absolute pseudoranges, we need to add an additional unknown when computing the so-called *coarse-time* navigation solution, as described in [18]. An alternative method is also presented in [1, Chapter 4].

(b) If $\mathbf{H}(\hat{v}_{\text{fine}}, \hat{\tau}_{N_{\text{scode}}}, \hat{p}, \hat{i}) < \gamma$, the satellite being searched is not present. The search for a new satellite must be restarted by coming back to step 1.
   end

   end

end

We see that in line with the snapshot philosophy, batches of samples are processed independently, so *no tracking stage is actually implemented*. This is consistent with the typical user test cases of snapshot receivers, where PVT is computed on-demand and batches of samples may not be consecutive. In these circumstances, we cannot guarantee the continuity of the observables at the PVT output, thus preventing the implementation of conventional closed-loop tracking techniques.

9.2.9   Single FFT Implementation

The double-FFT algorithm described in the previous section carries out the joint search of the time-delay, frequency shift and symbol transition by making extensive use of FFT operations. The implementation requires filling a multidimensional hypercube with the result of the correlation value for each tentative hypothesis being searched. Then, the hypercube is scanned to find the cell with the maximum value, thus providing an estimate of the time-delay, frequency shift and symbol transition position. The process involves a very significant amount of memory and a very extensive search, making this algorithm not suitable for direct implementation in low-cost or portable devices. Conventional software receiver implementations, not snapshot ones, prefer to rely on a single layer of FFT that is in charge of computing the correlation between the local replica and the received signal, only. They do not implement any symbol synchronization nor any refined frequency shift estimation at the acquisition stage because both are considered tasks to be carried out later at the tracking stage. It is for this reason that between the two FFT layers discussed in the previous section, only the first one covering the code correlation is widely implemented in conventional software receiver implementations. Nevertheless, there are some details related to how blocks of samples are processed through the first FFT that are worth being mentioned in this chapter. Particularly, when taking into account that we are focusing here on the block-processing of received signal samples.

The use of the FFT for code correlation was already introduced in Section 1.6.3 when discussing the step-by-step implementation of the full parallel search. The implementation of the correlation in the frequency domain is conceptually simple. It involves taking the FFT of both the input signal and the local code replica, conjugating the latter, multiplying both FFTs and finally doing the inverse FFT of such product. The interested reader can find more details for instance in [19, Chapter 7]. But subtleties appear when implementing the FFT in practice. One of them is regarding the FFT size, which should often be a power of 2 to achieve the fastest computational time. At least for the original implementation of the FFT algorithm. For software

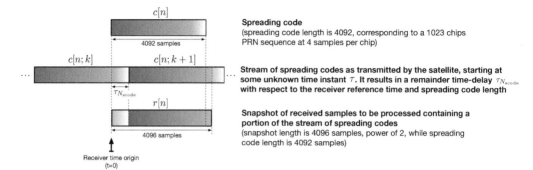

Figure 9.4 Example of a block of received signal samples with length slightly larger than the spreading code period

receivers running in a PC with MATLAB, this may not be a problem because the latest MATLAB versions already provide a wide set of efficient FFT algorithms for non–power-of-2 block sizes, by means of the open-source FFTW library. The only constraint to benefit from fast implementations is that the FFT size must have only small prime factors. But for software receivers running in a DSP or in dedicated hardware, the situation may be different because they may only afford the implementation of the original power-of-2-based FFT algorithm. In that case, using an FFT with a power-of-2 size may be the only way to benefit from a fast FFT implementation.

The problem when using a power-of-2 FFT size is that it often does not match with the actual length N_{scode} of the code replica that we need to convert to the frequency domain. To illustrate the situation, let us consider for simplicity a GPS L1 C/A signal and a sampling rate providing four samples per chip, that is, $F_s = 4R_c$ with $R_c = 1.023$ MHz. The code replica for this signal would have a length of $N_{\text{scode}} = 1,023$ chips $\times$ 4 samples/chip = 4,092 samples, but this is not a power-of-2. The nearest (and greater) power of 2 would be 4,096. So we may consider to use this value as the block size for implementing the code correlation using the FFT, that is, $N_{\text{FFT}} = 4,096$. Figure 9.4 represents how the received samples contained in a block size of $N_{\text{FFT}} = 4,096$ would look like. In the upper part, we see the original code of a given satellite with code length $N_{\text{scode}} = 4,092$ samples. In the central part, we see the stream of consecutive codes that are received from that satellite. Finally, in the bottom part, we see a block of $N_{\text{FFT}} = 4,096$ received samples that will be used for code correlation at the receiver using the FFT. As can be seen, the received samples are affected by a time shift $\tau_{N_{\text{scode}}}$ that corresponds to the fractional part of the propagation delay τ within one code period, N_{scode}. Note as well, that since the block length $N_{\text{FFT}} = 4,096$ is slightly larger than the code length $N_{\text{scode}} = 4,092$, we have slightly more than one full code period in a block of N_{FFT} received signal samples.

In Figure 9.4, we have seen how a block of $N_{\text{FFT}} = 4,096$ samples looks like. A similar situation occurs for the code replica, which needs to have the same FFT block size, $N_{\text{FFT}} = 4,096$ samples, even though the code length is just $N_{\text{scode}} = 4,092$ samples. In this case, we need to zero-pad the code replica from 4,092 to 4,096 samples. The result is illustrated in the upper left hand side of Figure 9.5, where we can see both the

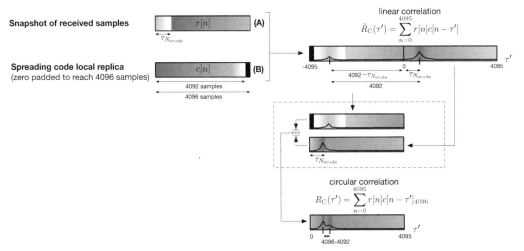

Figure 9.5 Result of the code correlation when implemented in the frequency domain using the FFT . In this example, there is a mismatch between the FFT size and the length of the code replica, being the former slightly larger than the latter.

block of N_{FFT} = 4,096 received samples and the block of N_{FFT} = 4,096 zero-padded code replica samples.

Now let us focus on these two blocks of N_{FFT} samples. If they were both correlated in the time domain, the result would lead to the linear correlation shown in the upper right hand side of Figure 9.5. Note that the result has a length equal to $2 \times N_{FFT} - 1 = 8,191$ samples, as expected for the linear correlation. But we are not interested in the time-domain correlation, but in the correlation implemented in the frequency domain by means of the FFT. So what we do is to take the N_{FFT}-size FFT of each block of samples, conjugate the FFT of the code replica and then multiply both FFTs. The result when coming back to the time domain by using the IFFT corresponds to the circular correlation, which is a *modular* version of the linear correlation. Such modular version can be understood by taking the linear correlation in the upper right hand side of Figure 9.5, splitting it into two blocks of N_{FFT} samples each, and then adding both blocks of samples together as illustrated in the central part of Figure 9.5. The result shown in the bottom part of Figure 9.5 corresponds to the circular correlation that would be obtained when implementing the correlation using the FFT.

It is interesting to note that due to the mismatch between the code length (N_{scode} = 4,092) and the FFT size (N_{FFT} = 4,096), two peaks do appear in the correlation obtained through the FFT. So care must be taken for selecting the correct one and discarding the aliased one. This situation, however, would not appear if both the FFT size and the length of the code replica were the same, $N_{FFT} = N_{scode}$. This is always possible in a MATLAB implementation because FFTs of any size can be implemented. In the worst case, when no optimizations were possible, the FFT algorithm would automatically switch to the DFT for the requested block size. The expected result would be obtained anyway just at the expense of an increase (maybe significant) of the computation time. In other computing devices, such as a DSP or dedicated hardware, it might happen that FFTs could only be implemented for power-of-2 block lengths, and then, the aforementioned discussion on the presence of two peaks should be borne in mind.

An alternative to the aforementioned implementation of the code correlation is to understand the problem as a block-convolution or a block-correlation one. That is, the problem of correlating an eventually infinite-length stream of samples with a finite-length block of samples. This is a problem that appears in many digital signal processing applications where a stream of incoming samples needs to be processed in real time. Implementing the conventional convolution or correlation in the time domain is unfeasible because one of the signals has actually infinite length. So one must resort to block-processing strategies. But care must be taken when performing the correlation in the frequency domain because we need the same result to be obtained as if the correlation had been obtained in the time domain instead. The two correlation peaks appearing in the bottom part of Figure 9.5 are something that we would like to avoid.

Specific methods to do so have been proposed in the literature such as the overlap-add and the overlap-save [20]. These methods chop the long stream of input samples into shorter blocks and perform the correlation via the FFT in each of these blocks, separately. Then, the individual results are put together again in a convenient way such that we obtain the same result as if we had done the linear correlation in the time domain on the long stream of input samples. An example is shown in Figure 9.6 for the overlap-save method. We are again interested in correlating the received samples with the code replica. The latter was shown in the previous example to have a code period of $N_{scode} = 4{,}092$ samples. Since we are correlating two blocks of samples (the one with the received samples and the one with code replica), the linear correlation has a length equal to twice the input block lengths, as observed in the upper right hand side of Figure 9.5. In order to account for that longer duration, we will now consider a code replica that is zero-padded up to twice the code period in order to leave some room for the aliased replicas that will appear in the circular correlation. This amounts at $2N_{scode} = 8{,}184$ samples, but since we are focusing on power-of-2 block lengths for efficiently implementing the FFT, we need a block size of $N_{FFT} = 8{,}192$ samples. The zero-padded code replica can now be observed in the upper left hand side of Figure 9.6, as well as the block of received samples, which both have a length of $N_{FFT} = 8{,}192$ samples.

The linear correlation of both blocks is shown in the central part of Figure 9.6 where one can see that the resulting length is twice that of the original signals, $2 \times N_{FFT} - 1 = 16{,}383$ samples. When the correlation is implemented via a 8,192-size FFT, the result leads to the circular correlation shown in the bottom part of Figure 9.6, which has a length equal to 8,192 samples as well. The overlap-save method then discards the second half of the result and keeps the first half only, which coincides with the correlation of the received signal with the code replica of one code period length. Since we keep 4,096 samples of the output correlation, for the next iteration,

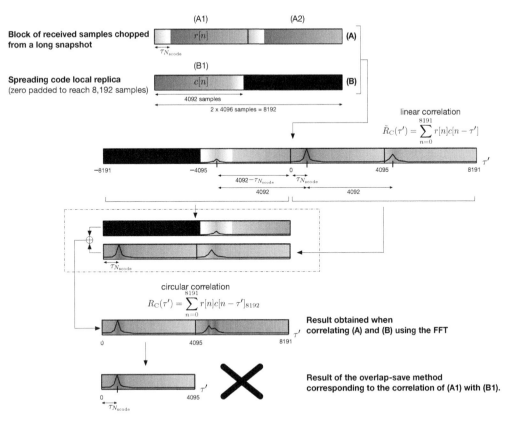

Figure 9.6 Illustration of the overlap-save method for performing the code correlation with received samples via the FFT, using blocks of samples of two code periods length (rounded up to a power-of-2 length).

we need to move forward the block of received samples by the 4,096+1 samples and start again with the steps illustrated in Figure 9.6 in order to obtain the next batch of 4,096 samples of output correlation.

Compared to the result in Figure 9.5, the correlation result that is obtained now in Figure 9.6 does not suffer from the presence of an aliased peak. Actually, the correlation we obtain with the overlap-save method coincides with the central part of the linear correlation, where we have the contribution of interest. The overlap-save method is actually implemented in the double-FFT algorithm described in Section 9.2.8, where pieces of two code periods length are considered (see steps 2.(b) and 2.(c).ii in Section 9.2.8).

9.3 Snapshot Positioning

So far, we focused on how to generate the satellite-to-receiver observations from a digital signal snapshot. Next, we focus on how to use these observations, mainly the

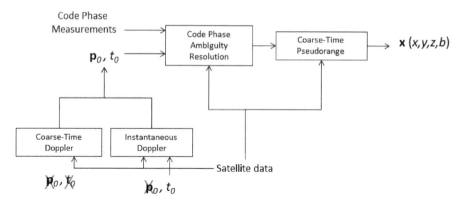

Figure 9.7 Snapshot PVT (Position, Velocity and Time) computation blocks.

code delay and the Doppler frequency observations, for something meaningful for the user: position and time. Our inputs to the positioning block are the satellite navigation data, the Doppler observations and the code-phase observations.[1] However, a snapshot receiver does not a priori know:

- The accurate time at which the observations are taken. This is generally obtained by demodulating the navigation data and synchronizing to a sequence transmitted by the satellite (see Section 1.8.1).
- The full pseudorange observation. Only the code delay, or code phase observation, that is, the point in time of the correlation peak within one primary code, is obtained from the signal processing block. That means that our input observation is a value between 0 and 1 ms for GPS L1 C/A, between 0 and 4 ms for Galileo E1 and so on, while the full signal time of arrival is between 66 and 90 ms approximately. We refer to this lack of information as *code-phase ambiguity*.

In order to solve the lacking information, we use Doppler observations. Unlike code delay observations, they are not subject to any ambiguity. They can provide a rough estimation of the position and time, accurate to some kilometers and some tens of seconds, respectively, which can be used to initialize the coarse-time pseudorange navigation algorithm, from which we can obtain the usual meterlevel accuracy.

Figure 9.7 shows the PVT computation blocks, to be used depending on the initial information available. If an initial position and time pair $(\mathbf{p}_0, t_0)$, accurate to around 150 kilometers and one minute respectively, is available, the PVT engine can use it to solve the code-phase ambiguity, compute a pseudorange and solve the coarse-time pseudorange navigation equations. If only an initial time t_0 is available, the instantaneous Doppler equations can be used to estimate $\mathbf{p}_0$. If neither $\mathbf{p}_0$ nor t_0 are available, the coarse-time Doppler equations can be used. These processes will be described in further detail in the next sections.

---

[1] The terms *code phase* and *code delay* are both used in this chapter. We use code phase in the context of *code-phase ambiguity*, which is the most used term.

9.3.1   Instantaneous Doppler Positioning

Positioning techniques based on the Doppler effect have been used since the beginning
of radionavigation. Initially, they were based on the observation of the variations in the
apparent signal frequency due to the relative movement of the transmitter with respect
to the receiver. By analyzing these variations over time, the receiver could determine
its position. The first satellite orbit determination method, applied on Sputnik in 1958,
was based on the Doppler effect. Some years later, and based on this idea, TRANSIT,
the first satellite positioning system and precursor of GPS, was also based on Doppler
variations, see [21]. In our case, as we are interested in snapshot positioning, we cannot
afford to wait and estimate the Doppler variations. We therefore focus on instantaneous
Doppler positioning and its variants.

Instantaneous Doppler Positioning with Reference Time
The instantaneous Doppler positioning equations can be obtained from the time
derivative of the standard pseudorange navigation equations in Section 1.10 [1]. For a
static or slow-moving receiver, the resulting system of equations is:

$$\Delta d = \begin{bmatrix} \Delta d^{(1)} \\ \vdots \\ \Delta d^{(i)} \\ \vdots \\ \Delta d^{(s)} \end{bmatrix} = \begin{bmatrix} \dot{e}^{(1)} & 1 \\ \vdots & \\ \dot{e}^{(i)} & 1 \\ \vdots & \\ \dot{e}^{(s)} & 1 \end{bmatrix} \begin{bmatrix} \Delta x \\ \Delta y \\ \Delta z \\ \Delta f_d \end{bmatrix} + \varepsilon, \tag{9.37}$$

where Δd is the update of the a priori Doppler measurement state, which is a column vector with the difference between the Doppler observations d and the predicted
Doppler observations $\hat{d}$, $\Delta d = d - \hat{d}$. s is the number of satellites, with index i. The
(1×3) vector $\dot{e}^{(i)}$ is the derivative in time of $e^{(i)}$, the receiver-to-satellite unit vector for
a given satellite i, whose expression is presented in Eq. (9.38), $[\, \Delta x \;\; \Delta y \;\; \Delta z \;\; \Delta f_d \,]^T$
is the update of the state vector, which estimates the receiver position (x, y, z) and the
frequency drift f_d, and ε is the error due to the noise and the linearization process.
The predicted Doppler observations $\hat{d}$ are based on the satellite positions at the reference time t_0 and the previous position, which for the first iteration is set to $\mathbf{p}_0$. If
no $\mathbf{p}_0$ is available, we can use the center of the polygon of the satellite projections
on Earth, although an initial position of $(0,0,0)$ will generally suffice. The system of
equations can be solved in a few iterations and through least-squares, as for standard
pseudorange equations.

   The receiver-to-satellite unitary vectors $e^{(i)}$ need to be derived in time, as they are
part of the pseudorange equation. The derivative in time of $e^{(i)}$ can be determined as
per Eq. (9.38). As the vector e is the division of the pseudorange vector by its module,
its time derivative depends on the derivative of each of them. Note that the superscript
i has been omitted.

$$\dot{e} = \frac{de}{dt} = \frac{d}{dt}\left(\frac{r}{r}\right) = \frac{1}{r^2}\left(\frac{dr}{dt}r - r\frac{dr}{dt}\right) = \frac{1}{r}\left(\dot{r} - e\dot{r}\right), \tag{9.38}$$

where $\dot{r}$ is the derivative in time of the range vector, that is, the derivative in time of the vector from the receiver position to the satellite position. This corresponds to the vector of satellite velocity relative to the receiver $\dot{r} = v^{(i)} - v_r$, with $v^{(i)}$ and v_r be the satellite and receiver velocity vectors, respectively. Notice that Doppler shift and range rate have inverse signs: When the satellite is approaching, and thus, the range is diminishing, the frequency is higher, and thus, the Doppler observation is positive:

$$\dot{\rho}^{(i)} = -(d^{(i)} - f_d)\lambda + \varepsilon, \qquad (9.39)$$

where $\dot{\rho}^{(i)}$ is the pseudorange rate of satellite i, the Doppler observation of the same satellite is $d^{(i)}$, λ is the wavelength of the carrier frequency (around 0.19 m for GPS L1 C/A at 1575.42 MHz) and ε represents the errors in the observation due to frequency misalignment and receiver noise.[2]

What we mean with slow-moving receiver is a receiver whose Doppler observations will not be highly impacted by its motion. Observation errors up to 40 Hz are still tolerable for our purposes, so a receiver at a speed of around 7.6 m/s (40 Hz×0.19 m for 1575.45 MHz of GPS L1 C/A or Galileo E1), or around 27 km/h, could still use these equations, depending on the accuracy of its Doppler estimation. Note that a 7.6 m/s speed would only be translated into a 40 Hz Doppler error when the velocity vector is aligned with the satellite line of sight, which is generally not the case. When the receiver is moving at a higher and unknown speed, the receiver velocity vector has to be estimated as part of the state vector as follows:

$$\Delta d = \begin{bmatrix} \dot{e}^{(1)} & e^{(1)} & 1 \\ & \vdots & \\ \dot{e}^{(i)} & e^{(i)} & 1 \\ & \vdots & \\ \dot{e}^{(i)} & e^{(i)} & 1 \end{bmatrix} \begin{bmatrix} \Delta x \\ \Delta y \\ \Delta z \\ \Delta v_x \\ \Delta v_y \\ \Delta v_z \\ \Delta f_d \end{bmatrix} + \varepsilon, \qquad (9.40)$$

where $[\Delta v_x \ \Delta v_y \ \Delta v_z]^T$ is the receiver velocity vector. Solving this system of equations requires at least seven satellite observations. In order to reduce the number of unknowns, we can have external estimations of height or velocity.

Instantaneous Doppler Positioning without Reference Time

When the snapshot time is unknown, we do not know where the satellites were at the time the observations were taken. The lack of reference time can be solved by adding an additional unknown to Eq. (9.37), as follows:

---

[2] Note that Eq. (9.39) includes the pseudorange rate, which also incorporates the clock drift, instead of the range rate. Note also that because of this sign difference between Doppler shift and range rate, the unitary vector in Eq. (9.37) is positive unlike in Eq. (1.149).

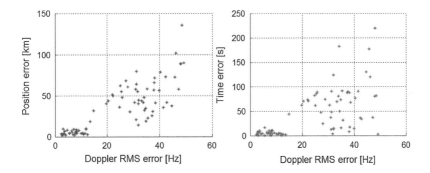

Figure 9.8 Position and time error of the coarse-time Doppler solution versus Doppler observation RMS error for 80 1-ms GPS L1 C/A snapshots, Danish GPS Center, Aalborg, 2010 [22] .

$$\Delta d = \begin{bmatrix} \dot{e}^{(1)} & 1 & -\ddot{\rho}^{(1)} \\ \vdots & & \\ \dot{e}^{(i)} & 1 & -\ddot{\rho}^{(i)} \\ \vdots & & \\ \dot{e}^{(s)} & 1 & -\ddot{\rho}^{(s)} \end{bmatrix} \begin{bmatrix} \Delta x \\ \Delta y \\ \Delta z \\ \Delta f_d \\ \Delta t_{c,d} \end{bmatrix} + \varepsilon, \tag{9.41}$$

where $-\ddot{\rho}^{(i)}$ denotes the satellite-i-to-receiver relative acceleration,[3] and $t_{c,d}$ is the new unknown: the coarse time between the observation time and the reference time. This closed-form solution allows computing position and time with an initial time error of more than an hour. This limitation is due to the assumption that the satellite relative acceleration is constant between the initial and the actual observation time, which is generally not the case, but is good enough for periods of time in that order.

For a dynamic receiver with unknown, nonnegligible velocity, the equation is extended to

$$5\Delta d = \begin{bmatrix} \dot{e}^{(1)} & e^{(1)} & 1 & -\ddot{\rho}^{(1)} \\ \vdots & & & \\ \dot{e}^{(i)} & e^{(i)} & 1 & -\ddot{\rho}^{(i)} \\ \vdots & & & \\ \dot{e}^{(s)} & e^{(s)} & 1 & -\ddot{\rho}^{(s)} \end{bmatrix} \begin{bmatrix} \Delta x \\ \Delta y \\ \Delta z \\ \Delta v_x \\ \Delta v_y \\ \Delta v_z \\ \Delta f_d \\ \Delta t_{c,d} \end{bmatrix} + \varepsilon. \tag{9.42}$$

Figure 9.8 shows the sensitivity of the coarse-time Doppler solution to errors in the Doppler observations. The figure shows a correlation between the Doppler observation quality and the position and time solution error. The method provides very good results

---

[3] We assume here constant receiver and satellite clock drifts.

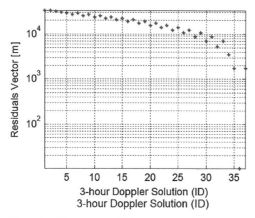

Figure 9.9 Time error of 3-hour coarse-time Doppler solutions over 4.5 days.

with good-quality observations when the Doppler RMS error is smaller than 15 Hz. However, very low-quality observations may provide an initial time-position pair that will not converge to the right pseudorange solution. Let us recall that our objective is to provide a rough initial position and time to initialize the coarse-time pseudorange equations.

Solving Long-Time Uncertainties

When the timing uncertainty of a snapshot is larger than a few hours, Eq. (9.41) or (9.42) will not converge, mainly due to the linearization errors of nonlinear equations: The assumption that the pseudorange rate changes linearly is only valid within a limited span of time. In order to estimate a solution over long periods of time without adding more unknowns, we can compute a solution for different times and study the residuals. For example, if the maximum initial time error is 1.5 hours, we can compute coarse-time Doppler solutions every three hours.

We can split the total time uncertainty into intervals of a given duration, define an initial time t_0 associated with each interval $T0_1, T0_2, T0_3, T0_4$ and $T0_5$ as per Figure 9.11, until the total time uncertainty is covered. For each initial time, the coarse-time Doppler navigation algorithm will provide a different time solution $T1_1, T1_2, T1_3, T1_4$ and $T1_5$. If the method converges to a plausible solution, the obtained observational residuals, in case of an overdetermined solution, will be below a certain threshold THR1, as is the case for $T1_2$, $T1_3$ and $T1_5$ in Figure 9.11. We then use the low-residual coarse-time Doppler navigation solutions to initialize a coarse-time pseudorange solution, which will converge to a correct and small-residual pseudorange solution below a certain residual threshold THR2, if the initial position and time were roughly correct. Figure 9.9 shows the module of the residual vector of the pseudorange solution initialized from plausible residuals smaller than THR1, over 4.5 days. The Doppler solutions are computed with initial times every three hours. We can see that there is only one solution (ID 36) with small enough residuals, the closest one having a residual vector module of around 2,000m, which seems large enough to be discarded.

Solving long-time uncertainties also requires overcoming the effect of GNSS orbital repeatability on Doppler positioning. GPS satellite orbits have a period of half a sidereal day, or almost 12 hours.[4] This means that instantaneous Doppler positioning can lead to a low-residual, plausible Doppler solution twice a day, if GPS-only observations are used. However, after half a sidereal day, the satellites will have completed their orbits and will be at the same positions with respect to an inertial reference frame, but the Earth will have turned 180°, so the low-residual solution will be found at a position with approximately the same latitude, but opposite longitude and half a day earlier or later. A low-residual solution at approximately the same latitude and longitude will be found one sidereal day later when the Earth will have performed a full 360° turn. This effect can be observed in Figure 9.12. It presents the results of the instantaneous Doppler positioning as per Eq. (9.37) over a period of 50 hours around the correct observation time. The left panel presents the Doppler observational residual vector. It shows that the residuals are small at the correct time, but also every 12 hours. The right panel shows the actual position error, which is small every sidereal day, as explained.

In order to resolve this ambiguity, the wrong solutions can be discarded at the later coarse-time pseudorange algorithm step. The application of ephemerides and clock corrections one or several days before or after will lead to large positioning errors that will allow us to discard the solution, as shown in Figure 9.9. If the receiver has observations from different GNSS such as GPS, GLONASS, Galileo or BDS, which have different orbital periods, the repeatability period will be extended.

9.3.2 Coarse-Time Pseudorange Positioning

We have explained how to compute, through instantaneous Doppler positioning methods, a good enough position that allows us to initialise our pseudorange-based equation system. However, we still have to solve the code-phase ambiguity, and we have to estimate an extra unknown, which is the time between our reference time from the Doppler solution and the observation time. This is explained in the next sections.

Code-Phase Integer Ambiguity Resolution

A conventional receiver, as described in previous chapters, computes pseudorange observations by subtracting the transmission time, which is known from the demodulated signal data, from the reception time. However, the receiver only extracts from the snapshot a code-phase observation that is a fraction of a full code. This is illustrated in a simplified way in Figure 9.10. In order to form the full pseudorange, the snapshot receiver needs to know the number of full PRN codes between the satellite and the receiver to solve the code-phase integer ambiguity. So the receiver needs to assign an integer number of codes $N^{(i)}$ to the fractional code phase of each satellite i. If there

---

[4] A sideral day is the time it takes the Earth to turn 360°, which is 11 hours, 56 minutes and four seconds.

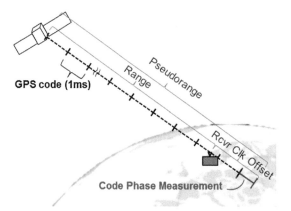

Figure 9.10 Representation of the satellite-to-receiver pseudorange, as the true range plus the receiver clock offset, and as the sum of integer 1-millisecond GPS codes plus the code-phase observation.

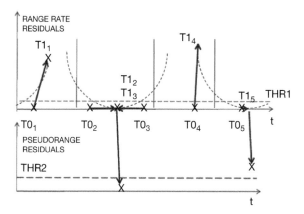

Figure 9.11 Residuals computation to determine the correct snapshot solution.

were no receiver clock bias, the $N^{(i)}$ could be easily estimated. However, due to the receiver clock bias – which is unknown at this stage – the problem becomes more complicated. To overcome this difficulty, the receiver may assign an $N^{(i)}$ for each satellite i and verify the residuals of a solution. If all $N^{(i)}$ are correct, the residuals will be in the order of a few meters. If not, there will be at least one pseudorange with an error of hundreds of kilometers (300 km for GPS L1 C/A, 1,200 km for Galileo E1), which has an impact on the residuals and the receiver can compute another solution.

A method to resolve the code-phase integer ambiguity is to perform the single differences between satellite pairs, so the unknown receiver clock bias is removed, as proposed in [1, Chapter 4]:

$$N^{(i)} = \text{round}(N^{(1)} + z^{(1)} - z^{(i)} + (\hat{r}^{(i)} - dt_t^{(i)}) - (\hat{r}^{(1)} - dt_t^{(1)})), \qquad (9.43)$$

where $N^{(i)}$ is the number of full codes of pseudorange $\rho^{(i)}$ for satellite i, $N^{(1)}$ is the number of full codes of satellite 1 a priori range, taken as a reference, $z^{(1)}$ is the observed code phase, or fractional pseudorange, of satellite 1, $z^{(i)}$ is the observed

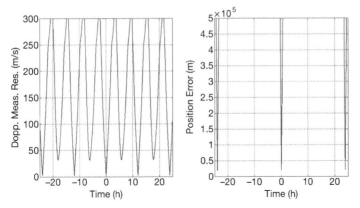

Figure 9.12 Instantaneous Doppler positioning, Danish GPS Center, 2010. The solutions are computed for every minute. Left panel: Doppler observational residuals. Right panel: position error.

code phase of satellite i, $\hat{r}^{(i)}$ is the a priori range of satellite i, $dt_t^{(i)}$ is the clock offset of satellite i, from the ephemeris, $\hat{r}^{(1)}$ is the a priori range of satellite 1 and $dt_t^{(1)}$ is the clock offset of satellite 1, also from the ephemeris. This method eliminates the integer rollover problem due to the receiver common bias.

Coarse-Time Pseudorange Positioning

Even if we solve the code-phase ambiguities and form the full pseudoranges, we still have an uncertainty in time between the observation time of the snapshot and the reference time coming from the Doppler solution or by our external reference, as shown in Figure 9.8. If we try to compute our position at a wrong time, given that the satellites are moving and their clocks are drifting, we will apply the wrong satellite positions and clock offsets, and our result will be erroneous. This problem can be solved by the coarse-time pseudorange navigation algorithm. This algorithm adds a fifth unknown, named *coarse time*, to the standard four component state vector of 3-D position and receiver clock bias [18]. The coarse-time observation equation is:

$$\Delta\rho = \mathbf{A}\Delta\mathbf{x}_{tc} + \boldsymbol{\varepsilon} = \begin{bmatrix} -\mathbf{e}^{(1)} & 1 & \dot{\rho}^{(1)} \\ \vdots & & \\ -\mathbf{e}^{(i)} & 1 & \dot{\rho}^{(i)} \\ \vdots & & \\ -\mathbf{e}^{(s)} & 1 & \dot{\rho}^{(s)} \end{bmatrix} \begin{bmatrix} \Delta x \\ \Delta y \\ \Delta z \\ \Delta b \\ \Delta t_{c,\rho} \end{bmatrix} + \boldsymbol{\varepsilon}, \tag{9.44}$$

where $\Delta\rho$ is a difference vector between the predicted $\hat{\rho}$ and the observed pseudoranges ρ, $(\Delta\rho = \rho - \hat{\rho})$, $\mathbf{A}$ is the matrix that relates the observations with the state vector $\Delta\mathbf{x}_{tc}$, that includes a new unknown for the coarse time, and $\dot{\rho}^{(i)}$ is the pseudorange rate of satellite i. $[\Delta x \quad \Delta y \quad \Delta z \quad \Delta b \quad \Delta t_{c,\rho}]^T$ is the update of the state vector $\Delta\mathbf{x}_{tc}$, which includes the receiver position (x, y, z), the receiver bias b and the coarse-time difference $t_{c,\rho}$ which is the actual observation time minus the estimated observation time. The observation and linearization errors are again represented by ε. Notice that $\dot{\rho}^{(i)}$ includes both the range rate and the satellite and receiver clock rates. Note also that Eq. (9.44) is similar to Eq. (9.41), but applied to pseudorange observations, with the difference that the coarse time t_c here can be up to around one minute, while in the Doppler case, $t_{c,d}$ can go beyond an hour. The accuracy of the coarse-time pseudorange solution will be similar to that of a standard pseudorange solution, with the difference that the additional unknown demands an additional observation and this may increase the DOP [1, Chapter 5].

9.4   Front End and Other Practical Information

9.4.1   Front End

The experimental results presented in this chapter were obtained using a front end formed by an integrated circuit SE4110L from SiGe semiconductors. Its features are presented in Table 9.1. It down-converts the signal to an intermediate frequency of $F_{IF} = 4.129$ MHz with a sampling frequency of $F_s = 16.368$ MHz [23].

9.4.2   Data Collection

The data were captured in Aalborg University (AAU), former Danish GPS Center, on March 2, 2010, 12:14 UTC, at a rooftop antenna of AAU's Danish GPS Center by O. Badia and T. Iacobescu. The snapshot sample provided lasts one second. GPS PRNs 4, 11, 12, 13, 17, 20, 23, 30, 31 and 32 were acquired in the Acquisition block and used for the position computation.

9.4.3   MATLAB Configuration and Functions

The snapshot receiver is located under the *snapshotgpsrx* folder. The receiver first reads the configuration parameters in the *InitSettings* function. This function includes the typical constants required for a GNSS receiver to work: the speed of light, π, and Earth parameters (gravitational constant, rotation rate, radius). These are fixed for all

Table 9.1 Configuration of front end for the GPS snapshot receiver example of this chapter.

Parameter	Value	Unit
Chipset	SiGe SE4110L	–
Sampling frequency	16.368	MHz
Intermediate frequency	4.129	MHz
Number of quantization bits	2	–
Down-converted signal type	Real	–

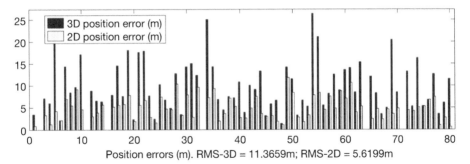

Position errors (m). RMS-3D = 11.3659m; RMS-2D = 5.6199m

Figure 9.13 Position error results, expressed as RMS-2D (horizontal) and RMS-3D for 80 contiguous snapshots from a sample taken on March 2, 2010, Danish GPS Center, Aalborg.

runs. Next, the GPS signal parameters are defined. As only the GPS L1C/A signal is processed, these parameters are fixed here, including carrier frequency (1,575.42 MHz), code length (1,023 chips) and code frequency (1.023 MHz). Another set of important parameters are the detection thresholds for acquisition (*acqThreshold*), Doppler positioning (*DoppTHR*) and pseudorange positioning (*CodeTHR*). They are defined heuristically for the samples used and may require some optimization for operational receivers (see Section 9.2.4). Other relevant configuration parameters are the acquisition integration time (set by default to 10 ms), and the amount of time/samples processed from the sample file.

If the acquisition is already performed, the position result of the snapshot receiver is instantaneous. If the acquisition is needed, it may take a few seconds to a few minutes, depending mostly on the configured integration time and number of snapshots. Therefore, the reader/programmer can quickly execute the code and understand how the functions are executed with the real data included with the software.

The parameters specific to a sample are set in the configuration (.cfg) file. These include the receiver sampling frequency (*samplingFreq*), intermediate frequency (*interFreq*), initial position and time, search period and execution mode: The receiver can run with a binary sample file (.bin or .snp), from which acquisition will be performed, or from an already acquired signal (.mat) from a previous execution. The configuration file also defines the number of acquisitions (and therefore position solutions) that will be generated. Note that the receiver is currently configured to read sample files captured at the intermediate frequency with a sample unit of ("int8") (*dataType* parameter). Also part of the configuration is the time interval between Doppler solutions (*doppInterval*). If too large, it may not find a solution that converges. This is now set to 3 hours. For the results shown, the start time in the current configuration is off by around 2 hours, but for the generation of plots in Figure 9.12 maximum time errors beyond 2 days were used. The code delay measurement was obtained through the parabolic interpolation shown in Figure 9.2. The least-square coarse-time pseudorange solution errors obtained are shown in Figure 9.13. The results show the higher error due to geometry in the vertical domain.

The receiver needs to download the navigation data for the period under analysis. This is performed by the *getFiles* and *loadRinex* functions, and store the ephemeris and clock information in structures *GPS* and *HEADER* for later use.[5]

Satellite data can be provided in many formats. The Receiver-INdependent EXchange (RINEX) format [25]. The RINEX format includes navigation data (in addition to observations, site, meteorological and clock data) in ASCII format. The MATLAB snapshot receiver accompanying this book uses RINEX V2.[6]

When configured to do so, the first step is to perform the signal acquisition, in the *acquisitionSnp* function. Currently, the GPS L1 C/A acquisition is performed for frequency steps of $\Delta v_t' = 500$ Hz, and $v_{t,\max} = 12$ kHz covering a range from $[-v_{t,\max}, +v_{t,\max}]$. Acquisition is performed through FFT. This means that, for each frequency bin, the signal is multiplied by the carrier for the sine and cosine components. The resulting baseband complex signal is transformed by the FFT, this is correlated (in the frequency domain) with the FFT of the spreading code and the IFFT is applied to the result. The result is equivalent to the convolution in time, which, if the signal is found, yields a peak when the code is correlated, that is, the code-phase measurement. When several integrations are performed (noncoherently, as the receiver does not know if there is a symbol transition), then the result is squared and added to the previous integration. This process is performed for each frequency bin. Once the acquisition is performed for all bins, if the signal is acquired, a peak is found. However, the code delay accuracy is limited by the sampling frequency. In order to refine the measurement, a parabolic interpolation is performed, as discussed in 9.2.5. When a satellite is found, its PRN is output through the screen. A comprehensive explanations of FFT-base acquisition are available in Section 1.6 and also before in this chapter.

As per the configuration, a coarse-time Doppler solution is calculated every 3 hours by the function *CTDoppler*. Once the Doppler estimation and geometry matrix are calculated, Eq. (9.41) is applied to solve the system. If the results of the residuals is below the threshold *DoppTHR* (or THR1 before), the solution is available to initialize the coarse-time pseudorange positioning solution. For each of the initial coarse-time Doppler initial solution, the standard, coarse-time pseudorange equation is computed in function *CTPRange*. The ionospheric delay is estimated according to GPS's Klobuchar model [27] and the tropospheric delay according to the Goad model [28], by *iono* and *tropo* functions, respectively. The function also includes the estimation of the integer part of the pseudorange, or code-phase ambiguity, according to van Diggelen's method [1, Chapter 4], Eq. (9.43). Once the ranges and geometry matrix are

---

[5] Options for download include servers from the International GNSS Service (IGS), with satellite data, atmospheric data and Earth rotation parameters of very high quality, available for postprocessing and in real time [24].

[6] Apart from RINEX files, the Radio-Technical Commission for Maritime services (RTCM) has developed other popular standards widely used. The RTCM SC-104 standard for D-GNSS provides binary messages which include satellite data. They were developed in the frame of D-GPS for maritime applications, but they are now adopted worldwide for many applications. Real-time RTCM messages from several stations worldwide are transmitted through Networked Transport of RTCM via Internet Protocol (NTRIP), see [26].

calculated, Eq. (9.44) is applied to solve the system. If the residuals module is below the threshold (THR2 before), the solution is considered as correct. Note that, in order to compute the residuals, plus the coarse time unknown, six satellites are needed. Note also that the above-described receiver is a simplified version of debugging-enabled version which, excluding the acquisition block, is provided under the folder *plots*. The reader can execute the *Main4Plots* function to generate the plots of the Danish GPS Center data capture used along the chapter (Figure 9.8).

References

[1] F. van Diggelen, *A-GPS, Assisted GPS, GNSS and SBAS*. Artech House, 2009.

[2] G. Lachapelle, H. Kuusniemi, D. T. H. Dao, G. Maccougan, and M. E. Cannon, "HSGPS Signal Analysis and Performance under Various Indoor Conditions," *Navigation: Journal of the Institute of Navigation*, vol. 51, no. 1, pp. 29–43, 2004.

[3] G. López-Risueño and G. Seco-Granados, "Measurement and Processing of Indoor GPS Signals Using One-Shot Software Receiver," in *Proceedings of ESA NAVITEC, Noordwijk, NL*, pp. 1–9, 2004.

[4] G. Seco-Granados, J. A. López-Salcedo, D. Jiménez-Baños, and G. López-Risueño, "Challenges in Indoor Global Navigation Satellite Systems," *IEEE Signal Processing Magazine*, vol. 29, pp. 108–131, March 2012.

[5] M. Manzano, J. Alegre-Rubio, A. Pellacani, G. Seco-Granados, J. López-Salcedo, E. Guerrero, and A. García, "Use of Weak GNSS Signals in a Mission to the Moon," in *Proceedings of ESA NAVITEC, Noordwijk, NL*, pp. 1–12, December 2014.

[6] V. Capuano, C. Botteron, J. Leclere, J. Tian, and Y. Wang, "Feasibility Study of GNSS as Navigation System to Reach the Moon," *Acta Astronautica*, vol. 116, pp. 186–201, 2015.

[7] M. H. Hayes, *Statistical Digital Signal Processing and Modeling*. John Wiley & Sons, 1996.

[8] M. L. Psiaki, "Block Acquisition of Weak GPS Signals in a Software Receiver," in *Proceedings of ION GPS, Salt Lake City, UT*, pp. 2838–2850, September 2001.

[9] D. Akos and M. Pini, "Effect of Sampling Frequency on GNSS Receiver Performance," *Navigation: Journal of the Institute of Navigation*, vol. 53, pp. 85–95, Summer 2006.

[10] I. Fernández-Hernández, *Snapshot and Authentication Techniques for Satellite Navigation*. PhD thesis, Aalborg University, Faculty of Engineering and Science, June 2015.

[11] G. López-Risueño, G. Seco-Granados, and A. García, "Evaluation of GPS Indoor Positioning Using Real Measurements and One-Shot Software Receiver," in *Proceedings of European Navigation Conference (ENC), Munich, NL*, pp. 1–9, 2005.

[12] G. López-Risueño and G. Seco-Granados, "C/N0 Estimation and Near-Far Mitigation for GNSS Indoor Receivers," in *Proceedings of IEEE Vehicular Technology Conference (VTC), Stockholm, Sweden*, pp. 2624–2628, 2005.

[13] E. Falletti, M. Pini, L. Lo Presti, and D. Margaria, "Assessment of Low Complexity C/N0 Estimators Based on M-PSK Signal Model for GNSS Receivers," in *Proceedings of IEEE/ION PLANS Conference, Monterey, CA*, pp. 167–172, 2008.

[14] J. A. López-Salcedo, J. M. Parro-Jiménez, and G. Seco-Granados, "Multipath Detection Metrics and Attenuation Analysis Using a GPS Snapshot Receiver in Harsh Environments," in *Proceedings of European Conference on Antennas and Propagation (EuCAP), Berlin, DE*, pp. 3692–3696, 2009.

[15] K. Borre, D. Akos, N. Bertelsen, P. Rinder, and S. H. Jensen, *A Software-Defined GPS and Galileo Receiver, Single Frequency Approach*. Birkhäuser, 2007.

[16] S. M. Kay, *Fundamentals of Statistical Signal Processing, Volume I: Estimation Theory*. Prentice-Hall, 1993.

[17] D. Jiménez-Baños, N. Blanco-Delgado, G. López-Risueño, G. Seco-Granados, and A. Garcia, "Innovative Techniques for GPS Indoor Positioning Using a Snapshot Receiver," in *Proceedings of ION GNSS, Fort Worth, TX*, pp. 2944–2955, 2006.

[18] B. Peterson, R. Hartnett, and G. Ottman, "GPS Receiver Structures for the Urban Canyon," in *Proceedings of ION GPS, Palm Springs, CA*, pp. 1323–1332, 1995.

[19] J. B.-T. Tsui, *Fundamentals of Global Positioning System Receivers: A Software Approach*. John Wiley & Sons, 2005.

[20] R. E. Blahut, *Fast Algorithms for Digital Signal Processing*. Addison-Wesley, 1985.

[21] P. Misra and P. Enge, *Global Positioning System: Signals, Measurements and Performance, Revised 2nd Edition*. Ganga-Jamuna Press, 2011.

[22] I. Fernández-Hernández and K. Borre, "Snapshot Positioning without Initial Information," *GPS solutions*, vol. 20, no. 4, pp. 605–616, 2016.

[23] O. B. Solé and T. I. Ioan, "Enhancement study of GPS snapshot techniques," *Aaborg University, Aaborg, Tesis*, 2011.

[24] IGS, "International GNSS Service–GPS Satellite Ephemerides Satellite and Television Station Clocks." www.igs.org/products, 2016.

[25] IGS. "RINEX Version 2" https://files.igs.org/pub/data/format/rinex2.txt.

[26] Radio Technical Commission for Maritime Services, Scientific Committee 104, *RTCM 10410.1 Standard for Networked Transport of RTCM via Internet Protocol (Ntrip)*, 2.0 with Amendment 1, June 28, 2011, ed., 2011.

[27] J. A. Klobuchar, *Global Positioning System: Theory and Applications*, Ch. Ionospheric Effects on GPS. American Institute of Aeronautics and Astronautics (AIAA), 1996.

[28] C. Goad and L. Goodman, "A Modified Hopfield Tropospheric Refraction Correction Model", in *Proceedings of the Fall Annual Meeting of the American Geophysical Union, San Francisco, CA*, 1974.

10 Acquisition and Tracking of BOC Signals

Elena Simona Lohan

10.1 Introduction

The present chapter deals with specific aspects of acquisition and tracking of signals using BOC modulations, which are characterized by a more complex inner structure compared to the traditional BPSK modulation. For instance, while a BPSK-modulated signal has most of its PSD concentrated at its center frequency, for a BOC modulation, the maximum values of the PSD are offset from the center frequency in a symmetric manner. Therefore, the peaks of the sidelobes that appear in the autocorrelation function (ACF) of a BOC signal create a pattern that makes it difficult to determine whether the tracking loop is tracking a left or a right peak of this correlation function. Hence the term *ambiguous*. Despite this limitation, BOC-modulated signals are essential for modernized GNSS signals. Due to their widespread adoption in Galileo, we will sometimes refer specifically to that system. However, BeiDou and GPS exploit this modulation type as well.

We start with an overview of the challenges in acquiring and tracking a BOC signal, by emphasizing the difference with BPSK modulation used in the traditional GPS L1 C/A signal. We then present a compact overview of the acquisition methods used more widely and which can be directly applicable to Galileo signals in the so-called *ambiguous* or conventional mode. Afterward, we focus on the Galileo-specific processing in acquisition, namely the *unambiguous* signal processing which takes into account the ambiguities of most of the proposed Galileo signals and tries to better cope with them. The next sections are dedicated to tracking: first focusing on an overview of generic tracking methods valid for any DS-SS system with and without multipath mitigation capabilities. Then, we discuss unambiguous tracking methods and multipath impact on tracking. This chapter is not offering an exhaustive review of all possible algorithms and related challenges in acquisition and tracking, but it tries to cover some of the main aspects, in such a way that the reader becomes familiarized with the basic challenges and the available solutions.

10.2 The Ambiguity Challenge in BOC-Modulated Signals

As described in Sections 1.6 and 1.7, the acquisition and tracking processes are based on correlating the incoming signal with a local replica of the pseudorandom

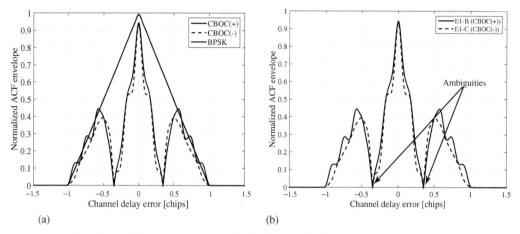

Figure 10.1 (a) ACF shapes for CBOC (Galileo E1-C pilot code and E1-B data code) and BPSK (GPS C/A code) for infinite bandwidth. (b) Ambiguities in the ACF of CBOC(−)- and CBOC(+)-modulated signals with 24 MHz double-sided front-end bandwidth.

code, typically using the same modulation as at the transmitter side or a simplified variant of it. Let us start with an example from Galileo where signals on the OS use a composite BOC (CBOC)-modulated signal. Then, the most straightforward processing, also called ambiguous or full BOC processing, correlates the incoming CBOC-modulated signal with a local CBOC-modulated signal replica. If we want to decrease the receiver complexity, we can correlate a CBOC-modulated signal with a sinBOC(1, 1)-modulated reference code as shown in [1]. Next, we compare the ACF of a CBOC(−)-modulated signal, used for Galileo E1 pilot signal, and of a CBOC(+)-modulated signal, used for Galileo E1 data signal, with the ACF of a conventional BPSK(1)-modulated signal used in GPS L1 C/A signals. The result is shown in Figure 10.1 for the case of infinite bandwidth, where we notice additional notches in the interval [−1, 1] centered at the maximum peak. These notches are also called *ambiguities*, and they are illustrated in more detail in Figure 10.2, this time for a 24.552 MHz double-sided bandwidth receiver limitation.[1] Such ambiguities pose two kinds of problems, according to whether we are in the acquisition or in the tracking stage.

In the *acquisition stage*, correlations are performed with different code phases of the local code replica. The time difference between one tentative code phase to the next one is typically called the time-bin step. For a CBOC(−)-modulated waveform, the notches are about 0.7 chip apart, as shown in Figure 10.2. If a too large time bin is used, it may happen that the correlations at various code phases of the local code replica occur close to the notches and therefore the main correlation peak may be missed. One solution is to decrease the time-bin step when scanning different code phases, but this translates into an increased acquisition time since more code phases to be tested. Alternatively, we can use the so-called unambiguous acquisition methods that are presented in this chapter.

---

[1] The term *ambiguity* must not be mixed up with the integer number of wavelengths between satellite and receiver antennas, or the integer number of milliseconds in travel time from satellite to receiver.

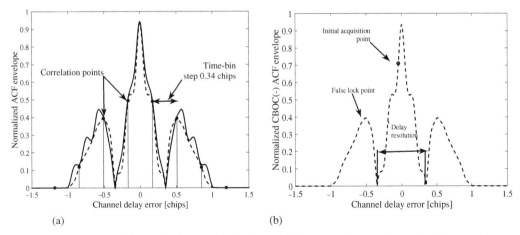

Figure 10.2 (a) Ambiguity problem in the acquisition stage for a too large time-bin step (i.e. larger than a quarter of the main ACF lobe width), in this case equal to half the main lobe width, 0.34 chips. (b) Example of false lock points in the tracking stage, CBOC(−) modulated signal, 24 MHz double-sided bandwidth.

In the *tracking stage*, we may face the problem of false lock peaks as illustrated in Figure 10.2. Indeed, the two additional sidelobes in the correlation function of a sinBOC(1, 1) or a CBOC-modulated signal have strong peaks, and the tracking process may converge to one of the false peaks. The problem is even more severe if we have higher order BOC modulations such as cosBOC(15, 2.5) with 22 additional side peaks to the main peak: 11 at each side around the main peak. The number of additional (or false) peaks within ±1 chip from the main (correct) peak equals $2N_B - 2$, N_B being the BOC modulation order. The peaks posing the greatest challenge are the two false peaks closest to the main peak.

In conventional methods, the ambiguities inherent to BOC signals are not dealt with and the receiver operates in the presence of ambiguities. This is the reason why conventional methods are also referred to as *ambiguous* methods herein. Since Galileo signals are basically DS-SS signals, any conventional or ambiguous DS-CDMA acquisition method can be applied for Galileo, with the only constraint of having a sufficiently small time-bin step as suggested in Table 1.2. In order to achieve a sufficiently high detection probability, a value equal to or smaller than 0.175 chip is usually recommended. Nevertheless, there are applications where there is no other chance but to compensate the presence of ambiguities in order to avoid the potential bias errors that could be incurred when falling into the wrong ACF peak. This involves new signal processing methods to be considered at both the acquisition and tracking, as discussed next.

10.3 Unambiguous Acquisition

Many unambiguous methods have been proposed in the literature in order to deal with the ambiguities of the BOC ACF, and they essentially follow one of these two approaches:

1. To convert through some filtering or other transforms, the ambiguous correlation-shape into a BPSK-like shape, thus removing notches and side peaks in the interval $[-1, 1]$ chip around the maximum correlation peak. These are the so-called *wide main lobe* correlation approaches.
2. To remove or diminish the sidelobes of an ambiguous correlation function and to keep only (or mainly) the main correlation lobe. These are the so-called *narrow main lobe* correlation approaches.

These two approaches will be introduced next. The readers interested in more detail can find a more comprehensive discussion in the survey on unambiguous acquisition methods for GNSS available in [2].

10.3.1   Wide Main Lobe Correlation Approach

The ideas within this approach can be divided into the following main algorithms:

1. *Betz & Fishman (B&F)* algorithm or sideband correlation, see [3], [4], processes individually each of the two spectral sidebands of the BOC signal as shown in Figure 10.3. The single-side version keeps only one of these bands, either upper or lower, when forming the decision statistics. This is the most frequent unambiguous acquisition method described in current GNSS literature.
2. *Martin & Heiries (M&H)* approach or BPSK-like techniques, see [5] and [6]: The block diagram is shown in Figure 10.4. In this approach, both main sidelobes of the received signal are filtered and correlated with a shifted version of the PRN reference code and up-sampled to the same rate as the incoming signal. The drawback of this method is that it is not working properly for odd BOC modulation orders N_B, but fortunately Galileo signals use even BOC modulation orders. For example, $N_B = 2$ for sinBOC(1, 1), and $N_B = 12$ for MBOC variants, see [7] and [8].
3. *Modified B&F method*, see [9]: It is based on shifting by $\pm \hat{a} R_c$ in frequency domain the received signal, with $\hat{a}$ an integer number, thus selecting the either the upper or lower main lobe, and correlating the resulting signal with the reference PRN code as shown in the upper plots (a) of Figure 10.5. The choice of $\hat{a}$ depends on the order of the BOC modulation, such as $\hat{a} = 1$ for sinBOC(1, 1) and CBOC.
4. *Modified M&H method*, see [9]: Its principle is shown in Figure 10.5, the middle signal spectrum (b). The difference with the modified B&F approach is that now we select both main lobes of the incoming signal, not only one of them. The performance of the modified M&H method is exactly the same as that of M&H method, but it has a lower complexity.
5. *Unsuppressed Adjacent Lobes (UAL) method*, see [9]: Its principle is shown in Figure 10.5, the lower signal spectrum (c). In order to decrease the implementation complexity, the filtering part is removed completely. Its

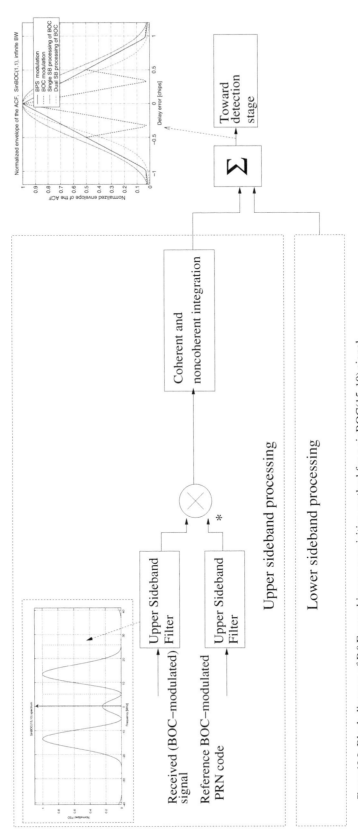

Figure 10.3 Block diagram of B&F unambiguous acquisition method for a sinBOC(15, 10) signal.

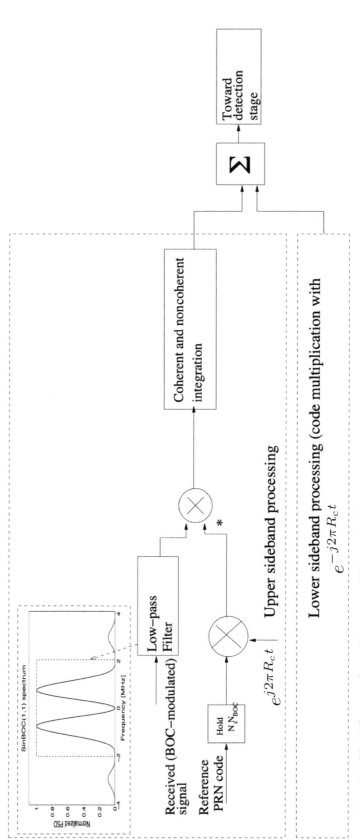

Figure 10.4 Block diagram of unambiguous acquisition method for a sinBOC(1,1) signal.

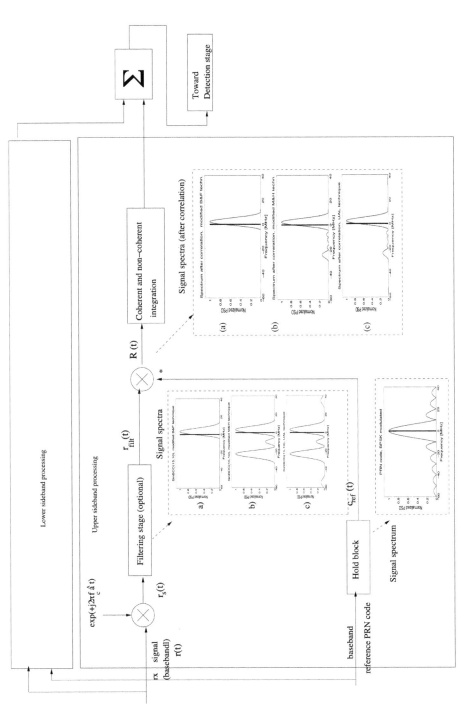

Figure 10.5 Block diagram of unambiguous acquisition methods of lower complexity than the B&F method: (a) modified B&F, (b) modified M&H and (c) UAL for a for a sinBOC(15,10) signal.

performance is slightly worse than the modified B&F and modified M&H approaches, but it offers the lowest complexity among all unambiguous acquisition approaches.

6. The contribution in [10] describes an algorithm based on filtering the correlation function with a 3-tap or 7-tap filter, adjusted to cancel the sidelobe threats. Two variants were provided: the first one where the sidelobes are cancelled with low complexity, but there is an unwanted frequency replica; the second one where a slightly higher-complexity filtering is used to cancel out also the unwanted frequency replicas. The impulse response $h(t)$ for the 3-tap filter used to process the square envelope of the ambiguous correlation in these two approaches are:

$$h(t) = \frac{1}{2}\delta(t + \tau^*) + \frac{1}{2}\delta(t) + \frac{1}{2}\delta(t - \tau^*), \tag{10.1}$$

while for the 7-tap filter is:

$$\begin{aligned} h(t) &= \frac{1}{4}\delta(t + 3\tau^*) + \frac{1}{2}\delta(t + 2\tau^*) + \frac{3}{4}\delta(t + \tau^*) + \delta(t) \\ &+ \frac{3}{4}\delta(t - \tau^*) + \frac{1}{2}\delta(t - 2\tau^*) + \frac{1}{4}\delta(t - 3\tau^*), \end{aligned} \tag{10.2}$$

where δ is the Dirac function and τ^* is the time spacing between the maximum correlation lobe and the first ambiguity. The second variant, Eq. (10.2), is able to eliminate the unwanted replica filtering at frequency equal to $1/2\tau^*$ created by the first filter.

7. The subcarrier shaping methods via *zero forcing* and *minimum mean square error* (MMSE), see [11], are prefiltering techniques that try to equalize the BOC filter effect. These algorithms also widen the main lobe of the correlation envelope, and they can be seen as extension of zero forcing and MMSE algorithms proposed in the communication context. Their main drawback is the high noise sensitivity.

As a summary, Figure 10.6 (a) illustrates the correlation shapes in absolute value when applying one of the B&F, M&H or UAL approaches presented above. Figure 10.6 (b) illustrated two additional unambiguous acquisition methods based on [10] (Benedetto1 and 2) together with the ambiguous BOC shape. The modified B&F and M&H shapes are very similar to the B&F and M&H ones. The dual-sideband and single-sideband approaches have similar normalized shapes, the difference stays in the nonnormalized power, which is about 1.5 dB lower in the single-sideband compared to dual-sideband approaches. A cosBOC(15,2.5) modulation was chosen in this example to illustrate better the widening of the main correlation lobe when we go from the ambiguous BOC (aBOC) to one of the unambiguous approaches.

10.3.2   Narrow Main Lobe Correlation Approaches

A wide main lobe, as described in the previous section, is highly suitable for fast unambiguous acquisition because it allows the use of a large time-bin step (and thus a faster

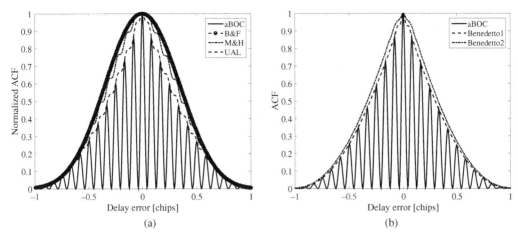

Figure 10.6 Absolute values of the correlation from: (a) B&F, M&H, UAL methods and the ambiguous BOC, where dual-sideband and single-sideband shapes are the same in the normalized scale; (b) from two versions of Benedetto's method and the ambiguous BOC.

computation) in the acquisition stage. Nevertheless, another form of getting rid of the ambiguities is to remove the sidelobes and keep a narrow main lobe. Such approaches have been investigated for acquisition by [12], [13], [14] and [15], as described below:

1. *General removing ambiguity via side-peak suppression* (GRASS), see [12], [13], is an unambiguous acquisition technique developed initially for generic sinBOC(kn, n) signals and later adapted to any BOC sine or cosine modulation, see [16]. It relies on correlating the incoming signals with both the reference BOC-modulated code and an auxiliary reference signal, and then combining noncoherently the resulting correlations. The resulting correlation shape will remove most of the side peaks, which create ambiguities, and it will keep a narrow main peak. However, spurious peaks around ±1 chips appear, as seen in Figure 10.7. Typically, the acquisition performance of GRASS is much lower than the acquisition performance of BPSK-like approaches presented in the previous section.

2. An *inhibition side-peak acquisition* (ISPA), see [14], is based on auxiliary signals in forming the receiver correlation functions. The main differences with GRASS stay in the way of forming the auxiliary functions and the way of combining them. The ISPA has, however, been studied in [14] only in terms of the ratio between main lobe and the strongest sidelobes; its detection probability in acquisition has not been studied, neither its robustness to noise.

3. *Correlation combination ambiguity removing technology* (CCART) [15] is a technique valid only for sinBOC(kn, n)-modulated signals, and it relies on the idea of combining an auxiliary correlation function with the autocorrelation function of sine-phased BOC(kn, n) signals. The auxiliary correlation function is obtained by correlating the incoming signal with the spreading code only. The resulting correlation function will be fully unambiguous, with no sidelobes, but

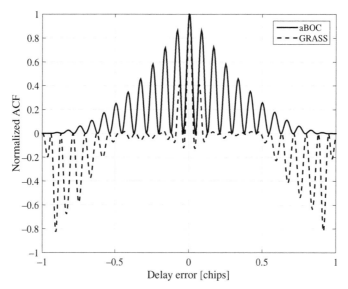

Figure 10.7 Absolute values of the correlation function from the GRASS method and the ambiguous BOC.

the width of the main lobe will be slightly higher that the width of the main lobe of an ambiguous BOC function. It was shown in [15] that CCART outperforms GRASS method in acquisition.

4. *Subcarrier phase cancellation* (SCPC), see [6], is based on the receiver correlations with two kind of local BOC codes: a sine-phase subcarrier and a cosine-phase subcarrier. The correlation envelope is widened. However, it was shown in [17] that SCPC acquisition performance is quite poor compared with other unambiguous acquisition approaches.

Several of the methods described so far have both a double-sideband and a single-sideband implementation, depending on whether both upper and lower bands are kept or only one of them, as well as combined noncoherently. Recent studies on unambiguous approaches on MBOC signals show that unambiguous methods can offer up to 4 dB enhancement in terms of C/N_0 in the acquisition process [18], provided that the receiver bandwidth is sufficiently wide. The best performance is achieved with B&F method which has also the highest complexity. For low receiver bandwidths, some studies showed that there is no benefit in using unambiguous acquisition, and the severe bandwidth limitation (typically 3–4 MHz double-sided bandwidth) acts in this way to alleviate the ambiguities, see [19].

A comparison between the acquisition performance of some of the aforementioned unambiguous approaches with full BOC acquisition is shown in Figure 10.8 for 81.84 MHz double-sided bandwidth, serial search and cosBOC(15,2.5) modulation. Clearly, if we can afford a very small time-bin step in the acquisition, which means a large complexity and large acquisition time, the ambiguous BOC approach works perfectly well, as seen in the left-hand plot. However, in realistic scenarios, when the

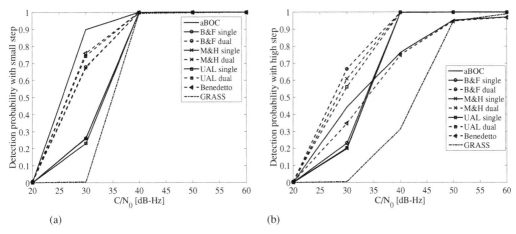

Figure 10.8 Detection probability for cosBOC(15, 2.5) modulation with ambiguous and unambiguous processing, hybrid search. Results for (a) a small time-bin step of 0.0063 chips and for (b) a high time-bin step of 0.5 chip.

acquisition speed is important, for example with a 0.5 chips time-bin step, the majority of the unambiguous approaches outperforms the ambiguous BOC case, as seen in the right plot of Figure 10.8. Exceptions are for narrow main lobe approaches such as GRASS, which are not suitable for acquisition, and also for Benedetto algorithm which works best at small time-bin steps. More examples regarding the performance of unambiguous approaches can be found for in [9], [18], [19], [17] and [16].

10.4 Ambiguous or Conventional Tracking Methods

Once unambiguous acquisition has been addressed, we now move to the tracking stage comprising both code and carrier tracking algorithms. Ambiguous (i.e. conventional) tracking methods will be considered first, and unambiguous tracking methods will be presented next in Section 10.5.

As already introduced in Chapter 1, once the acquisition stage is completed, the receiver enters into tracking mode. Tracking usually is split into two parts corresponding to code tracking and carrier tracking, and a brief overview of the techniques available in the literature for these two stages will be presented next. This will help the reader to better understand the rationale behind specific techniques proposed for the unambiguous code tracking of BOC signals in Section 10.5. The interested reader can find more details about conventional tracking techniques in [20], [21], [22] and [23].

10.4.1 Code Tracking

Many of the traditional code tracking methods described here can also be used as a basis for the unambiguous tracking methods discussed in this chapter. A classification of code delay tracking methods is shown in Figure 10.9, and each class is discussed in

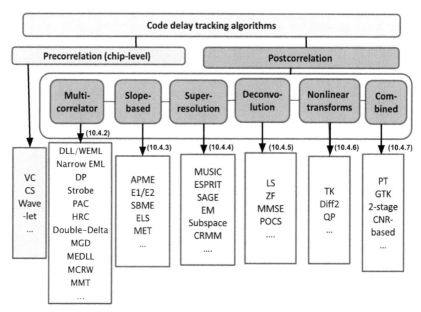

Figure 10.9 Classification of code-delay tracking methods. For each postcorrelation method, the number in brackets indicates the subsection where the method is discussed.

the following sections. Many of these methods have been derived to deal better with multipath propagation, and thus, they have some *multipath mitigation* properties as well. In this chapter, there is also a section dealing particularly with multipath effects and multipath mitigation in the context of GNSS code tracking.

Code tracking methods can be first divided into two subcategories: methods processing the received signal before the correlation with the local replica and methods relying on some form of correlation with the local replica, as shown in Figure 10.9. Even though the first category of methods is not widely adopted, an example corresponds to the vision correlator (VC) and the wavelet-based code tracking. The first one, VC, is a chip-level delay estimator where the multipaths are searched directly in the transitions of the received signal, before its correlation with the local replica. In order to deal with low signal levels of the spread signal, coherent integration is performed on several successive chips and an equivalent chip transition waveform is built and compared with some theoretical reference. The theoretical references are built for several multipath scenarios, and then, the best fit is selected. In [24], VC was shown to have better multipath resistance than the narrow correlators and double-delta correlators, which will be described in the next section. However, the results were shown only at rather high signal levels, and it is still an open issue whether the VC is able to deal with low and moderate carrier-to-noise ratios. Regarding wavelet-based code tracking, this is another chip-level code tracking methods studied for example in [25]. However, such a method has not been successfully adopted in the context of GNSS, no doubt due to its rather low robustness to noise.

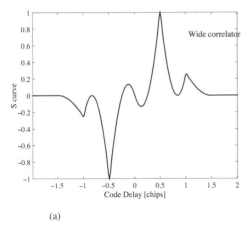

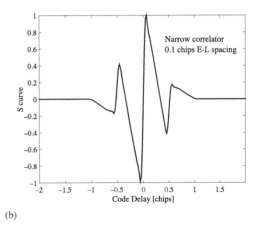

(a)   (b)

Figure 10.10 Examples of S-curves for sinBOC$(1,1)$-modulated codes for wide (a) and narrow (b) correlators. Single path channel with 24 MHz double-sided bandwidth.

The second category of code trackers, the one which relies on postcorrelation processing, is the one most widely adopted in practice, and it will be discussed in detail in the following subsections.

10.4.2   Maximum Likelihood or Multi-correlator Approaches

The idea behind any *Maximum Likelihood (ML)* estimator is to determine the parameters that maximize a likelihood function which is the joint PDF of the sample data. This estimation method does not require a priori information and assumes that the unknown parameters are constant over an observation period, typically hundreds of milliseconds or multiple seconds for high-sensitivity receivers, [26]. A straightforward implementation of an ML estimator is typically very complex, and suboptimal implementations are preferred.

1. A first example of low-cost ML implementation is given by the DLL implementation where the incoming signal is correlated with an Early (E), Prompt (P), and Late (L) version of the local replica of the PRN code, according to a certain initial delay estimate $\hat{\tau}$ coming from the acquisition stage. A tentative Doppler shift $\hat{v}_t$ is also employed in the local replica whose initial value also comes from the acquisition stage, but subsequent values come from the carrier phase and carrier frequency loops. According to the spacing δ_{EL} between early and late correlators, the DLL can be wide ($\delta_{EL} = 1$ chip) or narrow ($\delta_{EL} < 1$ chip). The narrow correlator concept was introduced in 1992 by [27] as a DLL variant better coping with noise. Since then it has proved to give moderate to good results also in the presence of BOC ambiguities and multipath. An example of the narrow correlators is the *narrow Early-Minus-Late (narrow EML)* code tracker, see [27].

   Examples of S- or discriminator curves for wideband correlator ($\delta_{EL} = 1$ chip) and narrowband correlator ($\delta_{EL} = 0.1$ chip) are shown in Figure 10.10 for sinBOC$(1,1)$ modulations and 0.1 chip cosBOC$(15,2.5)$ modulations. A double-sided bandwidth of 24 MHz was used in both cases with Butterworth filtering; the pull-in range (the linear region around the zero crossing) is smaller when we move from a larger time-bin step (wideband correlator) to a smaller time-bin step (narrow correlator). This implies a better resistance to multipaths, but also a higher likelihood of faster loss of lock in noisy situations. For cosBOC$(15,2.5)$, the wideband correlator is not sufficient, since the zero crossing at correct delay is barely visible.

2. If we introduce two additional early and late correlators, a so-called *very early* placed at $-\delta_{EL}$ from the prompt one and a *very late* placed at $+\delta_{EL}$ from the prompt one, we get the family of double-delta correlators: the *high resolution correlator (HRC)*, see [28], the *strobe correlators*, see [29], and *pulse aperture correlator (PAC)*, see [30]. An extension of the double-delta correlator concept has been presented in [31] under the form of *multiple gate delay (MGD)* structures. This structure allows the number and the spacing of the correlators to vary and can be optimized according to the estimated channel profile in order to achieve more robustness against multipaths. The optimization criterion in MGD is built according to the multipath errors. A least-squares based optimization of the S-curve has been proposed in [32]. The MGD structure uses weighting coefficients for forming the discriminator. The coefficients are chosen as to minimize the sum of squared errors.

3. The *Modified Correlator Reference Waveforms (MCRW)* concept introduced in [33] makes use of a modified replica code at the receiver to provide a cross-correlation function with inherent resistance to errors caused by multipath. An example of this family of MCRW is the gated correlator described in [28]. It consists of periodically blanking the received signal between code chip transitions. However, due to the modified reference correlation function, some losses in the SNR occur and in general the receiver becomes more sensitive to noise.

4. The *Multipath-Estimating Delay-Lock Loop (MEDLL)*, see [34], uses 6–10 correlators in order to determine accurately the shape of the multipath-corrupted correlation function. Instead of the discriminator function in Figure 10.10, a multipath estimation unit is incorporated in a loop. This unit determines via extensive ML search the best combination of LOS and multipath components (amplitudes, delays, phases and number of multipaths) which would have produced the measured correlation function. The *Reduced Search Space Maximum Likelihood (RSSML)* estimator described in [35] is an implementation of MEDLL of reduced complexity. It uses lookup tables.

5. The *Multipath Mitigation Technology (MMT)* introduced in [36] and studied later also in [37] is also based on a direct computation of log-likelihood functions, similar to MEDLL. However, unlike the MEDLL, a coordinate transformation is performed and the complexity of the search process is substantially reduced.

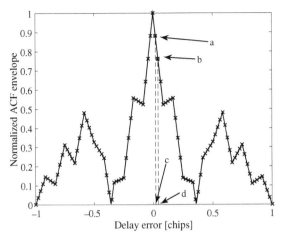

Figure 10.11 The principle of slope-based code tracking approaches.

10.4.3   Slope-Based Approaches

Slope-based approaches are based on the idea of using information from certain slopes of the ACF (usually close to the maximum ACF peak) and trying to infer multipath information based on such slope information. An example is shown in Figure 10.11 for a CBOC(+) ACF under infinite BW.

The slope in this example can be calculated as

$$\tan \theta = \frac{b - a}{d - c},$$ (10.3)

where a, b, c and d are the points shown in Figure 10.11. They depend on the chosen correlation values. Some other examples are listed below:

1. The *A Posteriori Multipath Estimation (APME)* as exposed in [38] relies on the a posteriori estimation of the multipath tracking error. It incorporates a narrow correlator plus an independent estimation module. The estimation module uses slope information according to the correlation function, with and without multipaths.

2. The *Early-Late Slope (ELS)* correlator, see [39] and [35], also known as multipath elimination technique (MET), is based on two correlator pairs at both sides of the correlation functions with parameterized spacing, similar to the MGD and HRC cases. Based on these correlation pairs, the slopes of the correlation function can be computed, and subsequently their intersection point can be found. A pseudorange correction is then applied. The MET is based on the symmetry of the autocorrelation function of the codes which is lost in the presence of multipath. Simulation results performed in [39] showed that the ELS correlator is outperformed by HRC and by the strobe correlators in multipath

environments. Improvements of ELS technique have recently been proposed in [6].

3. Another slope-based correlator structure is the Early 1/Early 2 (E1/E2) tracker, initially proposed in [40] and earlier analyzed in [39]. In the E1/E2 structure, two correlators are located on the early slope of the correlation function with an arbitrary spacing. The corresponding amplitudes are compared with the amplitudes of an ideal reference correlation function (no multipath), and some delay correction factor is computed based on the measured amplitudes and reference amplitudes.

4. A more recent Slope-Based Multipath Estimator, proposed in [41], attempts to compensate the multipath error contribution of a narrow correlator tracking loop by utilizing the slope information of an ideal normalized correlation function. The main difference with APME consists in the way of computing the multipath error contribution. For multipath environments, reported results are better than those of narrow correlators and HRC in multipath environments.

10.4.4   Subspace-Based Algorithms

Subspace-space algorithms involve a decomposition of the space spanned by the observation vector formed by the received signal samples, into the noise and signal subspaces. These algorithms use the orthogonality property between noise subspace and signal subspace in order to estimate the channel parameters. The subspace approaches involve eigenvector decomposition of high-order matrices. The main advantage of the subspace-based methods is their increased resolution in the parameter estimates. The main disadvantage is their increased complexity compared to other existing delay code trackers for DSSS signals. The best known subspace-based methods which have been employed in delay estimation of DSSS signals are the *Multiple Signal Classification (MUSIC)*, see [42], the Estimation of Signal Parameters via Rotational Invariance Techniques, see [42] and the Expectation Maximization (EM) algorithm, see [43]. The Space-Alternating Generalized Expectation-maximization algorithm described in [44], and [43] is a reduced complexity implementation of EM.

So far the subspace approach has been but little investigated in GNSS literature, mainly due to complexity and reported sensitivity to noise. Recent efforts in this direction are the studies of the complexity reduced multipath mitigation subspace techniques for GPS and Galileo in [45] and [46].

10.4.5   Deconvolution Approaches

Resolving the multipath components can be seen as a deconvolution problem in which we try to estimate the nonzero elements of an unknown gain vector, modeling the channel complex coefficients at each possible time delay, see [47]. The first nonzero component higher than a threshold will be the estimate of the LOS component. The classical solution of a deconvolution problem is obtained by a Least-Squares (LS) solution. However, the LS solution is highly sensitive to noise, and it is not well suited to GNSS signals, see [48]. Improved deconvolution approaches are the MMSE estimation like [35] and [48] and the projection onto convex sets estimation [49] and [50] and [48] and [51]. Deconvolution approaches typically suffer from low robustness to noise.

10.4.6 Non-linear Transform-Based Approaches

A different class of delay trackers or estimators is based on a nonlinear transformation of the correlation function. A nonlinear scheme based on the slope differential or second-order derivative (Diff2) of the correlation function has been proposed in [52]. This publication shows that this scheme has better multipath performance than narrow EML and strobe correlators.

Another nonlinear transform is offered by the nonlinear quadratic *Teager Kaiser (TK)* operator, first introduced for measuring the real physical energy of a system, see [53] and later applied to multipath delay estimation in DSSS systems, see [47] and in Galileo systems, see [48] and [54].

10.4.7 Combined Approaches

Combinations of two or several of the aforementioned algorithms have been studied in the literature with the scope of achieving better trade-offs between performance and complexity. For example, the peak tracking (PT) techniques described in [52] employ several nonlinear transforms of the correlation function and forms a pool of competitive peaks, among which the final LOS delay estimate is selected according to a weighting algorithm, thus achieving an increased performance in multipath environments. The *Generalized Teager Kaiser (GTK)* estimator combines the deconvolution algorithms with the TK nonlinear estimator, see [55]. It manages to outperform both TK and deconvolution stand-alone approaches.

The Differential Teager-Kaiser algorithm starts from TK and first-order derivatives and offers a new closed-loop algorithm for GNSS code tracking. A Fast Iterative Maximum Likelihood Algorithm combines the narrow correlator with a ML search, thus reducing the complexity of the final tracking algorithm, with respect to ML and MEDLL approaches, and still preserving a better multipath accuracy than the narrow correlator [26]. The work [56] focuses on ML estimators in conjunction with interpolation methods in order to increase the estimation accuracy.

10.4.8 Multipath Mitigation and Code Tracking

All the code tracking algorithms mentioned previously can cope with multipath to a certain extent. Typically, the trade-off is between complexity and performance, but sometimes even a simple structure such as the narrow correlator can outperform the

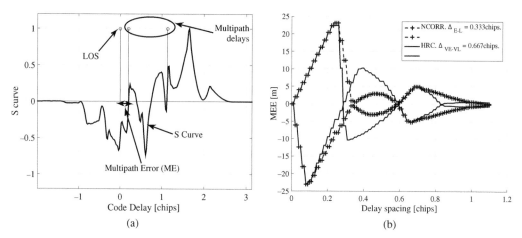

Figure 10.12 (a) Illustration of the computation of MEE in a three-path channel. (b) Example of MEE curves for narrow correlator and HRC, sinBOC(1, 1) signal, 24.552 MHz double-sided bandwidth, $\delta_{EL} = 0.04$ chip early-late spacing.

more complex ones, see [57] and [58]. More advanced multipath mitigation structures require *antenna arrays*, see [59]. The simplest way to analyze the performance of an algorithm in the presence of multipaths is to look at the Multipath Errors (ME) under some simplified channel assumptions – like very low or no noise, static channels and few paths – and to build so-called *multipath error envelopes (MEE)*, see [60], [57] or *Running Average Error (RAE)* curves, see [61].

The concept of multipath error is demonstrated in Figure 10.12, starting from an S-curve for a CBOC(-) modulated signal, narrow correlators with $\delta_{EL} = 0.08$ chip spacing and 24.552 MHz double-sided bandwidth, Butterworth filtering. In the absence of multipath on a single path channel, the S-curve would cross the zero axis at exactly zero delay error. In the presence of multipaths – one line of sight signal and two signals with multipath – the zero crossing and thus the delay estimation error is displaced, and we get a multipath error which depends on the channel profile. By comparing various MEs under similar channel profiles, we get the MEE. In case of two-channel paths, a typical analysis involves either the same phase (in-phase paths) or 180° shift (out-of-phase paths). The second path is typically assumed to be 3 dB or 6 dB weaker than the first one. The MEEs for these two extreme cases are built, and we may get plots similar to those shown in Figure 10.12(b).

Clearly, MEE shapes depend on the assumed channel profile and receiver bandwidth as well as on the delay tracker algorithm – or discriminator shape. A unified comparison between all code tracking algorithms, in terms of their performance in multipath scenarios, is currently lacking in the GNSS literature. A partial survey between several of the aforementioned algorithms can be found in [48] and [58].

10.5   Unambiguous Code Tracking

Unambiguous acquisition methods tend to destroy the main correlation peak, while removing ambiguities within one chip interval across the main peak. However, for code tracking purposes, we would like to preserve the narrow main correlation of BOC correlation, while removing or diminishing adjacent side peaks. A couple of unambiguous tracking approaches that widen the main lobe, instead of keeping it narrow, have been proposed in the literature. The next two subsections present the dichotomy in unambiguous code tracking, namely the unambiguous code tracking approaches based on widening the main correlation lobe and those based on keeping a narrow main lobe and removing or diminishing the sidelobes.

10.5.1   Wide Main-Lobe Correlation Approaches

The main unambiguous tracking approaches based on widening the main lobe are enumerated below:

1. *Astrium correlator*, see [62]: the idea is to use two independent, but cooperative loops: a phase (or frequency) lock loop (PLL or FLL) for carrier tracking and a subcarrier lock loop for subcarrier tracking. The main tracking is based on subcarrier tracking which can be interpreted in a similar way as the tracking of a BPSK signal with half of the subcarrier period as chip length.
2. Double estimator, also called *double estimator technique* (DET), see [63], is based on a 2D correlation function, computing the power levels in both carrier and subcarrier domains. The resulting 2D correlation function, shown for example for a cosBOC(10,5)-modulated signal in Figure 10.13, will depend on both the carrier phase and the subcarrier phase, and it can offer a more robust estimate of the carrier and subcarrier phases, which should be aligned. The code tracking error variance and the mean time to lose lock for DET were shown to be better than that of other unambiguous tracking algorithms in [63], but the trade-off is an increased complexity, as much more correlators are needed to perform the 2D correlation.
3. *Decimation double phase estimator* (DDPE), see [64], is a recently proposed unambiguous tracking procedure, similar with DET, but adding a decimation stage in the implementation in order to decrease the computational complexity. Performance-wise, DDPE is similar with DET, but it is able to reduce by up to 86 % the correlations needed at the receiver, compared with DET.

10.5.2   Narrow Main-Lobe Correlation Approaches

The principal narrow main-lobe unambiguous tracking approaches are enumerated below:

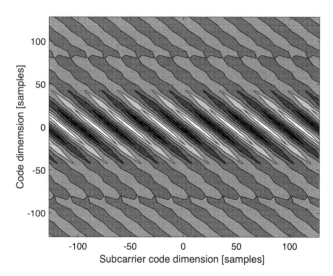

Figure 10.13 Illustration of the 2D correlation function as computed form the DET algorithm for a cosBOC(10,5)-modulated signal.

1. *ASPeCT*, see [65], is an innovative unambiguous tracking technique that keeps the sharp correlation of the main peak. It uses two correlation channels and thereby completely removing the side peaks from the correlation function in case of sinBOC(n, n)-modulated waveform. It was noted by [65] that the ASPeCT method could also be used for the signal acquisition process, allowing for a comparable sensitivity to the ambiguous acquisition scheme, but with the need of twice as many correlators in order to achieve the same mean-time-to-first-fix compared to an ambiguous acquisition scheme.

2. *Differential correlation* approaches, see [66], can be used to a certain extent in order to diminish the number of sidelobes or false peaks in the correlation envelope.

3. *Bump jumping* (BJ) technique is an unambiguous tracking technique which relies on five correlators: the prompt, an early, a late, a very early and a very late correlator. The very early and very late correlators are placed exactly on the side peaks which are the nearest to the main peak, and the decision is taken such that, if the very early or very late correlators are higher than the corresponding early or late correlators, the delay estimate is shifted to the maximum peak (very early or very late peaks). This ensures a better handling of the false lock situations.

4. *Sidelobe cancellation method* (SCM), see [67], is a method based on subtracting certain pulse shapes from the ambiguous BOC correlation in order to cancel out the undesired sidelobes. SCM can deal with any sine and cosine BOC-modulated waveform, and it removes to some extent the side peaks of the correlation envelope.
   An illustration of SCM principle is shown in Figure 10.14. A so-called subtraction waveform is generated based on the BOC-properties; this is done by

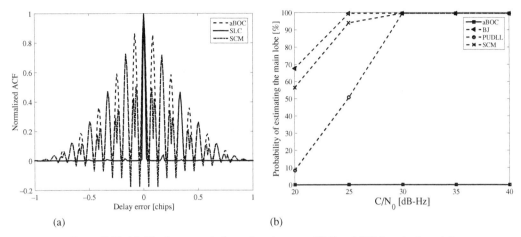

Figure 10.14 (a) Absolute correlation values between SLC and SCM methods and the ambiguous BOC. (b) Comparison of three unambiguous tracking methods with the ambiguous BOC in terms of estimating the correct main lobe of the correlation function when we start from an incorrect lobe, cosBOC(15,2.5) case.

covering the side peaks closest to the main peak. It is plotted as a dashed curve. The closed formula of the pulse to be subtracted for sinBOC and cosBOC modulations is given by [68] and [67]:

$$R_{\text{sub}}(\tau) = \sum_{i=0}^{N_B-1} \sum_{j=0}^{N_B-1} \sum_{k=0}^{N_{B_2}-1} \sum_{l=0}^{N_{B_2}-1} (-1)^{i+j+k+l}$$
$$\Lambda_{T_B}\left(\tau + i\frac{T_c}{N_B} - j\frac{T_c}{N_B} + k\frac{T_c}{N_B N_{B_2}} - l\frac{T_c}{N_B N_{B_2}}\right) \quad (10.4)$$

where $T_B = \frac{T_c}{N_B N_{B_2}}$ is the minimum BOC subinterval, $N_B = \frac{2n}{m}$ is the BOC modulation order and N_{B_2} is a flag equal to 1 for the sinBOC case and equal to 2 for the cosBOC case. For more details about this model, the reader is directed to [7].

This subtraction waveform needs first to be aligned to the maximum correlation peak and next subtracted in order to obtain a correlation function free of the highest false peaks. This is the reason that the SCM method cannot be used alone in multipath channels, but needs to be combined with some other multipath mitigation methods. Examples can be found in [67] where it is shown that SCM in combination with narrow correlators provides a good multipath performance while diminishing the threat of the false lock tracking points.

5. *Sidelobe cancellation* (SLC), see [69], is a method completely different from SCM, despite a similar name. It is based on correlations with several distinct auxiliary functions, according to the type of the BOC modulation, cosine or sine, and to the order of BOC modulation. The correlations with the set of the auxiliary functions will form the basic correlation functions which are then combined nonlinearly in order to cancel all the sidelobes in the correlation

function. Examples of the narrow main lobe correlations functions obtained after SLC and SCM processing are shown in Figure 10.14(a) for a cosBOC(15,2.5) signal. It can be seen that SCM approach cancels partly the odd-numbered side peaks, but it lets the even-numbered side peaks unaffected. On the other hand, SLC is able to cancel all the side peaks, at the cost of a large decrease in power; the power decrease is not seen in Figure 10.14(a) because the correlation envelopes are normalized for better clarity. The studies in [69] show that SLC is only useful at high and very high C/N_0 levels because it is very sensitive to noise. SLC is also very sensitive to residual frequency errors.

An example of comparing the probability of estimating the correct main lobe of three unambiguous tracking algorithms with the ambiguous tracking is shown in Figure 10.14(b) for a cosBOC(15,2.5)-modulated signal, one path fading channel, and 80.94 MHz double-sided receiver bandwidth. The correlation was performed with 4 ms coherent integration, 10 blocks of noncoherent integration and 1 second of smoothing of the discriminator curves. The initial delay estimate was assumed to be in the nearest incorrect peak of the correlation envelope. It is clear that the ambiguous BOC is not able to converge back to the correct peak, while the unambiguous tracking algorithms are able to converge back to do so.

10.6   Conclusions and Look Forward

Acquisition and tracking algorithms in Galileo and other modernized GNSS signals have to cope with the usual challenges in acquiring and tracking a DSSS signal over a wireless signal (multipath, dealing with acquisition of long codes and weak signal powers) but also with an additional challenge due to the utilization of BOC-class modulations. This challenge is related to the presence of ambiguities in the correlation envelopes and manifests itself differently in the acquisition and tracking stages. In the acquisition stage, one would like to preserve a wide correlation main lobe while removing the notches – while in the tracking stage the narrow main lobe is desirable to be preserved – at the same time as the additional sidelobes are removed. For these purposes, different unambiguous algorithms have been developed for the acquisition and tracking of BOC-modulated signals. The current existing ones in the GNSS literature were summarized in this chapter, and some illustrative examples were also shown. The performance of the unambiguous algorithms in the presence of multipath and fading channels is still a topic of intense research.

Appendix. General Formulation of BOC Signals

10.A   Introduction to BOC Signals

The motivation of this section is to further elaborate on the formulation behind BOC signals, which were succinctly introduced in Section 1.3.4. A general signal model will be presented herein that encompasses all modulations used nowadays in GNSS. It becomes a versatile tool for addressing the analysis of existing BOC signals as well as to facilitate the design of new signals.

As already discussed in Section 1.3.4, conventional BOC signals are characterized by two parameters. The first one is the subcarrier ratio with respect to the reference frequency $F_0 = 1.023$ MHz, $m = F_{sc}/F_0$ as defined in Eq. (1.39). The second one is the chip rate ratio with respect to the same reference frequency $F_0 = 1.023$ MHz, $n = R_c/F_0$ as defined in Eq. (1.40). In the sequel, we will define the parameter

$$N_1 = 2m/n, \tag{10.5}$$

so that when it becomes an integer, the sinBOC(m,n) and cosBOC(m,n) chip pulses are defined as follows:

$$p_{\text{sinBOC}}(t) = \text{sign}(\sin(N_1 \pi R_c t)), \tag{10.6}$$

$$p_{\text{cosBOC}}(t) = \text{sign}(\cos(N_1 \pi R_c t)) \tag{10.7}$$

which are nothing but a portion of the subcarrier waveform for $0 \le t < T_c$. Now let us define a rectangular pulse of unit amplitude and time support T_p as

$$p_{T_p}(t) = \Pi \left(\frac{t}{T_p} \right), \tag{10.8}$$

using the same notation already used in Eq. (1.10) for the rectangular pulse adopted in BPSK-modulated GNSS signals. Then, Eqs. (10.6)–(10.7) can alternatively be rewritten using Dirac's function δ and the convolution with $p_{T_p}(t)$, where the pulse duration is set to $T_p = T_c/N_1$. As shown in [7], this results in[2]

$$p_{\text{sinBOC}}(t) = p_{T_p}(t) * \sum_{i=0}^{N_1-1} (-1)^i \delta(t - iT_p) \tag{10.9}$$

$$p_{\text{cosBOC}}(t) = p_{T_p/2}(t) * \sum_{i=0}^{N_1-1} \sum_{k=0}^{N_2-1} (-1)^{i+k} \delta(t - iT_p - kT_p/2). \tag{10.10}$$

The cosine BOC modulation in Eq. (10.10) can actually be interpreted as a two-step BOC modulation: The signal is sine BOC modulated first, and then, the subchip is further split into two parts (due to $kT_p/2$).

In Section 10.B, we will show how the basic sine and cosine BOC modulations can be encompassed as a particular case of a more general BOC modulation referred to as complex double BOC, or CDBOC. Next, in Section 10.C, the case of multiplexed

---

[2] The $*$ symbol represents the convolution operator.

BOC (MBOC) modulation will be addressed, which is actually the modulation format used in modernized GPS L1C and Galileo E1 signals. Finally, the AltBOC modulation used in Galileo E5 will be introduced in Section 10.D. Several of the formulas, plots and M-scripts used therein are based on the work done by the author of this chapter. Actually, the important and remarkable paper [8] demonstrates that BPSK, QPSK, sinBOC, cosBOC and AltBOC can all be derived from the general family of *complex double-binary-offset-carrier modulations* (CDBOC) modulations. This new modulation allows a unified analysis of time and spectral properties of BOC signals. The method of analysis may be useful for future positioning signals as well. For the interested reader, further details on BOC signals including additional variants such as time multiplexed BOC (TMBOC) or CBOC is described in [70].

10.B   Complex Double BOC Modulation

The complex double BOC modulation model is the most generic model of BOC modulations, introduced in [8] to model in a unified way all the modulations used nowadays in GNSS. Let $p_{T_D}(t)$ be a rectangular pulse of unit amplitude and support

$$T_D = \frac{T_p}{N_2} = \frac{T_c}{N_1 N_2}, \tag{10.11}$$

for some integer N_2. Note that for $N_2 = 1$, we obtain the sinBOC modulation; for $N_2 = 2$, we get the cosBOC modulation and for $N_2 > 2$, we get a generic BOC modulation. We will refer to the latter as a two-step or double BOC(N_1, N_2, R_c) modulation, whose chip pulse is given by

$$p_{\text{DBOC}}(t) = p_{T_D}(t) * \sum_{i=0}^{N_1-1} \sum_{k=0}^{N_2-1} (-1)^{i+k} \delta(t - iT - kT_D). \tag{10.12}$$

We can continue with the derivation of the so-called complex double BOC (CDBOC) by using two signals $x_1(t)$ and $x_2(t)$, containing the data modulating symbols and the spreading codes to be pulse shaped with Eq. (10.12). To do so, let $u = \{1,2\}$ indicate each of these signals and $v = \{\text{Re}, \text{Im}\}$ their corresponding real or imaginary parts, so that $x_{u,\text{Re}} = \text{Re}\,[x_u(t)]$ and $x_{u,\text{Im}} = \text{Im}\,[x_u(t)]$. That is,

$$x_u(t) = x_{u,\text{Re}}(t) + j x_{u,\text{Im}}(t). \tag{10.13}$$

For each of these constituent signals, let the i-th complex data symbol be denoted by $d_u[i]$ (a pilot channel would have $d_u[i] = 1$ for all i), T_d the symbol period and $c_u[i,m]$ the m-th chip corresponding to the i-th symbol. The spreading factor is $F = T_d/T_c$, and the resulting data spread signal including data symbols and chips becomes

$$x_{u,v}(t) = \sum_{i=-\infty}^{+\infty} d_{u,v}[i] \sum_{m=1}^{F} c_{u,v}[i,m]\delta(t - iT_d - mT_c). \tag{10.14}$$

Next, we introduce four positive numbers $\{N_1, N_2, N_3, N_4\}$ such that $N_3 N_4$ is a divisor of $N_1 N_2$. The resulting CDBOC signal is modulated by $x_1(t)$ and $x_2(t)$ and results in the following general expression,

Table 10.1 Mapping of different CDBOC signals into GNSS modulations, based on the use of a binary complex signal (BCS) and $a = R_c/1.023$.

$x_1(t), x_2(t)$	N_1	N_2	N_3	N_4	Modulation
$x_1(t) = \text{BCS}, x_2(t) = 0$	1	1	—	—	BPSK
$x_1(t) = \text{BCS}, x_2(t) = 0$	> 1	1	—	—	$\text{sinBOC}(aN_1/2, a)$
$x_1(t) = \text{BCS}, x_2(t) = 0$	> 1	2	—	—	$\text{cosBOC}(aN_1/2, a)$
$x_1(t), x_2(t)$ distinct BCS	1	1	1	1	QPSK
$x_1(t), x_2(t)$ distinct signals	> 1	2	$N_3 = N_1$	1	$\text{AltBOC}(aN_1/2, a)$

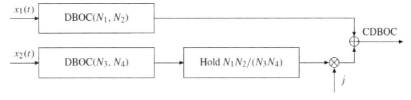

Figure 10.15 Block diagram of a CDBOC-modulated signal.

$$y_{\text{CDBOC}}(t) = x_1(t) * p_{T_D}(t) * \sum_{i=0}^{N_1-1} \sum_{k=0}^{N_2-1} (-1)^{i+k} \delta(t - iT_1 - kT_{12})$$

$$+ jx_2(t) * p_{T_D}(t) * \sum_{l=0}^{N_3-1} \sum_{m=0}^{N_4-1} \sum_{p=0}^{P-1} (-1)^{l+m} \delta(t - lT_3 - mT_{34} - pT_{12}),$$

$$\tag{10.15}$$

where

$$T_i = T_c/N_i, \quad T_{jk} = T_c/(N_j N_k). \tag{10.16}$$

The summation in Eq. (10.15) for the complex term is due to the need for equal subchip rates in the I- and the Q-arms, and it is represented in Figure 10.15 by the hold block. The subsample intervals after $\text{DBOC}(N_1, N_2)$ and $\text{DBOC}(N_3, N_4)$ processing are $T_c/(N_1 N_2)$ and $T_c/(N_3 N_4)$. Therefore, the difference P in subsample rates between the two branches in Figure 10.15 is compensated by the aforementioned hold block, which keeps the two branches at equal rate. The in-phase $x_1(t)$ and the quadrature $x_2(t)$ signals can be equal or distinct. Finally, for each branch, a DBOC modulation produces a complex signal that is sent to the channel.

In summary, Eq. (10.15) facilitates a unified analysis of all BOC modulations, a remarkable result quoted from [7]. This allow several GNSS signals to be modeled as special cases of Eq. (10.15), as summarized in Table 10.1. We may point to special cases such as sinBOC with $a = 1$ and $N_1 = 2$, which is $\text{sinBOC}(1,1)$. Similarly, cosBOC with $a = 2.5$ and $N_1 = 12$, which is $\text{cosBOC}(15, 2.5)$. Or finally, $a = 10$ and $N_1 = 3$, which leads to $\text{AltBOC}(15, 10)$. Figure 10.16 shows some special cases with $\text{sinBOC}(1, 1)$ and $\text{cosBOC}(1, 1)$ using $N = 2$, as well as $\text{sinBOC}(15, 2.5)$ and

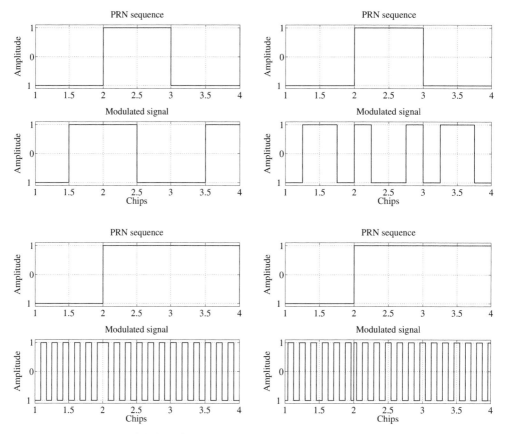

Figure 10.16 Modulation of a PRN sequence by $\sin$BOC$(1,1)$ (upper left), $\cos$BOC$(1,1)$ (upper right), $\sin$BOC$(15,2.5)$ (lower left) and $\cos$BOC$(15,2.5)$ (lower right).

$\cos$BOC$(15,2.5)$ using $N = 12$. The plots are generated by the M-script BOCDemo written by this chapter author.

10.B.1   Autocorrelation of CDBOC Signals

In order to assess the goodness of CDBOC signals, it is of interest to obtain their ACF, which is directly related to the accuracy these signals provide for ranging purposes. We can derive explicit expressions for the of the CDBOC-modulated signal $y(t)$, denoted herein by $R_{\text{CDBOC}}(t)$, by replacing Eq. (10.15) in the expression below:

$$R_{\text{CDBOC}}(\tau) = E\left[y(\tau) * y^*(-\tau)\right] = E\left[\int_{-\infty}^{\infty} y(t)y^*(t - \tau)\, dt\right], \tag{10.17}$$

where * denotes the conjugate operator. Assuming that the real and imaginary parts of the modulating signal $x_u(t)$ are independent and that the ACF for the spreading sequence is ideal, namely $R_{x_{u,v}}(\tau) = \delta(\tau)$, the ACF of the overall CDBOC signal results in,

$$R_{\text{CDBOC}}(\tau) = R_1(\tau) + R_2(\tau) \qquad\qquad \text{if } x_1(t) \neq x_2(t) \qquad (10.18)$$

$$R_{\text{CDBOC}}(\tau) = R_1(\tau) + R_2(\tau) + 2jR_{12}(\tau) \qquad \text{if } x_1(t) = x_2(t), \qquad (10.19)$$

where the partial correlation functions for the real and the imaginary components in Eq. (10.15), denoted by $R_1(\tau)$ and $R_2(\tau)$, as well as their cross correlation $R_{1,2}(\tau)$ turn out to be given by

$$R_1(\tau) = \sum_{i=0}^{N_1-1} \sum_{k=0}^{N_2-1} \sum_{i_1=0}^{N_1-1} \sum_{k_1=0}^{N_2-1} (-1)^{i+i_1+k+k_1} \triangle(\tau - (i - i_1)T_1 - (k - k_1)T_{12}) \quad (10.20)$$

$$R_2(\tau) = \sum_{l=0}^{N_3-1} \sum_{m=0}^{N_4-1} \sum_{l_1=0}^{N_3-1} \sum_{m_1=0}^{N_4-1} \sum_{p=0}^{P-1} \sum_{p_1=0}^{P-1} (-1)^{l+l_1+m+m_1} \triangle(\tau - (l - l_1)T_3$$
$$- (m - m_1)T_{34} - (p - p_1)T_{12}) \qquad (10.21)$$

$$R_{12}(\tau) = \sum_{i=0}^{N_1-1} \sum_{k=0}^{N_2-1} \sum_{l=0}^{N_3-1} \sum_{m=0}^{N_4-1} \sum_{p=0}^{P-1} (-1)^{i+k+l+m} \triangle(\tau - iT_1 + lT_3 - kT_{12} + mT_{34} + pT_{12}),$$
$$(10.22)$$

with $\triangle$ a triangular pulse of support $2T_{12}$, centered at 0. Equations (10.20)–(10.22) are valid for nonzero modulating signals. If the modulating signal is $x_2(t) = 0$, then $R_{\text{CDBOC}}(\tau) = R_1(\tau)$. The expressions may appear complex, but they simply express addition of delayed triangular pulses and can be easily implemented in MATLAB. See for example Fct_idealACF_generic_BOC.

All autocorrelation functions for BOC signals have a profile with several peaks, as already discussed in this chapter. Therefore, it is important to make sure that the channel is tracking the main peak in order to avoid selecting a wrong or false peak. Additional correlators are needed to measure the profile at half a subcarrier phase from the prompt correlator at either side. If one output value of the very-early and very-late correlators is higher than the central point correlation, the channel is tracking a side peak and corrective action is taken. Other techniques exist, too. Some examples of ACF for typical BOC signals are represented in Figure 10.17 for Galileo and GPS candidates for BOC modulations. One can notice how the main lobe is narrower for larger a, thus yielding a higher resolution in tracking the pseudorange and a smaller error in the presence of multipath.

The improved behavior of BOC signals can also be regarded in terms of the early and late correlation function, as discussed when addressing the tracking stage of a GNSS receiver. It consists on the difference between the ACF value at an early and a late correlation lag, typically plus/minus one half of the chip period. By taking a look at the results in Figure 10.18, the following conclusions can be drawn:

1. Both discriminator functions are linear around the center of the ACF. The linear regions extend from −0.25 to 0.25 chip code offset.

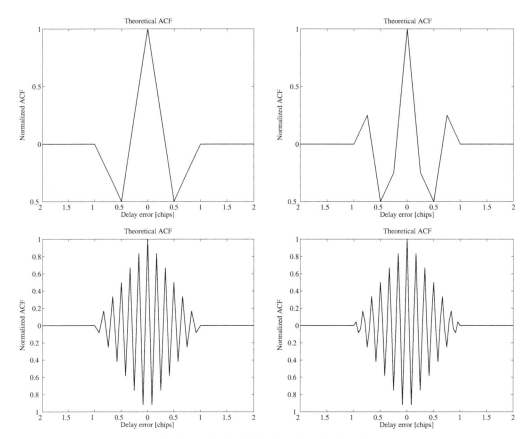

Figure 10.17 Normalized ACF for the following BOC signals: $\text{sinBOC}(1,1)$ ($N_1 = 2, a = 1$) and $\text{cosBOC}(1,1)$ ($N_1 = 2, a = 1$) in upper row and $\text{sinBOC}(15,2.5)$ ($N_1 = 12, a = 2.5$) and $\text{cosBOC}(15,2.5)$ ($N_1 = 12, a = 2.5$) in bottom row.

2. The slope of the BOC discriminator in the linear region is three times the slope of the C/A discriminator. The C/A code discriminator output is used to adjust the code NCO to align the code phase with the incoming signal; this adjustment will succeed for tracking errors less than 1.25 chips.

3. The C/A discriminator is stable in the entire region where the discriminator curve is nonzero and the DLL will converge. The BOC discriminator has stable regions next to the linear region as well, but tracking errors in the outer regions. Absolute errors between 5/8 and 5/4 chip will cause the DLL to diverge and lose lock.

Finally, Figure 10.19 shows the degradation incurred at the ACF of BOC(1,1) signals for a limited bandwidth. As can be seen, the peak value becomes less than one because not all power is available in the signal. Part of the power is blocked by the bandlimiting. For $b = 1$, the bandlimiting results in a slight rounding off at the edges of the ACF. For $b = 0.5$, the frequencies lower than twice the square wave frequency are stopped by the filter. This results in oscillations outside the

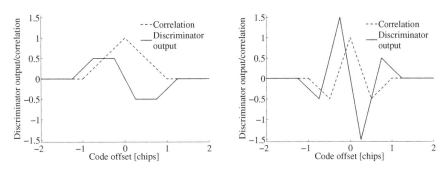

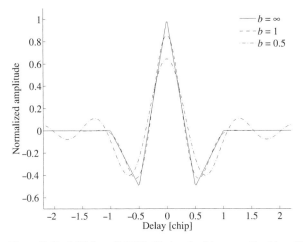

Figure 10.18 Early minus late correlation functions (discriminator function) for C/A code (left panel) and BOC(1, 1) (right panel). The correlator spacing is $d = 0.5$ chip.

Figure 10.19 ACF for a BOC(1, 1) signal with normalized bandlimits $b = 0.5$, 1 and with ∞ bandwidth.

chip length region. In the case of multipath, this could lead to undesirable sidelobe effects.

10.B.2 Power Spectral Density of CDBOC Signals

The PSD of CDBOC signals is obtained as the Fourier transform of the ACF in Eq. (10.17), resulting in:

$$S_{\text{CDBOC}}(f) = \frac{1}{T_c} S_1(f) |H(f)|^2 \qquad \text{if } x_1(t) = x_2(t)$$

(10.23)

$$S_{\text{CDBOC}}(f) = \frac{1}{T_c} \left(S_1(f) |H_{\text{upper}}(f)|^2 + S_2(f) |H_{\text{lower}}(f)|^2 \right) \qquad \text{if } x_1(t) \neq x_2(t),$$

(10.24)

where $S_1(f)$ and $S_2(f)$ are the PSD of the data-modulating signals $x_1(t)$ and $x_2(t)$ in Eq. (10.15), which are assumed to be independent when $x_1(t) \neq x_2(t)$. The factor $1/T_c$

normalizes the signal power over infinite bandwidth. $H(f)$ is the equivalent transfer function of the block diagram in Figure 10.15, where $H_{\text{upper}}(f)$ and $H_{\text{lower}}(f)$ are the corresponding transfer functions of the two branches. The overall transfer function $H(f)$ can be found from Figure 10.15 and Eq. (10.15) as

$$H(f) = P_{T_{12}}(f)(H_{12}(f) + H_{34}(f)H_{\text{hold}}(f)), \tag{10.25}$$

where $P_{T_{12}}(f) = T_{12}\text{sinc}(fT_{12})$ is the Fourier transform of a rectangular pulse with time support T_{12} and $\text{sinc}(x) = \sin(\pi x)/x$. In turn, the transfer function $H_{12}(f)$ of the DBOC(N_1, N_2) modulation in the upper branch of Figure 10.15 is given by

$$H_{12}(f) = \sum_{i=0}^{N_1-1} \sum_{k=0}^{N_2-1} (-1)^{i+k} e^{j2\pi f(iT_1+kT_{12})}$$

$$= \left(\frac{1 - (-1)^{N_1} e^{-j2\pi fT_c}}{1 + e^{-j2\pi fT_1}}\right)\left(\frac{1 - (-1)^{N_2} e^{-j2\pi fT_1}}{1 + e^{-j2\pi fT_{12}}}\right). \tag{10.26}$$

By replacing 1 by 3 and 2 by 4, we obtain a similar expression for H_{34} for the lower branch of Figure 10.15. Finally, the transfer function for the hold block in Figure 10.15 becomes

$$H_{\text{hold}}(f) = \sum_{p=0}^{P-1} e^{j2\pi fpT_{12}} = \left(\frac{1 - e^{-j2\pi fT_c}}{1 - e^{-j2\pi fT_{12}}}\right). \tag{10.27}$$

Furthermore,

$$H_{\text{upper}}(f) = P_{T_{12}}(f)H_{12}(f) \tag{10.28}$$

$$H_{\text{lower}}(f) = P_{T_{12}}(f)H_{34}(f)H_{\text{hold}}(f). \tag{10.29}$$

Substituting Eqs. (10.25)–(10.29) into Eqs. (10.24) and (10.23) yields the final expressions of the PSD for the CDBOC-modulated waveforms, depending on the inputs $S_1(f)$ and $S_2(f)$. If we assume that the sequences $x_1(t)$ and $x_2(t)$ are built of zero-mean and uncorrelated chips, then $S_1(f)$ and $S_2(f)$ are constants and equal to the sequence power, thus not influencing the shape of the output PSD.

A result of the subcarrier modulation is a split of the classical BPSK spectrum into two symmetrical parts with no remaining power on the carrier frequency. The product is a symmetric split spectrum with two main lobes shifted from the carrier frequency by the subcarrier frequency. Figure 1.2 nicely illustrates this fact by comparing the GPS C/A and Galileo BOC(1, 1) signals, which share the same L1 band, but still the two signals can be separated.

Finally, and for the sake of completeness, Figure 10.20 depicts the PSDs for the BOC modulations specified in Figure 10.16.

10.C   Multiplexed BOC Modulation

In Chapter 1, it was already discussed that ionospheric delay and multipath effects often are today's dominant error sources for any GNSS. In this context, the idea of introducing multiplexed binary offset carrier (MBOC) modulation was to remedy the

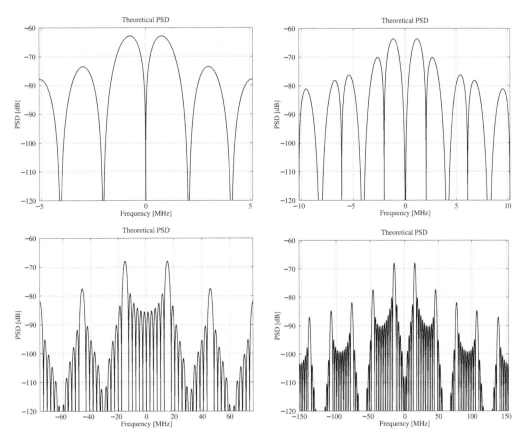

Figure 10.20 PSDs for a sinBOC(1,1) ($N_1 = 2$, $a = 1$) (upper left), cosBOC(1,1) ($N_1 = 2$, $a = 2$) (upper right), sinBOC(15,2.5) ($N_1 = 12$, $a = 2.5$) (lower left) and cosBOC(15,2.5) ($N_1 = 12$, $a = 2.5$) (lower right).

multipath degradation. The MBOC modulation was introduced for E1 and the modernized L1C signals, and it is formally defined as a weighted combination of sinBOC(1,1) and sinBOC(6,1). This offers flexibility to receiver designers since simple receivers may still be able to track one of these two BOC signals only, typically sinBOC(1,1), thus experiencing a modest degradation compared to the whole MBOC. The values 10/11 and 1/11 were selected by Galileo Joint Undertaking for weighting the two constituent BOC signals of MBOC. The combined PSD becomes

$$S_{\mathrm{MBOC}}(f) = \frac{10}{11} S_{\mathrm{BOC}(1,1)}(f) + \frac{1}{11} S_{\mathrm{BOC}(6,1)}(f), \tag{10.30}$$

where $S_{\mathrm{BOC}(m,n)}(f)$, or equivalently $S_{\mathrm{BOC}(N_1)}$ using N_1 defined in Eq. (10.5), is the following normalized PSD,

$$S_{\mathrm{BOC}(m,n)}(f) = S_{\mathrm{BOC}(N_1)}(f) = \frac{1}{T_c} \left(\frac{\sin(\pi f \frac{T_c}{N_1}) \sin(\pi f T_c)}{\pi f \cos(\pi f \frac{T_c}{N_1})} \right)^2. \tag{10.31}$$

MBOC has better tracking properties than sinBOC(1, 1) because of the sinBOC(6, 1) component, since higher-order BOC signals do provide better tracking properties due to their larger mean square bandwidth. The advantage of using MBOC is that it also has a better spectral separation from GPS C/A codes. The disadvantage is that using MBOC is more complex, and the acquisition is slightly more difficult compared to BOC(1,1).

The MBOC modulation is defined in the frequency domain, and it allows for several implementations in the time domain. We shall describe two such implementations, CBOC and TMBOC, currently used for GPS and Galileo, respectively:

1. The CBOC modulation is a weighed sum or difference of sinBOC(1,1) and sinBOC(6,1); nowadays, we have both CBOC(+) for the Galileo E1-B data and CBOC(−) for the Galileo E1-C pilot signal. The first part is passed through a hold block in order to match the rate of the sinBOC(6,1) part. Lohan et al. [8] indicated a method to derive the waveform, resulting in

$$s_{CBOC(\pm)}(t) = w_1 \, s_{sinBOC(1, 1),held}(t) \pm w_2 \, s_{sinBOC(6, 1)}(t)$$

$$= w_1 \sum_{i=0}^{N_1-1} \sum_{k=0}^{(N_2/N_1)-1} (-1)^k \, c\left(t - i\frac{T_c}{N_1} - k\frac{T_c}{N_2}\right) \pm w_2 \sum_{i=0}^{N_2-1} (-1)^i \, c\left(t - i\frac{T_c}{N_2}\right),$$

$$(10.32)$$

where $N_1 = 2$ and $N_2 = 12$. The weights w_1 and w_2 are set to $w_1 = \sqrt{10/11}$ and $w_2 = \pm\sqrt{1/11}$ with $w_1^2 + w_2^2 = 1$. In Eq. (10.32), the function $c(t)$ is the pseudorandom code including all data symbols, so the model applies to both data and pilot channels. Let b_n be the n-th code symbol (in a pilot signal $b_n = 1$ for all n), $c_{m,n}$ the m-th chip of the n-th symbol, E_b the code symbol energy and F the spreading factor or number of chips per code symbol. Then, we have

$$c(t) = \sqrt{E_b} \sum_{n=-\infty}^{\infty} b_n \sum_{m=1}^{F} c_{mn} p_{T_2}(t - nT_c F - mT_c). \qquad (10.33)$$

Figure 10.21 depicts the CBOC-modulated signal with weight $w_1 = \sqrt{10/11}$ using binary spreading symbols. This figure is plotted by a MATLAB code based on Eq. (10.33).

2. TMBOC modulation divides the signal into blocks of N code symbols. M of these N are sinBOC(1,1)-modulated, while $N - M$ are sinBOC(6,1)-modulated. This splitting applies to both data and pilot channels.

Following the derivations in [8], the TMBOC waveform can be expressed as

$$y_{TMBOC}(t) = \sqrt{E_b} \sum_{n \in S} b_n \sum_{m=1}^{F} c_{mn} \sum_{i=0}^{N_1-1} \sum_{k=0}^{(N_2/N_1)-1} (-1)^i p_{T_2}\left(t - i\frac{T_c}{N_1} - k\frac{T_c}{N_2}\right)$$

$$+ \sqrt{E_b} \sum_{n \notin S} b_n \sum_{m=1}^{F} c_{mn} \sum_{i=0}^{N_2-1} (-1)^i p_{T_2}\left(t - i\frac{T_c}{N_2}\right). \qquad (10.34)$$

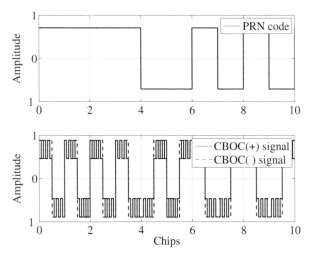

Figure 10.21 Example of composite BOC modulation with $w_1 = \sqrt{10/11}$.

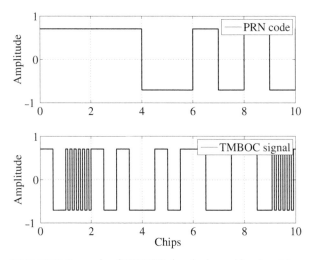

Figure 10.22 Example of TMBOC signal where chips 1 and 9 are sinBOC(6, 1)-modulated, while all other chips are sinBOC(1, 1)-modulated.

where S is a set of chips or spreading symbols that are sinBOC(1, 1) modulated, with M symbols out of N in the total set. A sample of TMBOC signal is shown in Figure 10.22, produced by the M-code Plot_MBOCwaveforms. The choice of M and N depends on the split between data and pilot power. Many solutions satisfy Eq. (10.30). If for example the data symbols are sinBOC(1, 1) modulated, a power fraction x_p is placed on the pilot channel and the remaining $1 - x_p$ power fraction on the data channel, then $\frac{M}{N} = 1 - \frac{1}{11x_p}$. The number 11 originates from the definition in Eq. (10.30).

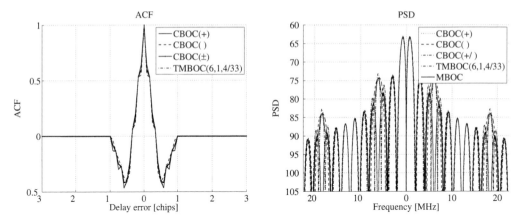

Figure 10.23 ACF and PSD for some examples of MBOC signals.

10.C.1   Autocorrelation of MBOC Signals

The autocorrelation $R_{\mathrm{MBOC}}(\tau)$ of an MBOC-modulated signal is obtained as

$$R_{\mathrm{MBOC}}(\tau) = E\left[y_{\mathrm{MBOC}}(\tau) * y_{\mathrm{MBOC}}^*(-\tau)\right] = E\left[\int_{-\infty}^{\infty} y_{\mathrm{MBOC}}(t)y_{\mathrm{MBOC}}^*(t-\tau)\,dt\right].$$

(10.35)

Assuming ergodic signals with zero mean and unitary variance, some algebra leads to the following result:

$$
\begin{aligned}
R_{\mathrm{MBOC}}(\tau) = E_b\Bigg(& w_1^2 \sum_{i=0}^{N_1-N_1-(N_2/N_1)-(N_2/N_1)-1} \sum_{i_1=0}^{} \sum_{k=0}^{} \sum_{k_1=0}^{} (-1)^{i+i_1}\triangle\left(\tau - (i-i_1)\frac{T_c}{N_2} - (k-k_1)\frac{T_c}{N_2}\right) \\
& + w_2^2 \sum_{i=0}^{N_2-1} \sum_{i_1=0}^{N_2-1} (-1)^{i+i_1}\triangle\left(\tau - (i-i_1)\frac{T_c}{N_2}\right) \\
& + w_1 w_2 \sum_{i=0}^{N_1-1} \sum_{i_1=0}^{N_2-1} \sum_{k=0}^{(N_2/N_1)-1} (-1)^{i+i_1}\triangle\left(\tau - i\frac{T_c}{N_1} - (k-i_1)\frac{T_c}{N_2}\right) \\
& + w_1 w_2 \sum_{i=0}^{N_1-1} \sum_{i_1=0}^{N_2-1} \sum_{k=0}^{(N_2/N_1)-1} (-1)^{i+i_1}\triangle\left(\tau + i\frac{T_c}{N_2} + (k-i_1)\frac{T_c}{N_2}\right)\Bigg),
\end{aligned}
$$

(10.36)

where $\triangle$ is the triangular pulse of time support $2T_2$, centered at 0 and with maximum amplitude 1. Recall that $w_2 = \sqrt{M/N}$ for TMBOC signals and $N_1 = 2$, $N_2 = 12$ for the MBOC signals defined in Eq. (10.30).

10.C.2   Power Spectral Density of MBOC Signals

The PSD for MBOC modulation was our starting point as described in Eq. (10.30). Figure 10.23 depicts the PSD for some examples of MBOC signals such as CBOC and TMBOC. The Main_MBOC calls the functions Fct_idealACF_MBOC and Fct_PSD_MBOC. Clearly, both MBOC- and TMBOC- modulated signals move away the signal energy from the carrier, and thus, they reduce the level of interference with the traditional BPSK-modulated signal used by the C/A GPS signal. Also, if we compare CBOC with TMBOC ACF waveforms, we can see that TMBOC ACF is smoother than the CBOC ACF, pointing out toward a slightly better noise robustness. On the other hand, the main peak of the CBOC ACF is slightly narrower than for the TMBOC ACF, pointing out toward slightly better multipath mitigation capability of CBOC than of TMBOC. Last but not least, the PSD figures of CBOC and TMBOC show the need of a much wider receiver bandwidth than for BPSK case in case we want to capture enough spectral energy of the signal. As a rule of thumb, at least 20 MHz double-sided bandwidth is needed for efficient processing of CBOC and TMBOC signals.

10.D   Alternate BOC Modulation

Any AltBOC-modulation modulates four real signals (or two complex signals) onto the two phases of orthogonal subcarriers [71]. Assuming that the two complex signals are denoted by $x_1(t)$ and $x_2(t)$, the non-constant envelope (NCE) AltBOC(m,n)-modulated signal can be written as derived in [7] and [8]:

$$y_{\text{AltBOC}(m,n),\,\text{NEC}}(t) = x_1(t) * p_{T_B}(t) * \sum_{i=0}^{N_B-1} (-1)^i \delta(t - iT_B)$$

$$+ x_2(t) * p_{T_B/2}(t) * \sum_{k=0}^{1} \sum_{i=0}^{N_B-1} (-1)^{i+k} \delta(t - iT_B - k\frac{T_B}{2}),$$

(10.37)

where $T_B = T_c/N_B$ and $N_B = 2m/n$.

10.D.1   Autocorrelation of AltBOC Signals

Examples of an ACF for an AltBOC-modulated signal with a nonconstant envelope are shown in Figure 10.24, and the following expression can be found in the two references just mentioned above.

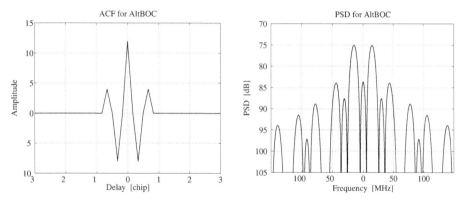

Figure 10.24 ACF and PSD for the AltBOC(15,10) signal.

$$R_{\text{AltBOC}(15,10),\text{NCE}}(\tau) = \sum_{i=0}^{2}\sum_{k=0}^{2}\sum_{i_1=0}^{1}\sum_{k_1=0}^{1}(-1)^{i+i_1+k+k_1}$$

$$\times \delta(\tau - i\frac{T_c}{3} - i_1\frac{T_c}{6} - k\frac{T_B}{6} + k_1\frac{T_B}{6}) * \Lambda_{T_c/6}(\tau) * R_s(\tau)$$

$$+ \sum_{l=0}^{2}\sum_{m=0}^{2}\sum_{p=0}^{1}\sum_{p_1=0}^{1}(-1)^{l+m}$$

$$\times \delta(\tau - l\frac{T_c}{3} - m\frac{T_c}{6} - p\frac{T_B}{6} + p_1\frac{T_B}{6}) * \Lambda_{T_c/6}(\tau) * R_s(\tau),$$

$$(10.38)$$

where $\Lambda_{T_c/6}(\tau)$ is a triangular pulse of width $T_c/6$.

10.D.2   Power Spectral Density of AltBOC Signals

The PSD of the NCE AltBOC(15,10) can be computed as described in the two papers just quoted, resulting in:

$$S_{\text{AltBOC}(15,10)\text{NCE}}(f) = \frac{4}{T_c}\left(\frac{\sin(\pi f\frac{T_c}{6})\cos(\pi f T_c)\tan(\pi f\frac{T_c}{3})}{\pi f}\right)^2$$

$$\left(\frac{S_{b_1}\,e^{j2\pi f T_c}}{\cos^2(\frac{\pi f T_c}{6})} + \frac{S_{b_2}\,e^{j2\pi f T_c}}{\sin^2(\frac{\pi f T_c}{6})}\right),$$

$$(10.39)$$

where $S_{b_1}\,e^{j2\pi f T_c}$ and $S_{b_2}\,e^{j2\pi f T_c}$ are the PSDs of the transmitted discrete-time symbol trains $\{b_{1n}\}_n$ and $\{b_{2n}\}_n$, corresponding to the signal waveforms $x_1(t)$ and $x_2(t)$. S_{b_1} is the PSD of a SinBOC(10,5) signal and S_{b_2} is the PSD of a CosBOC(10,5) signal; maybe you can simply replace these in Eq. (1.34) and then we don't need to redefine the $\{b_{1n}\}_n$ and $\{b_{2n}\}_n$

In order to avoid the issue of a NCE, Ries et al. [72] proposed a constant-envelope method, where the subcarrier waveforms are chosen such that the sum and difference of complex values always lie on the unit circle in the complex plane. The PSD of the *constant envelope* AltBOC(15,10) has been derived in [73], under the assumption of $S_{b_1}\,e^{j2\pi f T_c} = 1$, $S_{b_2}\,e^{j2\pi f T_c} = 1$. The result is:

$$S_{\text{AltBOC}(15,10),\text{CE}}(f) = \frac{4}{T_c}\left(\frac{\cos(\pi f T_c)}{\pi f\cos(\pi f\frac{T_c}{3})}\right)^2$$

$$\times \left(\cos^2\left(\frac{\pi f T_c}{3}\right) - \cos\left(\frac{\pi f T_c}{3}\right) - 2\cos\left(\frac{\pi f T_c}{3}\right)\cos\left(\frac{\pi f T_c}{6}\right) + 2\right). \quad (10.40)$$

An AltBOC waveform is a particular case of a CDBOC modulation. Figure 10.24 shows the ACF and PSD for AltBOC(15,10). It is created by the M-code Main_AltBOC which calls Fct_idealACF_AltBOC and Fct_PSD_AltBOC.

References

[1] E. S. Lohan, "Analytical Performance of CBOC-Modulated GALILEO E1 Signal Using Sine BOC(1,1) Receiver for Mass-Market Applications," in *Proceedings of ION-PLANS 2010, Palm Springs, CA*, 2010.

[2] E. S. Lohan, D. A. de Diego, J. A. Lopez-Salcedo, G. Seco-Granados, P. Boto, and P. Fernandes, "Unambiguous Techniques Modernized GNSS Signals: Surveying the Solutions," *IEEE Signal Processing Magazine*, vol. 34, pp. 38–52, September 2017.

[3] J. W. Betz, "Binary Offset Carrier Modulations for Radionavigation," *Navigation*, vol. 48, pp. 227–246, 2002.

[4] P. Fishman and J. Betz, "Predicting Performance of Direct Acquisition for the M-Code Signal," in *Proceedings of ION NMT, Anaheim, CA*, pp. 574–582, 2000.

[5] N. Martin, V. Leblond, G. Guillotel, and V. Heiries, "BOC(x,y) Signal Acquisition Techniques and Performances," in *Proceedings of ION GPS/GNSS 2003, Portland, OR*, pp. 188–197, 2003.

[6] V. Heiries, D. Oviras, L. Ries, and V. Calmettes, "Analysis of Non Ambiguous BOC Signal Acquisition Performance," in *Proceedings of ION GNSS, Long Beach, CA*, 2004.

[7] E. S. Lohan, A. Lakhzouri, and M. Renfors, "Binary-Offset-Carrier Modulation Techniques with Applications in Satellite Navigation Systems," *Wireless Communications and Mobile Computing*, vol. 7, pp. 767–779, 2006.

[8] E. S. Lohan, A. Lakhzouri, and M. Renfors, "Complex Double-Binary-Offset-Carrier Modulation for a Unitary Characterization of GALILEO and GPS Signals," *IEE Proceedings – Radar, Sonar and Navigation*, vol. 153, pp. 403–408, 2006.

[9] E. S. Lohan, A. Burian, and M. Renfors, "Low-Complexity Acquisition Methods for Split-Spectrum CDMA Signals," *International Journal of Satellite Communications and Networking*, vol. 26, pp. 503–522, 2008.

[10] F. Benedetto, G. Giunta, E. S. Lohan, and M. Renfors, "A Fast Unambiguous Acquisition Algorithm for BOC-Modulated Signals," *IEEE Transactions on Vehicular Technology*, vol. 62, no. 3, pp. 1350–1355, 2013.

[11] P. Anantharamu, D. Borio, and G. Lachapelle, "Sub-carrier Shaping for BOC Modulated GNSS Signals," *EURASIP Journal on Advances in Signal Processing*, vol. 2011, no. 1, p. 133, 2011.

[12] Z. Yao, M. Lu, and Z. Feng, "Unambiguous Sine-Phased Binary Offset Carrier Modulated Signal Acquisition Technique," *IEEE Transactions on Wireless Communications*, vol. 9, no. 2, pp. 577–580, 2010.

[13] Z. Yao and M. Lu, "Side-Peaks Cancellation Analytic Design Framework with Applications in BOC Signals Unambiguous Processing," in *Proceedings of the ION NMT*, pp. 775–785, August 2011.

[14] F. Liu and Y. Feng, "A New Acquisition Algorithm with Elimination Side Peak for All BOC Signals," *Mathematical Problems in Engineering*, vol. 2015 Article ID 140345, 9 pages, 2015.

[15] L. Wen, X. Yue, D. Zhongliang, J. Jichao, and Y. Lu, "Correlation Combination Ambiguity Removing Technology for Acquisition of Sine-Phased BOC(kn,n) Signals," *China Communications*, vol. 12, pp. 86–96, April 2015.

[16] F. Shen, G. Xu, and D. Xu, "Unambiguous Acquisition Technique for Cosine-Phased Binary Offset Carrier Signal," *IEEE Communications Letters*, vol. 18, pp. 1751–1754, October 2014.

[17] H. Wang, Y. Ji, H. Shi, and X. Sun, "The Performance Analysis of Unambiguous Acquisition Methods for BOC(m,n) Modulated Signals," in *Proceedings of International Conference on Wireless Communications, Networking and Mobile Computing (WiCOM)*, Wuhan, China, pp. 1–4, September 2011.

[18] M. F. Samad and E. S. Lohan, "MBOC Performance in Unambiguous Acquisition," in *Proceedings of ENC-GNSS*, Naples, Italy, May 2009.

[19] E. S. Lohan, "Limited Bandwidths and Correlation Ambiguities: Do They Co-exist in GALILEO Receivers?," *Positioning Journal*, pp. 14–21, 2011.

[20] E. D. Kaplan and C. J. Hegarty, *Understanding GPS - Principles and Applications.* Artech House, Inc., 2006.

[21] B. W. Parkinson and J. J. Spilker Jr., eds., " Global Positioning System: Theory and Applications," *Progress in Astronautics and Aeronautics Volume 163.* American Institute of Aeronautics and Astronautics, Inc., 1996.

[22] P. Misra and P. Enge, *Global Positioning System: Signals, Measurements and Performance, Revised 2nd Edition.* Ganga-Jamuna Press, 2011.

[23] K. Borre, D. Akos, N. Bertelsen, P. Rinder, and S. H. Jensen, *A Software-Defined GPS and Galileo Receiver, Single Frequency Approach.* Birkhäuser, 2007.

[24] P. Fenton and J. Jones, "The Theory and Performance of Novatel Inc.'s Vision Correlator," in *Proceedings of the ION GNSS, Long Beach, CA*, 2005.

[25] S. Burley, M. Darnell, and P. Clark, "Enhanced pn Code Tracking and Detection Using Wavelet Packet Denoising," in *Proceedings of the IEEE-SP International Symposium on Time-Frequency and Time-Scale Analysis, Pittsburgh, PA* , pp. 365–368, IEEE, 1998.

[26] M. Sahmoudi and R. Landry, "Probabilities and Multipath: Mitigation Techniques Using Maximum-Likelihood Principles," *Inside GNSS*, 2008.

[27] A. J. V. Dierendonck, P. C. Fenton, and T. Ford, "Theory and Performance of Narrow Correlator Spacing in a GPS Receiver," *Journal of the Institute of Navigation*, vol. 39, pp. 265–283, June 1992.

[28] G. McGraw and M. Braasch, "GNSS Multipath Mitigation Using Gated and High Resolution Correlator Concepts," in *National Technical Meeting of ION, San Diego, CA*, pp. 333–342, 1999.

[29] L. Garin and J. Rousseau, "Enhanced Strobe Correlator Multipath Rejection for Code & Carrier," in *Proceedings of the ION GPS-97, Kansas City, MO*, pp. 559–568, 1997.

[30] J. Jason, P. Fenton, and B. Smith, "Theory and Performance of the Pulse Aperture Correlator", Tech. Rep., Novatel, Alberta, Canada, 2004.

[31] H. Hurskainen, E. S. Lohan, X. Hu, J. Raasakka, and J. Nurmi, "Multiple Gate Delay Tracking Structures for GNSS Signals and Their Evaluation with Simulink, Systemc and VHDL," *International Journal of Navigation and Observation*, 2008.

[32] T. Pany, M. Irsigler, and B. Eissfeller, "Optimum Coherent Discriminator Based Code Multipath Mitigation by S-curve Shaping for BOC(n,n) and BPSK signal," in *Proceedings of ENC-GNSS, Munich, Germany*, 2005.

[33] A. Zhdanov, V. Veitsel, M. Shodzishsky, and J. Ashjaee, "Multipath Error Reduction in Signal Processing," in *Proceedings of ION GPS 1999, Nashville, TN*, pp. 1217–1223, 1999.

[34] D. J. R. van Nee, "The Multipath Estimating Delay Lock Loop," in *IEEE Second International Symposium on Spread Spectrum Techniques and Applications, Yokohama, Japan*, pp. 39–42, 1992.

[35] M. Z. H. Bhuiyan, E. S. Lohan, and M. Renfors, "A Reduced Search Space Maximum Likelihood Delay Estimator for Mitigating Multipath Effects in Satellite-Based Positioning," in *Proceedings of 13th International Association of ION, Stockholm, Sweden*, 2009.

[36] H. Saarnisaari, "ML Time Delay Estimation in a Multipath Channel," in *IEEE 4th International Symposium on Spread Spectrum Techniques and Applications*, vol. 3, pp. 1007–1011, 1996.

[37] R. Weill, "Multipath Mitigation Using Modernized GPS Signals: How Good Can It Get?" in *Proceedings of ION GPS 2002, Portland, OR*, pp. 493–505, 2002.

[38] J. M. Sleewaegen and F. Boon, "Mitigating Short-Delay Multipath: A Promising New Technique," in *Proceedings of ION GPS 2001, Salt Lake City, Utah*, 2001.

[39] M. Irsigler and B. Eissfeller, "Comparison of Multipath Mitigation Techniques with Consideration of Future Signal Structures," in *CDROM Proceedings of ION-GNSS, Portland, OR*, September 2003.

[40] P. Mattos, "Acquiring Sensitivity to Bring New Signal Indoor," *GPS World*, vol. 5, pp. 28–33, May 2004.

[41] M. Z. H. Bhuiyan, E. S. Lohan, and M. Renfors, "A Slope-Based Multipath Estimation Technique for Mitigating Short-Delay Multipath in GNSS Receivers," in *Proceedings of IEEE ISCAS, Paris, France*, 2010.

[42] A. Jakobsson, A. L. Swindlehurst, and P. Stoica, "Subspace-Based Estimation of Time Delays and Doppler Shifts," *IEEE Transactions on Signal Processing*, vol. 46, no. 9, pp. 2472–2483, 1998.

[43] F. Antreich, O. Esbri-Rodriguez, J. A. Nossek, and W. Utschick, "Estimation of Synchronization Parameters Using SAGE in a GNSS-Receiver," in *Proceedings of ION GNSS, Long Beach, CA*, 2005.

[44] A. Fessler and A. O. Hero, "Space-Alternating Generalized Expectation-Maximization Algorithm," *IEEE Transactions on Signal Processing*, vol. 42, pp. 2664–2677, 1994.

[45] I. Groh and S. Sand, "Complexity Reduced Multipath Mitigation in GNSS with the Granada Bit- True Software Receiver," in *IEEE/ION Position Location and Navigation Symposium (PLANS), Monterey, CA*, 2008.

[46] I. Groh, *Efficient Time-Variant Synchronization in Spread-Spectrum Navigation Receivers*. PhD thesis, Universitas Miguel Hernández, Elche, Spain, 2011.

[47] E. S. Lohan, R. Hamila, A. Lakhzouri, and M. Renfors, "Highly Efficient Techniques for Mitigating the Effects of Multipath Propagation in DS-CDMA Delay Estimation," *IEEE Transactions on Wireless Communications*, vol. 4, pp. 149–162, January 2005.

[48] E. S. Lohan, A. Lakhzouri, and M. Renfors, "Feedforward Delay Estimators in Adverse Multipath Propagation for Galileo and Modernized GPS Signals." *EURASIP Journal on Applied Signal Processing*, vol. 2006, pp. 1–19 2006.

[49] Z. Z. Kostic, M. I. Sezan, and E. L. Titlebaum, "Estimation of the Parameters of a Multipath Channel Using Set-Theoretic Deconvolution," *IEEE Transactions of Communications*, vol. 40, pp. 1006–1011, 1992.

[50] D. Skournetou, A. H. Sayed, and E. S. Lohan, "A Deconvolution Algorithm for Estimating Jointly the Line-of-Sight Code Delay and Carrier Phase of GNSS Signals," in *Proceedings of ENC-GNSS, Naples, Italy*, 2009.

[51] K. Dragūnas, "Indoor Multipath Mitigation," in *IEEE International Conference on Indoor Positioning and Indoor Navigation*, pp. 578–584, 2010.

[52] M. Z. H. Bhuiyan, *Analyzing Code Tracking Algorithms for GALILEO Open Service Signal*. MSc dissertation, Tampere University of Technology, 2006.

[53] J. F. Kaiser, "On a Simple Algorithm to Calculate the Energy of a Signal," in *Proceedings of IEEE International Conference on Acoustics, Speech, and Signal Processing (ICASSP), Alburquerque, NM*, pp. 381–384, 1990.

[54] D. de Castro, J. Diez, and A. Fernandez, "High Resolution Multipath Mitigation Technique Based on the Teager-Kaiser Operator for GNSS Signals," in *Proceedings of ION GNSS 2007, Fort Worth, TX*, 2007.

[55] R. Hamila, A. Lakhzouri, E. S. Lohan, and M. Renfors, "A Highly Efficient Generalized Teager-Kaiser-Based Technique for LOS Estimation in WCDMA Mobile Positioning," *EURASIP Journal on Advances in Signal Processing*, vol. 2005, no. 5, p. 685032, 2005.

[56] N. Blanco-Delgado, *Signal Processing Techniques in Modern Multi-constellation GNSS Receivers*. PhD thesis, Technical University of Lisbon, 2011.

[57] M. Z. H. Bhuiyan, E. S. Lohan, and M. Renfors, "Code Tracking Algorithms for Mitigating Multipath Effects in Fading Channels for Satellite-Based Positioning," *EURASIP Journal on Advances in Signal Processing*, vol. 2008, p. 88, 2008.

[58] M. Z. H. Bhuiyan and E. S. Lohan, "Advanced Multipath Mitigation Techniques for Satellite-Based Positioning Applications," *International Journal of Navigation and Observation*, vol. 2010, pp. 1–15, 2010.

[59] G. Seco-Granados, *Antenna Arrays for Multipath and Interference Mitigation in GNSS Receivers*. PhD dissertation, Universitat Politecnica de Catalunya, 2000.

[60] J. K. Ray, *Mitigation of GPS Code and Carrier Phase Multipath Effects Using a Multi-antenna System*. PhD thesis, University of Calgary, 2000.

[61] J. A. Avila-Rodriguez, S. Wallner, and G. Hein, "MBOC: The New Optimized Spreading Modulation Recommended for GALILEO E1 OS and GPS L1C," in *Proceedings of ESA NAVITECH 2006, Noordwijk, the Netherlands*, 2006.

[62] F. M. Schubert, J. Wendel, M. Söllner, M. Kaindl, and R. Kohl, "The Astrium Correlator: Unambiguous Tracking of High-Rate BOC Signals," in *Proceedings of IEEE/ION Position Location and Navigation Symposium (PLANS), Monterey, CA*, pp. 589–601, 2014.

[63] J. Ren, W. Jia, H. Chen, and M. Yao, "Unambiguous Tracking Method for Alternative Binary Offset Carrier Modulated Signals Based on Dual Estimate Loop," *IEEE Communications Letters*, vol. 16, pp. 1737–1740, November 2012.

[64] T. Feng, "Decimation Double-Phase Estimator: An Efficient and Unambiguous High-Order Binary Offset Carrier Tracking Algorithm," *IEEE Signal Processing Letters*, vol. 23, pp. 905–909, July 2016.

[65] O. Julien, C. Macabiau, M. Cannon, and G. Lachapelle, "ASPeCT: Unambiguous Sine-BOC(n,n) Acquisition/Tracking Technique for Navigation Applications," *IEEE Transactions on Aerospace and Electronic Systems*, vol. 43, pp. 150–162, 2007.

[66] A. Schmid and A. Neubauer, "Differential Correlation for Galileo/GPS Receivers," in *Acoustics, Speech, and Signal Processing, 2005. Proceedings. (ICASSP'05). IEEE International Conference on*, vol. 3, pp. iii–953, IEEE, 2005.

[67] A. Burian, E. S. Lohan, and M. Renfors, "Efficient Delay Tracking Methods with Side-lobes Cancellation for BOC-Modulated Signals," *Journal on Wireless Communications and Networking*, vol. 2007, 2, 2007.

[68] A. Burian, E. S. Lohan, and M. Renfors, "BPSK-Like Methods for Hybrid-Search Acquisition of GALILEO Signals," in *Proceedings of ICC 2006, Turkey*, 2006.

[69] M. Navarro-Gallardo, G. Lopez-Risueno, M. Crisci, and G. Seco-Granados, "Analysis of Side Lobes Cancellation Methods for BOCcos(n, m) Signals," in *2012 6th ESA Workshop on Satellite Navigation Technologies and European Workshop on GNSS Signals and Signal Processing (NAVITEC)*, pp. 1–7, Dec 2012.

[70] E. S. Lohan and M. Renfors, "On the Performance of Multiplexed-BOC (MBOC) Modulation for Future GNSS Signals," in *Proceedings of European Wireless Conference, Paris, France*, 2007.

[71] G. Artaud, L. Lestarquit, and J. L. Issler, "AltBOC for Dummies or Everything You Always Wanted to Know About AltBOC," in *Proceedings of ION GNSS, Savannah, GA*, 2008.

[72] L. Ries, F. Legrand, L. Lestarquit, W. Vigneau, and J. L. Issler, "Tracking and Multipath Performance Assessments of BOC Signals Using a Bit-Level Signal Processing Simulator," in *Proceedings of ION GPS, Portland, OR*, pp. 1996–2010, 2003.

[73] E. Rebeyrol, C. Macabiau, L. Lestarquit, L. Ries, J. L. Issler, and M. L. Boucheret, "BOC Power Spectrum Densities," in *Proceedings of ION National Technical Meeting, San Diego, CA*, 2005.

11 SDR Front Ends, Platforms and Setup

José A. López-Salcedo, Gonzalo Seco-Granados
and Ignacio Fernández-Hernández

11.1 Introduction

This chapter complements the rest of the book with an overview of the related hardware that is needed to gather live GNSS signals and experiment with the companion software receiver. First, we present a brief overview of SDR front ends, with focus on those used to gather the samples provided with this book, but also including other models available in the market. Later, we present current trends in technologies in SDR platforms and applications. Finally, we discuss some practical considerations on the required experimentation setup that the reader would need to gather GNSS samples.

11.2 SDR Front Ends

The front end is the hardware element in charge of conditioning the RF received signals in order to filter the out-of-band noise and to down-convert the signal of interest to either an intermediate frequency or to baseband. In digital receiver implementations, the front end is also in charge of the analog to digital conversion, and thus, it provides at its output a stream of digitized signal samples, as discussed in Section 1.4.3. The purpose of the present section is twofold. First, to provide a brief review of the front ends used in previous chapters of the book along with their main configuration parameters. Second, to present the existing market of low-cost SDR front ends, which has facilitated the access to GNSS samples in a simple and affordable way.

11.2.1 Main Features of the SDR Front Ends Used in This Book

Two different front ends have been considered in this book. The first one is the Stereo front end developed by Nottingham Scientific Limited [1], which was used to gather the samples processed in Chapters 2–7. The second one was the SdrNav40 developed by OneTalentGNSS [2], which was used to gather the dual-frequency samples processed in Chapter 8. Both devices make use of commercial integrated circuits (ICs) readily available in the market, and they embed them into a dedicated board together with a microcontroller, input/output ports and a power supply.

Table 11.1 Main specifications and configuration parameters of the RF front ends used in previous chapters of the book.

	Stereo		SdrNav40		SiGe
	Front end 1	Front end 2	Front end 1	Front end 2	Front end 1
IC	MAXIM 2769B	MAXIM 2112	MAXIM 2112	MAXIM 2120	SE4110L
Input frequency range (MHz)	1550–1610	925–2175	925–2175	925–2175	1570–1580
Bandwidth (MHz)	4.2	10	20.46	20.46	4
Sampling frequency (MHz)	26	26	27.456	27.456	16.368
Intermediate frequency (MHz)	{0, 6.39, 6.5}	{0}	0.132	0.418	4.129
Quantization bits	2	3	8	8	2

The SdrNav40 incorporates both the MAXIM 2112 [3] and the MAXIM 2120 [4] ICs into the same board, providing two options for implementing the filtering and down conversion operations. These two ICs are actually analog tuners so their outputs need to be digitally converted. This is done with the MAXIM 19505 ADC where the stream of digitized samples is finally obtained. The Stereo front end also provides two different options, either using the same MAXIM 2112 followed by an ADC or just the MAXIM 2769B [5] that already incorporates an ADC into the same chip. The main parameters of both the Stereo and SdrNav40 front ends are summarized in Table 11.1.

It is interesting to note that the MAXIM 2769B used in the Stereo front end was specifically optimized for GNSS, having an input frequency range covering the upper L-band from 1,550 to 1,610 MHz. This is where GPS L1, Galileo E1 and GLONASS G1 signals are transmitted. These features were extended to cover GPS L2 and L5, Galileo E5 and E6, as well as BeiDou B1, B2 and B3 in the multi-band version MAXIM 2771 [6]. The MAXIM 2120 and MAXIM 2120, on the other hand, are actually TV tuners targeting satellite digital video broadcasting (DVB-S). However, their flexibility, performance and wide input frequency range have made them very popular among the SDR community for a myriad of different applications far beyond satellite television.

11.2.2 Alternative SDR Front Ends in the Market

In the recent years, there has been an increasing interest in the SDR paradigm motivated by the flexibility provided by software to reconfigure hardware resources transparently to the user [7]. Furthermore, the unprecedented advances in semiconductors and the cost reduction due to economies of scale have democratized the access to the RF domain for the general public. Many initiatives have joined the arena, and at the present moment, there are dozens of multipurpose SDR front ends available in the market ready to be used by radio enthusiasts, researchers and students. This is a radically different situation to the one existing when *A Software-Defined GPS and Galileo Receiver: A Single-Frequency Approach*, the preceding book, written more than 15 years ago. As a result, users have now available a plethora of SDR front ends that could be used to gather their own GNSS samples.

Probably one of the most popular SDR front ends to do so is the RTL-SDR. It is a low-cost device that incorporates the Realtek RTL2832U HDTV decoder based on the R820T2 tuner from RafaelMicro [8]. The input frequency range is quite wide covering from 24 to 1,766 MHz, thus including all GNSS bands. However, its very limited bandwidth (2.4 MHz) circumscribes its operation to GPS L1 signals only. But other alternatives are available in the market at the expense of a trade-off between performance and cost.

Choosing the right SDR front end is not a difficult task, and some steps are briefly discussed next. The first one is to make sure that the front end is able to tune the L-band where GNSS signals are transmitted. This involves a range of frequencies ranging from ∼1.1 GHz to ∼1.6 GHz, depending on the desired GNSS signal to be processed. The requirement is already taken for granted when considering an RF front end specifically designed for GNSS. However, it must be borne in mind when selecting a general purpose SDR front end from those already available in the market. Many of them are targeting different broadcast services such as amateur radio, digital television, and it could be the case that they can only be tuned at frequency bands different to those used by GNSS.

Once it is confirmed that the SDR front end can tune the L-band, the next step is to check that the bandwidth is wide enough to let the GNSS signal of interest to pass by without an excessive degradation. The wider the bandwidth, the better the position accuracy we will obtain, and the smaller the power losses and distortion incurred when truncating the power spectral density of the received signal. Nevertheless, the bandwidth is linked to the sampling frequency through the well-known Nyquist criterion, as already discussed in Section 1.4.3. So a larger bandwidth does also involve a larger sampling rate and thus a faster data rate at the output of the front end. Many low-cost commercial SDR front ends are not able to provide such high data rates. But even if they did so, we are not always interested in gathering GNSS signals with the widest possible bandwidth and the largest possible sampling rate. For GPS L1 C/A signals, for instance, it is often enough for most amateur experiments to work with at least 2 MHz bandwidth and a sampling rate on that order, or just a bit higher to compensate for the transition band of the front-end bandpass filter. For Galileo E1, it is enough with 4 MHz bandwidth and 8 MHz for GLONASS.

The aforementioned bandwidths can be accommodated by most SDR front ends, but limitations may be faced when choosing a very low-cost device. For instance, the RTL-SDR provides a stable sampling rate up to 2.4 MHz, only. This value is enough for sampling GPS L1 C/A signals with a small bandwidth, for example, 2 MHz,[1] but

---

[1] Recall that bandwidth is considered here in the bandpass sense, so a 2 MHz bandwidth for GPS L1 C/A involves a spectrum that once down-converted to baseband ranges from [−1, +1] MHz.

not enough for sampling Galileo signals. If a higher sampling rate is required, then an alternative SDR front end must be found whose ADC allows for that. Some examples are shown in Table 11.2 where the most relevant SDR front ends at the time of writing this book are listed. The sampling rates shown in this table correspond to the maximum value provided by each front end, since many of them allow the sampling rate to be configured by software. When this is the case, it is always advisable to adjust the sampling rate to the one strictly required in our application in order to avoid excessive overhead of data transmission, data storage and data processing at the user's computer. The sampling rate is one of the key parameters of the ADC of each SDR front end. The other key parameter is the number of quantization bits for each output sample. Most SDR front ends provide at least 8 bits per sample, which is more than enough for most GNSS applications as discussed in Section 1.4.3.

Finally, the last but not least element to be checked is the front-end oscillator. This element generates a pure sinusoidal signal that is used as a reference signal for the frequency synthesizers and local oscillators of the front end. The quality of this reference signal is therefore of the utmost importance for gathering GNSS signals and subsequently, for processing them, as already discussed in Section 1.6. Unfortunately, practical oscillators are affected by two main degradations. On the one hand, the generated signal is far from being pure sinusoid because it is affected by phase noise, and thus, it exhibits rapid and random frequency variations in the short-term. On the other hand, external effects such as temperature variations and ageing introduce a deterministic frequency variation in the long-term.

The impact of long-term frequency stability is mainly at the acquisition stage of the GNSS receiver, where it is perceived as an additional frequency offset on top of the satellite Doppler to be estimated. The larger this offset, the larger the search space and therefore, the larger the acquisition time. Fortunately, such offset varies slowly, so it can be estimated and compensated for at the receiver. Actually, some SDR front ends provide additional software routines to calibrate for the oscillator long-term frequency instability. They usually do so by processing signals from nearby cellular base stations whose nominal frequencies are known in advance. This process should be repeated from time to time in order to account for the possible variations that might occur when the aforementioned conditions significantly change.

In contrast, nothing can be done to compensate for the short-term frequency stability except for using a better oscillator. This often means choosing a different SDR front end, but sometimes, an external oscillator can be used if the SDR front end has a built-in port allowing to do so. The oscillator short-term stability has a significant impact in many stages of the GNSS receiver. At the acquisition stage, it prevents the implementation of long coherent integration, and therefore, it constrains the acquisition sensitivity, which plays a key role in high-sensitivity GNSS applications. It is worth mentioning that, as a rule of thumb, and for a fixed total integration time, the sensitivity increases by 1.5 dB for each twofold increase in frequency stability [9]. At the tracking stage, rapid variations of the oscillator phase noise hinder carrier tracking and may even cause the receiver to lose lock when using narrow loop filter bandwidths. But even if lock is not lost, both short- and long-term clock variations are superimposed to the user's dynamics thus biasing the position, velocity and time estimates.

Table 11.2 List with some of the SDR front ends available in the market at the time of writing this book.

Device	Bandwidth (MHz)	Freq. band (MHz)	Sampling rate (Msps)	ADC bits	Freq. stab. (ppm)	Rx chann.	Bias-tee	Ext. clock	Cost
RTL-SDR v3	2.4	0.5–1766	2.4	8	±0.5	1	Yes	No	$
Airspy Mini	6	24–1700	10	12	±0.5	1	Yes	No	$
RSP1A	10	0.01–2000	10	14	±0.5	1	Yes	No	$
Adalm Pluto	20	0.325–3800	61.44	12	±25	1	Yes	No	$
Lime SDR mini	30	10–3500	30.72	12	±4	1	No	Yes	$
Kerberos SDR	2.4	0.5–1766	2.4	8	±0.5	4	No	No	$
BladeRF 2.0μ x4	56	47–6000	61.44	12	±1	2	Yes	Yes	$$
HackRF One	20	1–6000	20	8	±20	1	Yes	Yes	$$
RSPduo	10	0.001–2000	10	14	±0.5	2	Yes	Yes	$$
Lime SDR	56	0.1–3800	61.44	12	±4	2	No	No	$$
USRP B200mini-i	56	70–6000	61.44	12	±2	1	No	Yes	$$$
ADRV9364-Z7020	56	70–6000	61.44	12	±2	1	No	Yes	$$$

So having a good enough clock for the application at hand is always advisable. OCXO oscillators offer a very good short-term stability but a not so good long-term stability. CSAC oscillators, on the other hand, provide a not so good short-term stability, but a very good long-term stability. The problem is that good clocks such as OCXO and CSAC are expensive, and this is at odds with the low-cost philosophy of many SDR front ends. This is the reason why most SDR front ends are mounting TCXO instead, which provides not a very good performance at neither short- nor long-term stability, but acceptable for most amateur applications.

Choosing the right oscillator involves checking its short- and long-term stability. The former is often expressed through the spectral mask of the phase noise power spectral density, but from a more practical point of view, it can also be measured through the Allan variance, $\sigma_y^2(T_A)$. This is a unit-less metric that provides information on the frequency deviation of an oscillator over a period of time T_A, typically from 0.01 to 100 seconds. For instance, for an oscillator with nominal frequency F_0 Hz and Allan variance $\sigma_y^2(T_A)$, the actual frequency of the output oscillator sinusoid has a standard deviation of $F_0\sigma_y(T_A)$ Hz when observed over a T_A seconds period. The long-term stability is often decoupled into temperature stability and ageing stability by many oscillator manufacturers, expressed in parts per million or parts per unit (e.g. 1 ppm or 10^{-6}). But when it comes to SDR front ends, most of the datasheets dismiss the distinction between short- and long-term stability and they just provide instead a single number for the overall frequency stability of the oscillator, as indicated in Table 11.2.

In summary, Table 11.2 provides a list of features for the most relevant SDR front ends available at time of writing this book. The interested reader will find in this table a guiding source to select the front end that best fits a target application. Any of these front ends can be used to gather GNSS samples that can later be processed with the MATLAB routines accompanying this book, provided that the proper parameter configuration is set on either side. The present section provided just a brief overview of SDR front ends and their main constituent elements. Interested readers can find more details about the front-end RF components in the preceding book, [10, Chapter 4], as well as in more recent publications such as [11, Chapter 6] or [12, Chapter 14].

11.3   SDR Platforms, Architectures and Applications

Most of the existing GNSS receivers are implemented in hardware based on proprietary designs from manufacturers such as Javad Positioning, Trimble, NovAtel, Thales, Topcon or Septentrio and chip manufacturers like Broadcom, ST Microelectronics, Qualcomm, Texas Instruments or u-blox, among many other. The latter provide miniaturized chipsets at a very reduced cost thanks to massive production and economies of scale. However, they all provide very limited access and control of the internal receiver parameters, which becomes a drawback for research activities dealing with algorithm prototyping, testing and also for the development of custom applications.

As discussed throughout the book, the use of an SDR device together with a GNSS software receiver became an appealing alternative to cope with the stringent limitations of GNSS hardware receivers. The underlying principle of the SDR philosophy is to move the digital signal processing as close as possible to the antenna, in such a way that almost all the tasks to be carried out by the receiver can be implemented in software. This provides extreme flexibility to the designer, who can easily reconfigure and upgrade the receiver without the constraints and limitations of hardware design. As discussed in the previous section, the minimum required hardware is a GNSS antenna and a simple front end capable of providing the digitized samples to a software receiver, which is nothing but a piece of software running on a computer. The main advantages of GNSS software receivers are that they facilitate the implementation and testing of advanced signal processing algorithms which are difficult to implement in hardware; they provide full control of the receiver parameters; and they allow the postprocessing of the received signal samples, thus enabling the possibility to apply record-and-replay analyses and to test different algorithm configurations under the same set of received signal samples.

On the other hand, the main disadvantage may already have been experienced by the reader after running our accompanying MATLAB receivers: except in the snapshot case, it takes several minutes to perform all the operations until the first position fix is available. Current software receiver implementations intend to optimize processing time and energy consumption to get closer to the performance of hardware receivers.

The first reported GNSS software receivers were developed during the 1990s [13] where one of the main challenges was the implementation of the acquisition stage using the FFT algorithm [14]. Later on, different GNSS software receivers were presented by introducing novel features such as real-time processing and compatibility beyond GPS, by incorporating Galileo and EGNOS signal processing, see [15], [16] and [10]. Nowadays, GNSS software receivers are widely deployed across research institutions, and some of them are also commercially available such as the SX3 (formerly SX-NSR) by Ifen GmbH, ARAMIS by IP Solutions, the software receiver by Galileo Satellite Navigation Ltd (GSN) or the open multiconstellation software receiver by M3 Systems [17] just to mention a few. Some others are used for research and educational purposes at their own institutions, such as the GSNRx software receiver developed by the PLAN group at the University of Calgary [18], the ipexSR at Institute of Geodesy and Navigation, University FAF[19], the N-FUELS at LINKS (formerly ISMB) [20], the snapshot GNSS receiver at UAB [21] and many others at different institutions researching in GNSS. A number of open-source software receivers are also available such as the GNSS-SDRlib by Taroh Suzuki [22], the fastGPS by Morgan Quigley and Pieter Abbeel at Stanford University [23], the GNSS-SDR project [24] and the single-frequency GNSS software receiver supporting the predecessor of this textbook [10], all of them providing access to the signal processing tasks of the GNSS receiver.

There are other GNSS software platforms available whose focus is on the processing of GNSS measurements, assumed to be available as an input. RTKLIB [25] and PPPWizard [26] are good examples of that. They provide free software for high-accuracy applications, based on carrier phase measurement processing and carrier phase ambiguity estimation, but they are beyond the scope of this book.

Multi-core Processing Units

In many emerging GNSS-based applications, traditional application-specific integrated circuits (ASIC) of hardware receivers have been gradually substituted by more flexible, transparent, configurable and low-cost implementations based on SDR. This allows a higher degree of flexibility for testing and validating GNSS signal processing algorithms without going through the rigid and long-term design flow of ASIC development. Moreover, and since the signal specification of future GNSS systems is still evolving, SDR allows the designer to rapidly reconfigure the receiver algorithms in order to accommodate any required changes.

However, the problem with traditional SDR implementations is that they pose many limitations to cope with the high computational demand of GNSS signals, particularly when real-time operation is sought. This is the case when dealing with either field programmable gate arrays (FPGA) controlled by digital signal processors (DSP), ARM central processing units (CPU) or general purpose processors (GPP). Their computational speed is ultimately limited by physical barriers due to voltage leakage across internal chip components and heat dissipation limits, thus causing a clock speed wall that prevents designers from improving the computing performance without bound. In order to circumvent this limitation, new strategies have been devised such as the use of hyper-threading technology for executing parallel threads within a single core, and later on, the replication of several processing cores within the same chip leading to the multicore CPU, see [27].

Multicore technology is currently widespread deployed and 2–8 cores can easily be found in most mass-market CPUs, thus becoming a kind of de facto standard in current desktop and laptop computers as well as in smartphones. Professional CPUs, on their side, can offer several tenths of cores in a single chip following a trend that is rapidly evolving. At the present time, up to 56 parallel cores can be found for instance in the Intel Xeon Platinum family. This certainly paves the way to paralleling many of the signal processing tasks to be carried out in a GNSS software receiver and thus make possible the real-time operation of advanced GNSS signal processing, see [28]. Taking full advantage of multicore technology requires a twofold approach. First, a divide-and-conquer strategy must be applied to split all the computationally intensive GNSS signal processing tasks into several simple operations. Second, the repetitiveness of many GNSS signal processing operations must also be taken into account and whenever it is possible, launch each of these tasks into a separate core. This applies for instance to the case of repeating the acquisition stage for each of the satellites, each of the frequencies and each of the constellations to be searched. This is schematically illustrated in Figure 11.1 where each box would be associated to a processing core.

Graphical Processing Units

Graphical processing units (GPU) make one more leap forward in the race to parallel as many GNSS signal processing tasks as possible into a single device, thus making

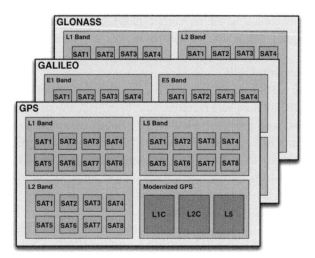

Figure 11.1 Illustration of the independent threads that could be launched in a multicore SDR-based platform for a multiconstellation and multifrequency snapshot GNSS receiver.

feasible the implementation of real-time GNSS receivers. One of the extensive operations carried out within a GNSS receiver is the correlation between the received signal and the local code replica. This operation is fundamental for despreading the received signal and thus being able to estimate the code and carrier parameters, which are needed to compute the user's position. In hardware implementations, it is common to embed hundreds of thousands of correlators leading to massive parallel correlation architectures, but this same approach becomes unfeasible in software implementations. A much more efficient approach is to carry out the correlation in the frequency domain, since this can easily be done in software using the efficient FFT algorithm, as already described throughout the book. A step forward is to move the execution of the FFT to a GPU which is a dedicated graphics rendering device specializing in the massive implementation of efficient FFT processors.

GPUs were originally designed to cope with the very high computational demands of video applications in personal computers and game consoles. Since image processing techniques are essentially based on FFT operations, GPUs turn out to be very well suited for implementing FFT-based correlation algorithms of GNSS software receivers, as well. The main advantage of using GPUs is that they offer formidable computational power in a very dense multicore architecture, as it is the case for instance of the Nvidia GeForce RTX3090 GPU, offering more than 10,000 cores running in parallel and providing up to 35 TFlops (Tera-floating point operations per second) in 32-bits float precision. This unparalleled computational power exceeds by large the one currently offered by multicore general purpose CPUs. But this comes at the expense of having to migrate and optimize the code to be run into the GPU using native GPU language (CUDA or OpenCL) and then having to interface between the local CPU and the GPU for I/O operations. Several GPU-based GNSS software receivers have been reported in the existing literature [29], [30] and [31], even though one of their limitations is the latency due to the transfer of data from the local CPU and disk to the GPU unit. Actually, this has been reported as one of the bottlenecks of GPUs for the implementation of real-time signal processing, which drastically limits the potential advantages of GPU in front of multicore CPUs, see [32] and [33]. In practice, this is an issue that needs to be carefully addressed in order to assess the actual impact in the application to be implemented.

Migration to Cloud Infrastructures

The unprecedented advances in very large scale integration (VLSI) have made possible the widespread deployment of multicore devices which open the door to a new paradigm in the implementation of software GNSS receivers. This trend, however, is at odds with the miniaturization, cost and power constraints of many applications which prevent the adoption of advanced computing devices at the user terminal. This is the case of GNSS-enabled personal devices (smart watches and car navigators) or the advent of miniaturized sensors and trackers with applications in the Internet of Things (IoT). In the latter case, the size and power constraint of these devices make completely unfeasible to implement barely any GNSS signal processing task.

This problem can benefit from the concept of cloud computing, a well-established paradigm nowadays based on offering computing services through the Internet. Cloud computing opens the door for the migration of all the computational tasks of GNSS receivers to a remote platform, thus substituting a dedicated computing device by the nearly unlimited resources, scalability, distributed computing and high-performance of a cloud infrastructure. Following this approach, the user terminal would only need to collect a batch of RF samples and send them through a communication link to the cloud infrastructure where the GNSS signal processing would actually take place. This situation is schematically depicted in Figure 11.2 where a set of heterogeneous GNSS front ends send their gathered samples to the cloud and the results of the GNSS signal processing are sent back to the user or to a control center.

This new paradigm dramatically reduces the power consumption of the user's terminal, facilitates the implementation of sophisticated signal processing techniques such as authentication and spoofing detection, enables the seamless upgrade for the compatibility with new GNSS signals and grants access to restricted services without the need of having a decryption key at the user's terminal. Some cloud-based GNSS receivers are currently available in different flavors depending on whether the RF samples are truly processed in the cloud, or locally at the user's terminal, in which case the result of such local processing is what is actually sent to the cloud. Some examples are the products available by companies such as Baseband Technologies [34], RxNetworks [35] or the prototype developed by Universitat Autònoma de Barcelona [36] and licensed to Loctio [37].

Applications

The GNSS market is currently dominated by hardware GNSS receivers due to their low cost and reliable performance. This is the result of the more than three decades of operation and optimization of such devices, as well as to the unbeatable economies

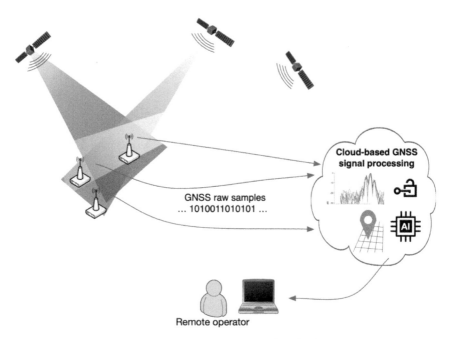

GNSS raw samples
... 1010011010101 ...

Figure 11.2 Illustration of the cloud GNSS receiver paradigm for the case where the RF samples of the gathered GNSS signals are sent to the cloud.

of scale in VLSI. GNSS software receivers imply a radically different paradigm by relying on a software code that runs in a multipurpose processor. It is for this reason that for the time being, they have been circumscribed to specific areas such as research and education where the main interest is on the code itself. GNSS software receivers are an attractive tool for developing and testing new algorithms (e.g. multipath, interference or spoofing detection/mitigation), as well as an indispensable tool for newcomers, offering full transparency on how a GNSS receiver works.

Most of the new GNSS signal processing techniques unveiled in the past decade are the result of experimentation with GNSS software receivers. Actually, snapshot GNSS receivers did appear as a result of such experimentation. They have become a paradigm shift for the GNSS receiver industry, which mostly relies on hardware receivers that operate continuously, following the conventional GNSS architecture with acquisition and tracking. However, snapshot-based GNSS receivers are attractive for many of the new emerging applications of GNSS, where a technological leap forward is required to circumvent the limitations of conventional GNSS receiver architectures. Some fields of industrial and commercial applications for snapshot-based GNSS receivers are described below:

– Weak signal applications, such as those dealing with GNSS positioning in indoor environments, see [38], or GNSS navigation in Space missions.
– High-dynamics applications, such as the case of onboard GNSS receivers in unmanned aerial vehicles, launchers and applications subject to severe dynamics such as radio occultation [39].

- Location-based services, where position fixes are requested by the user at the time the service is requested, as with many smartphone applications for social networking, geo-marketing and advertising, or in emergency calls.
- INS/GNSS integration, where the tight pull-in ranges of conventional tracking architectures significantly limit the integration of external data coming from inertial sensors [40].
- Miniaturized, ultra low-power trackers attached to objects or animals that are sporadically switched on to find out their location.
- IoT where uniquely identified devices are interconnected and may require position information as well as a precise time reference. This information could be obtained from their own gathered GNSS samples and the support of a cloud computing platform for carrying out the required GNSS signal processing. The use of a large population of IoT sensors opens the door for crowd-sourcing and signal integrity monitoring in a given geographical area.

GNSS software receivers not operating in snapshot mode are thus following the same architecture as conventional GNSS hardware receivers. In a practical application where the GNSS signals need to be permanently monitored, a GNSS hardware receiver may offer a more efficient implementation nowadays. However, with the advent of cloud computing and future network infrastructures such as 5G and eventually 6G, providing low-latency and ultra-fast-speed communication links, the widespread operation of software GNSS receivers may soon become a reality. Indeed, many telecom operators are pushing to migrate most of their current hardware infrastructure into a software-based one. Concepts like software-defined networks (SDN) are currently a reality, and this trend for networking hardware is likely to embrace GNSS hardware as well. For instance, the GNSS hardware receivers currently mounted in many cellular base stations for providing precise (i.e. GNSS) network time synchronization could be migrated into GNSS software receiver implementations running in the operator global SDN. Therefore, it is expected that software GNSS receivers may play a increasing role in the coming decades.

11.4   Experimentation Setup to Gather GNSS Samples

The MATLAB software accompanying this book can readily be used to process GNSS samples gathered by the reader using an SDR front end. But for this to be possible, some additional elements are needed first as shown in Figure 11.3 where the experimentation setup is schematically illustrated. It can be seen that a GNSS antenna and eventually a bias tee are needed at the input of the front end, while a host computer is required at the output of the front end. The host computer is needed to control and command the front end, since the latter is not a stand-alone component, as well as to store the stream of binary samples coming out of it.

At the input of the front end, a GNSS antenna is needed to capture the GNSS signals propagating through the air. This antenna should be compatible with the frequency

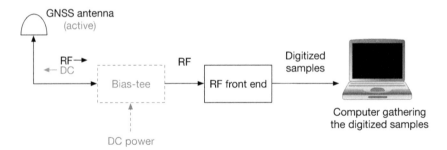

Figure 11.3 Setup for gathering GNSS samples using an SDR front end.

bands to be processed, so for gathering dual-frequency samples and thus being able to run the software from Chapter 8, a dual-frequency L1/L5 GNSS antenna is needed. Otherwise, a conventional L1 antenna is enough. Many GNSS antenna manufacturers have an antenna selection tool in their website to facilitate the user to find the best candidate antenna for the application at hand. Different configurations are possible depending on the frequency band, the mounting type (e.g. OEM/embedded, surface mount either adhesive or magnetic, survey mount), the RF connector type (SMA, TNC, N, IPEX) and the cable length. GNSS antennas are also classified into active and passive antennas. Actives ones are named that way because they include a low-noise amplifier (LNA) to amplify the received signals, which is an active element that needs to be powered. This means that power supply needs to be provided to active antennas, which is done by means of a bias tee as shown in Figure 11.3. A bias tee is a three-port component that takes the input DC from one port and delivers it to one of the two other ports, only, where the active antenna must be connected. The remaining port is the one to be connected to the front end. This remaining port has a capacitor to block DC so that only the RF signal reaches the front-end. Otherwise, the front-end components could be damaged.

The use of an active GNSS antenna is always advisable provided the very weak power levels of the received GNSS signals. Furthermore, once the GNSS signals are received at the antenna, they must be transferred to the front end through a coaxial cable and some additional losses will be incurred as well. These losses can be signif-icant if the antenna and the front end are far apart, as in the case where the antenna is placed on top of a building and the front end is placed indoors, in a room of that build-ing. One may wonder why to care about a few meters-length cable when the received signals have successfully been able to propagate through tens of thousands of kilome-ters. The reason is that propagation losses through a wired medium are much higher than those through radio transmission. Actually, the former depends exponentially on the cable length, while the latter depends just on the square of the traveled distance.

The user may be familiar with coaxial cables used in residential TV distribution systems, but these cables are not suitable for GNSS applications. First, because they often have an impedance of 75 Ω while most RF components used in GNSS have an impedance of 50 Ω. Keeping the same impedance along the whole RF chain is impor-tant to avoid reflections that may significantly degrade the signal reception. Second, because TV distributed signals are transmitted at low frequencies, on the order of a few hundreds of MHz, while GNSS signals impinging on the antenna and propagating through the cable to the front end do so at frequencies of GHz, thus one order of magnitude larger. This involves much larger propagation losses that can even reach more than 100 dB per 100 meters. It is for this reason that the type of coaxial cable should be carefully chosen for cable lengths on the order of several meters. Standard coaxial cables for GNSS applications such as the RG-58 or RG-214 provide ~30 dB of attenuation per 100 meters, while this value goes down to ~15 dB per 100 meters for new generation RF coaxial cables such as the LBM/LMC-400. For shorter cable lengths of just a few meters or even less than a meter, the impact of using a high- or low-loss coaxial cable may be negligible, though.

Active GNSS antennas are the most widely used in practice, but there might be certain applications where passive GNSS antennas might be of interest as well. The latter do not contain any LNA inside, so they just provide the GNSS signals as they impinge onto the radiating element. If we are working in outdoor clear sky conditions and the front end is just next to the antenna, using a passive antenna may be fine. This is often the case in miniaturized devices that have both the antenna and front-end integrated into the same board, and they get rid of the LNA of antenna to save power and thus to extend the battery lifetime. Finally, apart from being active or passive, there are several other features that advanced users might be interested in such as the radiation pattern, polarization properties, noise figure and phase center, but they are out of the scope of the present text. The interested reader is referred to [41, 42] for more details, as well as to the survey of commercially available GNSS antennas that is periodically published by the magazine GPS World [43].

References

[1] Nottingham Scientific Limited (NSL), "Stereo Front-End Datasheet." www.nsl.eu.com/datasheets/stereo.pdf. Accessed: May 25, 2019.

[2] SdrNav40 – OneTalent GNSS. www.onetalent-gnss.com/ideas/software-defined-radio/sdrnav40. Accessed: May 25, 2019.

[3] Maxim Integrated, "MAX2112, Complete, Direct-Conversion Tuner for DVB-S2 Applications." https://datasheets.maximintegrated.com/en/ds/MAX2112.pdf. Accessed: April 2, 2020.

[4] Maxim Integrated, "MAX2120, Complete, Direct-Conversion Tuner for DVB-S and Free-to-Air Applications." https://datasheets.maximintegrated.com/en/ds/MAX2120.pdf. Accessed: April 2, 2020.

[5] Maxim Integrated, "MAX2769B, Universal GPS Receiver." https://datasheets.maximintegrated.com/en/ds/MAX2769B.pdf. Accessed: April 2, 2020.

[6] Maxim Integrated, "MAX2771, Multiband Universal GNSS Receiver." https://datasheets.maximintegrated.com/en/ds/MAX2771.pdf. Accessed: April 2, 2020.

[7] R. W. Stewart, L. Crockett, D. Atkinson, K. Barlee, D. Crawford, I. Chalmers, M. McLernon, and E. Sozer, "A Low-Cost Desktop Software Defined Radio Design Environment Using MATLAB, Simulink, and the RTL-SDR," *IEEE Communications Magazine*, vol. 53, no. 9, pp. 64–71, 2015.

[8] S. Cass, "A $40 Software Defined Radio," *IEEE Spectrum*, vol. 3, pp. 22–23, 2013.

[9] F. van Diggelen, *A-GPS, Assisted GPS, GNSS and SBAS*. Artech House, 2009.

[10] K. Borre, D. Akos, N. Bertelsen, P. Rinder, and S. H. Jensen, *A Software-Defined GPS and Galileo Receiver, Single Frequency Approach*. Birkhäuser, 2007.

[11] J. B.-T. Tsui, *Fundamentals of Global Positioning System Receivers: A Software Approach*. John Wiley & Sons, 2005.

[12] J. W. Betz, *Engineering Satellite-Based Navigation and Timing: Global Navigation Satellite Systems, Signals, and Receivers*. Wiley IEEE Press, 2016.

[13] D. M. Akos, *A Software Radio Approach to Global Navigation Satellite System Receiver Design*. PhD thesis, Ohio University, 1997.

[14] D. Van Nee and A. Coenen, "New Fast GPS Code-Acquisition Technique Using FFT," *Electronics Letters*, vol. 28, no. 2, pp. 158–160, 1991.

[15] D. Akos and J. B. Tsui, "Design and Implementation of a Direct Digitization GPS Receiver Front End," *IEEE Transactions on Microwave Theory and Techniques*, vol. 44, no. 12, pp. 2334–2339, 1997.

[16] D. Akos, P.-L. Normark, P. Enge, A. Hansson, and A. Rosenlind, "Real-Time GPS Software Radio Receiver," in *Proceedings of ION National Technical Meeting, Long Beach, CA*, pp. 809–816, 2001.

[17] M3 Systems, "Open Multiconstellation GNSS Receiver." www.ni.com/pl-pl/innovations/case-studies/19/developing-an-open-multiconstellation-gnss-receiver-for-education.html. Accessed: September 20, 2021.

[18] M. Petovello, C. O'Driscoll, G. Lachapelle, D. Borio, and H. Murtaza, "Architecture and Benefits of an Advanced GNSS Software Receiver," *Positioning*, vol. 1, pp. 66–78, December 2008.

[19] C. Stöber, M. Anghileri, A. S. Ayaz, D. Dötterböck, I. Krämer, V. Kropp, J.-H. Won, B. Eissfeller, D. S. Güixens, and T. Pany, "ipexSR: A Real-Time Multi-Frequency Software GNSS Receiver," in *Proceedings of ELMAR-2010, Zadar, Croatia*, pp. 407–416, 2010.

[20] E. Falletti, D. Margaria, M. Nicola, G. Povero, and M. Troglia, "N-FUELS and SOPRANO: Educational Tools for Simulation, Analysis and Processing of Satellite Navigation Signals," in *Proceedings of IEEE Frontiers in Education Conference (FIE), Oklahoma City, OK*, pp. 1–6, 2013.

[21] SPCOMNAV, Universitat Autònoma de Barcelona, "Snapshot HS-GNSS Software Receiver." http://spcomnav.uab.es/docs/projects/datasheet_HSGNSS-SPCOMNAV.pdf. Accessed: September 20, 2021.

[22] T. Suzuki, "GNSS-SDRLIB." https://github.com/taroz/GNSS-SDRLIB. Accessed: September 20, 2021.

[23] S. Gleason, M. Quigley, and P. Abbeel, "A GPS Software Receiver," in D. Gebre-Egziabher and S. Gleason (eds.), *GNSS Applications and Methods, Technology and Applications Series*, Ch. 5, Artech House, 2017.

[24] GNSS-SDR, "An Open Source Global Navigation Satellite Systems Software-Defined Receiver." https://gnss-sdr.org/. Accessed in September 20, 2021.

[25] T. Takasu, "RTKLIB: An Open Source Program Package for GNSS Positioning." www.rtklib.com/. Accessed: September 20, 2021.

[26] CNES, "The PPP-WIZARD Project. Precise Point Positioning with Integer and Zero-Difference Ambiguity Resolution Demonstrator." www.ppp-wizard.net/. Accessed: September 20, 2021.

[27] A. C. Sodan, J. Machina, A. Deshmeh, K. Macnaughton, and B. Esbaugh, "Parallelism via Multithreaded and Multicore CPUs," *Computer*, vol. 43, no. 3, pp. 24–32, 2000.

[28] T. Humphreys, J. A. Bhatti, T. Pany, B. M. Ledvina, and B. W. O'Hanlon, "Exploiting Multicore Technology in Software-Defined GNSS Receivers," in *Proceedings of ION GNSS Conference, Savannah, GA*, pp. 1–13, 2009.

[29] T. Hobiger, T. Gotoh, J. Amagai, Y. Koyama, and T. Kondo, "A GPU Based Real-Time GPS Software Receiver," *GPS Solutions*, vol. 14, pp. 207–216, 2010.

[30] T. Pany, B. Riedl, and J. Winkel, "Efficient GNSS Signal Acquisition with Massive Parallel Algorithms Using GPUs," in *Proceedings of ION International Technical Meeting of the Satellite Division, Portland, OR*, pp. 1889–1895, 2010.

[31] J. Seo, C. Yu-Hsuan, D. S. De Lorenzo, S. Lo, P. Enge, D. Akos, and J. Lee, "A Real-Time Capable Software-Defined Receiver Using GPU for Adaptive Anti-Jam GPS Sensors," *Sensors*, vol. 1, no. 11, pp. 8966–8991, 2011.

[32] V. W. Lee, C. Kim, J. Chhugani, M. Deisher, D. Kim, A. D. Nguyen, N. Satish, M. Smelyanskiy, S. Chennupaty, P. Hammarlund, R. Singhal, and P. Dubey, "Debunking the 100X GPU vs. CPU Myth: An Evaluation of Throughput Computing on CPU and GPU," in *Proceedings of International Symposium on Computer Architecture (ISCA), Saint-Malo, France*, pp. 451–460, 2010.

[33] C. Gregg and K. Hazelwood, "Where Is the Data? Why You Cannot Debate CPU vs. GPU Performance without the Answer," in *Proceedings of IEEE International Symposium on Performance Analysis of Systems and Software (ISPASS), Austin, TX*, pp. 134–144, 2011.

[34] "Snapshot Positioning Technology." http://basebandtech.com/snapshot-receiver/. Accessed: August 5, 2016.

[35] "Ultra-Sensitive GNSS Receiver." http://rxnetworks.com/location-io/ultra-sensitive-gnss-receiver/. Accessed: August 5, 2016.

[36] V. Lucas-Sabola, G. Seco-Granados, J. A. López-Salcedo, J. A. García-Molina, and M. Crisci, "Cloud GNSS Receivers: New Advanced Applications Made Possible," in *Proceedings of ICL-GNSS, Barcelona, Spain*, pp. 1–6, 2016.

[37] Loctio. www.loctio.com. Accessed: September 20, 2021.

[38] G. Seco-Granados, J. A. López-Salcedo, D. Jiménez-Baños, and G. López-Risueño, "Challenges in Indoor Global Navigation Satellite Systems," *IEEE Signal Processing Magazine*, vol. 29, pp. 108–131, March 2012.

[39] W. J. Hurd, J. I. Statman, and V. A. Vilnrotter, "High Dynamic GPS Receiver Using Maximum Likelihood Estimation and Frequency Tracking," *IEEE Transactions on Aerospace and Electronic Systems*, vol. 23, no. 4, pp. 425–437, 1987.

[40] F. van Graas, A. Soloviev, M. Uijt de Haag, and S. Gunawardena, "Closed-Loop Sequential Signal Processing and Open-Loop Batch Processing Approaches for GNSS Receiver Design," *IEEE Transactions on Selected Topics in Signal Processing*, vol. 3, no. 4, pp. 571–586, 2009.

[41] Gerald J. K. Moernaut and D. Orban, "GNSS Antennas: An Introduction to Bandwidth, Gain Pattern, Polarization and All That," *GPS World*, vol. 2, pp. 42–48, 2009.

[42] B. Rama Rao, W. Kunysz, R. Fante, and K. McDonald, *GPS/GNSS Antennas*. Artech House, 2013.

[43] GPS World, "GNSS Antenna Survey." www.gpsworld.com/resources/gps-world-antenna-survey/. Accessed: April 2, 2020.

12 SDR MATLAB Package

M. Zahidul H. Bhuiyan, Stefan Söderholm, Giorgia Ferrara, Sarang Thombre, Heidi Kuusniemi, Ignacio Fernández-Hernández, Padma Bolla, José A. López-Salcedo, Elena Simona Lohan and Kai Borre

12.1 Introduction

This chapter presents the MATLAB GNSS SDR package attached to this book. It shows how to download and run the code. It links the different parts of the code with the preceding chapters, so the reader can easily relate theory and practice.

The MATLAB GNSS SDR package, including all the receivers, samples and other files, is available at the following link: https://www.cambridge.org/gnss.

There are four parts in the SDR package attached to this book:

- The FGI-GSRx, a multi-GNSS receiver able to process GPS, GLONASS, Galileo, BeiDou and NAVIC, and which is the main part of the package. It supports Chapters 2–7.
- The GPS dual-frequency receiver, which processes GPS L1 C/A and L5 signals as per Chapter 8.
- The Snapshot receiver, which processes GPS L1 C/A as per Chapter 9.
- The functions that are used to generate the figures of this book, in particular in Chapters 1 and 10.

In the next sections, we explain in more detail each part of the package. In the previous chapter (Chapter 11) we proposed some GNSS front ends that the readers can purchase and generate their own data captures.

12.2 Multi-GNSS SDR Receiver (FGI-GSRx)

12.2.1 Download and Installation

In order to run the FGI-GSRx, you need a current version of MATLAB®. The FGI-GSRx was developed using MATLAB® R2016b. Thus, any newer version should also be compatible. If not already installed download and install MATLAB® using the install guide from Mathworks website [1]. In order to use the GSRx, the MATLAB Communications toolbox is required.

You can also download the latest version of the FGI-GSRx project from following link: https://github.com/nlsfi/FGI-GSRx.

Download the following files and folder:

- `FGI-GSRx.zip`
- `FGI-GSRx Example MATLAB DataFiles.zip`
- `Raw IQ Data` *(Folder)*

After you have downloaded the files, extract the two zip folders to your directory of choice and move the "`Raw IQ Data`" folder there as well. Now you can open MATLAB® and navigate to said directory.

At first you should add the folders to MATLAB® path by either right-clicking the folder and pressing "`Add to Path → Selected Folder and Sub folders`."

12.2.2   Execution

This section presents an overview on how to execute the FGI-GSRx. Further details are provided in the user manual downloaded with the code.

To run the receiver with default configurations use the MATLAB® editor to navigate to "`/FGI-GSRx/param/defaultReceiverConfiguration.txt`" and open it. In this file, you will find multiple path information given (e.g. line 34 or line 63). These are the data files to load or the RF files, respectively. In order for the program to find these files, copy the lines in the `.txt` file containing data paths to a newly created `.txt` file and change the paths to the paths on your local machine. Once you have created the file containing the right data paths, you should be able to run the receiver.

Running the receiver in default configuration mode can be achieved by simply navigating to "`/MATLAB/FGI-GSRx/main`" and calling "`gsrx(path/to/file.txt)`", whereas file.txt is the file we created in the previous section.

Running the receiver in nondefault configuration is done in a similar fashion as in default parameters. In order to create a configuration file familiarize yourself with the parameters. Only the parameters you want to change need to be included in the personalized configuration file.

Folder Structure
The following folders are included in the FGI-GSRx:

- Acquisition (*acq*) folder contains all scripts related to the acquisition of all signals.
- Correlation (*corr*) contains all scripts related to signal correlation.
- Frame decoding (*frame*) contains all scripts related to data frame decoding.
- Geo (*geo*) folder contains all the scripts related to coordinate transformation.
- Ionex (*ionex*) contains all the scripts related to the handling of ionex files.
- Ionopsheric (*iono*) folder contains all the scripts related to ionospheric models.
- Least Square Estimation (*lse*) contains all the scripts related to the least square navigation filter.
- Main (*main*) folder contains the main execution scripts
- Modulation (*mod*) folder contains all the scripts related to modulation of the signals.
- Navigation (*nav*) folder contains all the scripts related to navigation.

– Observation (*obs*) folder contains all the scripts related to observation handling.
– Parameter (*param*) folder contains all the scripts related to the parameter system.
– Plot (*plot*) folder contains all the scripts related to plotting of data.
– Radio front end (*rf*) folder contains subfolders for all supported radio frontends. Each subfolder contains configuration files for the various supported front ends.
– Satellite (*sat*) folder contains all the scripts related to satellite orbit calculations.
– Statistics (*stats*) folder contains all the scripts related to statistical analysis of the outputs.
– Time (*time*) folder contains all the functions related to time conversion.
– Tracking (*track*) folder contains all the scripts related to tracking signals.
– Troposhere (*tropo*) folder contains all the scripts related to the tropospheric models.
– User interface (*ui*) folder contains all the scripts related to the output of data to command line.
– Utility (*utils*) folder contains utility script needed in various parts of the receiver.

General Operation

The main function is called "gsrx.m." It calls some necessary functions mentioned in Table 12.1 in order to carry out the GNSS receiver functionalities.

The user must run the main function gsrx.m with a desired parameter file given as string input to the function like gsrx("MyReceiverConfigurationParameters.txt"). An example parameter file is given with the software. The parameter input format has to be followed for proper processing of the FGI-GSRx multi-GNSS receiver.

Acquisition

The acquisition block searches and acquires satellite signals one signal at a time. After the search is completed for one signal acquisition, it will start for the next signal. After all signals have been searched for the results are handed over to tracking and finally navigation. The acquisition block can use all available assistance data to aid the process. Plots for all found signals are generated and a combined histogram plot showing searched and found signals. The acquisition block is executed via a function call:

```
acqData = doAcquisition(param)
```

Acquisition functions are listed in Table 12.3. The acquisition data are stored in a structure called "acqData." There is one structure for each found signal (e.g. acqResults.gpsl1, acqData.gale1b). Each structure has information concerning the signal acquired (e.g. nrObs, duration, signal, channel) as well as statistics from the acquisition phase (e.g. peakMetric, peakValue, variance, baseline, SvId: [1x1 struct], codePhase, carrFreq). Table 12.2 presents an example of acquisition parameter configuration for Galileo E1B signal.

Tracking

The tracking block uses the output from the acquisition and correlates the incoming signal with signal replicas to produce the raw measurements from all signals and

Table 12.1 List of necessary functions called from the main script.

Function name	Tasks performed	Example code
gsrx()	The main function to carry our multi-GNSS receiver processing	gsrx("default.txt")
readSettings()	Read user-defined parameter files	settings = readSettings(varargin1);
generateSpectra()	Generate spectra and bin distributions	generateSpectra(settings);
doAcquisition()	Acquire signals	acqData = doAcquisition(settings);
plotAcquisition()	Plot the acquisition results	plotAcquisition(acqData.gale1b,settings, "gale1b"); %Example showed only with Galileo E1B signal, "gale1b"
doTracking()	Track signals	trackData = doTracking(acqData, settings);
plotTracking()	Plot the tracking results	plotTracking(trackData, settings);
generateObservations()	Generate observations	obsData = generateObservations(trackData, settings);
doFrameDecoding()	Decode navigation data from the ephemeris	[obsData, ephData] = doFrameDecoding(obsData, trackData, settings);
doNavigation()	Perform navigation on the processed observables	[obsData,satData.navData] = doNavigation(obsData. settings, ephData);
calcStatistics()	Calculate receiver statistics	statResults = calcStatistics(navResults);

Table 12.2 Example of acquisition parameters.

File, parameter name, value	Explanation
gale1b,acqSatelliteList,[1:30],	Specify what GPS satellites to search for [PRN numbers]
gale1b,nonCohIntNumber,2,	Number of noncoherent integration rounds for signal acquisition
gale1b,cohIntNumber,1,	Coherent integration time for signal acquisition [ms]
gale1b,acqThreshold,9,	Threshold for the signal presence decision rule
gale1b,maxSearchFreq,6000,	Maximum search frequency in one direction
gale1b,modType,'CBOC',	Can take input either of the two modulation types: "CBOC" or "SinBOC"

Table 12.3 Acquisition functions.

Function name	Tasks performed	Example code
doAcquisition()	Attempts to acquire signal for each enabled GNSS constellation (i.e. "gpsl1" or "gale1b")	acqData = doAcquisition(settings);
initAcquisition()	This function initializes the acquisition structure in a favorable way	acqResults = initAcquisition(allSettings);
getDataForAcquisition()	Reads 100 milliseconds (ms) of data for acquisition: The FGI-GSRx receiver does not attempt any reacquisition. Therefore, this is the only data that is used for acquisition	[pRfData,sampleCount] = getDataForAcqui-sition(signalSettings, msToSkip, 100);
acquireSignal()	Generates codes, performs the modulation and up sampling before doing the FFT-based correlation. Use FFT-based acquisition technique for acquiring the satellites from the enabled constellation	acqResults.(signal) = acquireSignal(pRfData, signalSettings);
searchFreqCodePhase()	This function performs the parallel code phase search acquisition for a given satellite signal	results = searchFreqCode-Phase(upSampledCode, signalSettings, pRfData, PRN);
plotAcquisition()	Plot the acquisition results	plotAcquisition(acqData.gale1b, settings, "gale1b"); %Example showed only with Galileo E1B signal, "gale1b"

Table 12.4 Tracking functions.

Function name	Tasks performed	Example code
doTracking()	Main tracking function which in turn calls other associated tracking functions in order to carry out the code tracking and the carrier tracking successfully	trackData = doTracking(acqData, settings);
initTracking()	Initializes tracking channels from acquisition data	trackResults = initTracking(acqResults, allSettings);
GNSSCorrelation()	Performs code and carrier correlation	trackResults.(signal) = GNSSCorrelation(trackResults.(signal), channelNr);
GNSSTracking()	Performs state-based tracking for the received signal	trackResults.(signal) = GNSSTracking(trackResults.(signal), channelNr);
showTrackStatus()	Prints the status of all track channels to the command window	showTrackStatus(trackResults, allSettings, loopCnt);
plotTracking()	Plots all tracking-related results for each tracked channel. Plots also the C/N0 of all satellites in each constellation	plotTracking(trackData, settings);

satellites. The processing is done for all found signals in parallel, and the results are sent to navigation. Tracking functions are listed in Table 12.4. The tracking block is executed via a function call:

```
trackData = doTracking(acqData, param)
```

Data Decoding and Navigation

The data decoding block uses the output from the tracking and converts the processed tracking data to ephemeris for different available GNSS constellation. It finds the preambles first and then starts decoding the ephemeris for that constellation.

The navigation block has three main tasks: to convert the measurements to observations, to calculate the satellite positions and velocities, and finally to calculate the position of the user. Each element of the *obsData*, *satdata* and *navData* data structure contains data for one signal. The *param* structure contains the parameters for one signal. Navigation functions are listed in Table 12.5.

12.2.3 Data Management

The receiver architecture has been designed so that the intermediate data from acquisition and tracking can be saved, and processing can start from any presaved data file.

Table 12.5 Navigation functions.

Function name	Tasks performed	Example code
doNavigation()	This is the main loop for navigation. It utilizes receiver observables, satellite-specific information, correction information and offers a navigation solution based on the user configuration settings	[obsData,satData,navData] = doNavigation(obsData, settings, ephData);
initNavigation()	This function calculates some initial values needed for the main navigation loop	[obsData, nrOfEpochs, startSampleCount, navData, samplesPerMs] = initNavigation(obsData, allSettings);
applyObservationCorrections()	This is the main function to apply all available corrections to the specific observable	obsData = applyObservationCorrec-tions(allSettings, obsData, satData, navData, corrInputData);
checkObservations()	This function decides which current observations are going to be used in the navigation computation	obsData = checkObserva-tions(obsData, satData, allSettings, navData);
getNavSolution()	Calculates Least Square Estimation (LSE)-based navigation solutions for the measured receiver observables	[obsData, satData, navData] = getNavSolution(obsData, satData, navData, allSettings);
showNavStatus()	Prints the status of navigation to the command window	showNavStatus(allSettings, currMeasNr, navData, obsData, satData);

A set of user parameters for the multi-GNSS receiver processing are read from a text file. The parameters specify the configurations of the IF data which include sampling frequency, data type and bandwidth for further signal processing. If requested by the user, the acquisition is executed using the IF data stored in the file system, and the results are stored to the memory on the computer. Optionally, if the acquisition has been carried out before and the already stored acquisition data can be retrieved from the file system, the user can choose to bypass the acquisition process by setting appropriate parameter in the user parameter file. The result is the same regardless of which approach the user takes. The acquisition output is passed on to the tracking stage. The same options are available for tracking. If tracking has been carried out and the already stored tracking data can be retrieved from the file system, the user can choose to bypass the tracking. The result from tracking is then passed on to position computation. Figure 12.1 shows the functional blocks of the multisystem receiver highlighting the feature that allows the user to skip acquisition and tracking and use prestored results.

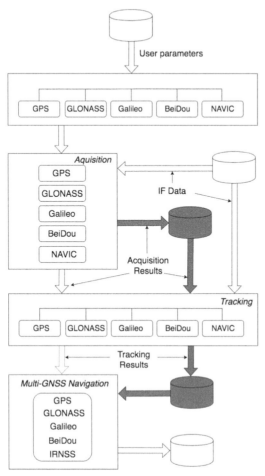

Figure 12.1 Functional blocks in the multisystem receiver. The dark gray parts indicate the option to use prestored output from acquisition and tracking.

12.2.4 Parameter System

The flexibility of modifying different receiver parameters is one of the main advantages of a software-defined implementation. In FGI-GSRx, the parameter system is based on text files. All default values are kept in a default parameter file, and each user can have their own parameter file depending on their specific needs. Therefore, changing parameters do not really require any detailed changes to the actual underlying code. This flexibility makes it very easy to test different algorithms with many different data sets and in a way that can be easily reproduced later. The multi-GNSS receiver is therefore a very useful platform for researchers, as it enables them to independently develop and to test new algorithms in each stage of the receiver chain from raw IF samples to PVT solution. FGI-GSRx is a very valuable tool also for whomever is simply interested in learning how a GNSS receiver operates. The user can configure different parameters related to the acquisition, tracking and position calculation modules, as well as parameters specifying the front-end configuration used to store the IF data. Given the possibility to set center frequency, sampling frequency, sample size, data type and bandwidth, the software-defined receiver can be used with different front ends. Furthermore, it is possible for the user to enable/disable certain GNSS signals and to specify the number of milliseconds to process in the data file.

12.3   Dual-Frequency GNSS Software Receiver (DF-GSRx)

12.3.1   General Overview

The dual-frequency GNSS software receiver is developed based on the open source GPS L1 single frequency software receiver published by Prof. Kai Borre. The software receiver for dual-frequency GNSS software receiver was developed by Padma Bolla under the supervision of Prof. Kai Borre at Samara National Research University, Samara, Russia. In this manual, an overview of the DF-GSRx is presented to the readers by taking GPS L1/L5 dual-frequency signals as a specific example. This dual-frequency receiver software can be modified and used for processing dual-frequency signals from any global and regional satellite navigation systems. Furthermore, Part-2 of this code has dual-frequency Navigation solution using carrier-smoothed pseudorange observations processed at 20ms signal integration time.

Download and Installation

In order to run the DF-GSRx, you need a latest version of MATLAB. The DF-GSRx was developed using MATLAB release version R2015b on a PC running the Mac operating system. In order to run the DF-GSRx, the MATLAB Communications toolbox is required.

Installation of the code is done by downloading the main folder and associated sample data files as listed below,

– DF-GSRx.zip
– DF-GSRx data files.zip

The MATLAB code can be placed anywhere in the PC. Before executing it is recommended that the path of the DF-GSRx is added to the MATLAB, environment. Now you can open the MATLAB and navigate to the directory where the DF-GSRx folder is stored.

Execution

This section presents an overview on how to execute the DF-GSRx. Detailed information can be found in the user manual provided along with the code. The receiver can be executed in two modes : default configuration mode and user configuration mode.

– In default configuration mode, navigate the MATLAB editor to the folder "MATLAB/DF-GSRx/" and execute the main function "DFSRx.m". This process will read the default receiver parameters through a function named "settingsLx.m" for each of the subscripted frequency channels and process the digital samples stored in the .dat files.

Table 12.6 List of necessary functions called from the main script.

Function name	Tasks performed	Example code
DFSRx()	The main function to carry out dual-frequency receiver signal processing	DFSRx()
SettingsL1()	Read L1 user-defined parameters files	settingsL1 = SettingsL1_1ms();
SettingsL5()	Read L5 user-defined parameters files	settingsL5 = SettingsL5_1ms();
DFAcquisition()	Acquire two-frequency signals	[acqResultsL1,acqResultsL5] = DFAcquisition(settingsL1, settingsL5);
plotAcquisition()	Plot the acquisition results for each channel	plotAcquisition(acqResultsL1, settingsL1);
DFTracking()	Track two-frequnecy signals	[trackResultsL1,trackResultsL5] = DFTracking(channelL1, settingsL1, channelL5, settingsL5);
plotTracking()	Plot the tracking results of each channel	plotTracking(trackResultsL1, settingsL1);
Data_Decoder()	Decode navigation message from the Tracking loop output	[eph,TOW,subFrameStart, activeChnList] = Data_Decoder(trackResultsL1, settingsL1);
DFNavigation()	Perform navigation from the processed two-frequency channel observables	[navSolutionsL1L5,epochs] = DFNavigation(trackResultsL1, trackResultsL5,settingsL1, settingsL5);

- In user configuration mode, user can modify the receiver parameters in "settingsLx.m." Also, user can process newly recorded two-frequency signals by adding the appropriate file path and defining the signal format in "settingsLx.m" for each subscribed frequency channel.

Folder Structure

Along with the other major functions listed in Table 12.6, the following folders are included in the DF-GSRx:

- include (*include*) folder contains all the scripts related to PRN code generation, tracking loop filter coefficient generation, parity check and data decoding for navigation solution and so on
- geoFunctions (*geoFunctions*) folder includes all the scripts related coordinate transformation and navigation solution.

General operation

The dual-frequency software receiver code is executed by the main function "*DFSRx.m.*" Subsequently, main function initiates some necessary function calls listed in Table 12.6. The software receiver can be configured through a set of design parameters listed in a file named, "*SettingsLx.m.*" Otherwise, the receiver will automatically use the default configuration parameters in the file. The intermittent results from each of the receiver functions can be stored in "mat" file, for further use in algorithm

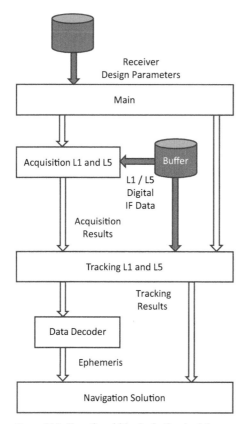

Figure 12.2 Functional blocks in the dual-frequency software receiver.

experimentation. As an example, acquisition and tracking loop output results stored in "mat" files are provided to the user along with the software. These sample files can be utilized by the user to test the specific receiver functions by skipping the traditional way of running the entire receiver code.

The DF-GSRx consists of five different functional blocks: DFSRx, DFAcquisition, DFTracking, Data decoding and DFNavigation. The input and output for these blocks are defined in Table 12.6. After the main functional block is executed, the acquisition, tracking, data decoding and navigation are processed in sequence, for the received two-frequency signals. The execution flow is shown in Figure 12.2.

12.3.2   Main function

The main function "DFSRx();" is executed first in the process. The block has two main tasks: reading the configuration settings files and obtaining receiver design parameters. The main functional block also generates plots in the time and frequency domain as well as the histogram plot for the bit distribution. The receiver design parameter settings for each frequency channel are defined separately for receiver signal processing at acquisition, tracking and navigation blocks. The user can set the design parameters based on customized signal processing algorithms developed in the receiver. The settings parameters are defined as it was described in the open source single-frequency software receiver via a function call:

SettingsLx();

The receiver design parameters are used to configure the number of signal processing functions in the dual-frequency software receiver.

12.3.3 Acquisition Block

The acquisition block searches and acquires satellite signals transmitted in the two-frequency channels. Once the acquisition is done, the results are handed over to tracking and finally navigation. Acquisition plots for all the detected signals in each frequency channel are generated. The acquisition block is executed via a function call:

[acqResultsL1,acqResultsL5] = DFAcquisition(settingsL1,settingsL5);

12.3.4 Tracking Block

The tracking block uses the output from the acquisition and correlates the incoming signal with signal replicas to produce the raw measurements from all signals and satellites. The tracking process is done for all detected two-frequency signals in parallel, and the results are sent to data-decoding and navigation process. The tracking block is executed via a function call:

[trackResultsL1,trackResultsL5] = DFTracking(channelL1,settingsL1,channelL5,settingsL5);

12.3.5 Data-Decoding Block

The data-decoding block uses the output from the tracking and converts the processed tracking data to ephemeris for GPS L1/L5 satellite constellation. It finds the preambles first to identify the subframe start and then start decoding the ephemeris for that constellation. The two-frequency signals are transmitted coherently from the same satellite; hence, the data decoding from one of the two channels is sufficient to decode the ephemeris of the satellite. The data decoding block is executed via a function call:

[eph,TOW,subFrameStart,activeChnList] = DataDecoder(trackResultsL1, settingsL1);

12.3.6 Navigation Block

The navigation block in dual-frequency receiver has four main tasks:

- to convert the tracking loop code and carrier phase to range and range-rate observations

- to generate ionosphere-free pseudorange and range-rate observations
- to calculate the satellite positions and velocities
- finally, to calculate the position of the receiver.

The ionosphere-free code pseudorange observations are further carrier smoothed using ionosphere-free range-rate observations to improve the observation precision. These observations from visible satellite signals along with the satellite position information are used to compute the receiver position, velocity and time by making use of least square algorithm. The navigation solution is computed via function call:

[navSolutionsL1L5,epochs] = DFNavigation(trackResultsL1,trackResultsL5, settingsL1,settingsL5);

12.3.7   Part-2

Dual-frequency software receiver, computes Navigation solution using carrier-smoothed pseudorange observations processed at 20ms signal integration time. The GNSS signal tracking at 20ms signal integration needs bit synchronisation, hence, Part-2 of this code can be enabled after completing signal tracking at 1ms integration.

12.4   Snapshot GPS L1 C/A Receiver

12.4.1   Snapshot Receiver Overview

This subsection describes the MATLAB GPS L1 C/A Snapshot Receiver. Its acquisition block is based on the GPS L1 MATLAB receiver provided with [3]. The positioning block is based on the algorithm using Doppler measurements for a coarse position and code phase measurements for a precise position, as described in Section 9.3.

12.4.2   Installation

Unpack snaprx.zip. It will create the folders *snapsamples*, with the samples, and *snapshotgpsrx*, with the MATLAB code. The code can be placed anywhere. Before executing it is recommended that the path of the Snapshot receiver is added to the MATLAB environment.

12.4.3   Configuration

The configuration file defines:

- The receiver intermediate and sampling frequencies.
- The initial position and time, and time error. The initial position for all the available snapshots is considered as unknown and set to the Earth center. The initial time is around some hours off, for the given sample files. If set to 0, the receiver will search within the day of initial time. If set to "n," the receiver will search in an interval of the initial day +/– "n" days.

– The acquisition type, and the input file: "0" implies that the acquisition is already performed and stored in a *.mat* file. "1" means that the acquisition has to be performed from a samples file. Currently, the two options are available, with one of them commented ("%").
– The receiver stores the acquisition in a file whose name is that of the configuration file, plus the suffix "*_Acqresults.mat*," in the *"results"* folder. The samples are stored in the *"../samples"* folder.
– The number of acquisitions: If higher than one, the receiver will calculate several positions from several acquisitions, from either the *.mat* or the samples file.

12.4.4 Execution

Run *Main.m*, with or without a configuration file. If no configuration file is introduced, the user will be prompted to choose one. Some configuration files are available in the *"config"* folder, corresponding to several snapshots/acquisitions.

The receiver gets the satellite ephemerides and clocks from RINEX files. It first checks if the files are already available in the "ephemeris" folder for the configured time window and, if not, it downloads them from NASA server:

https://cddis.nasa.gov/archive/gnss/data/daily/

For this operation, an internet connection is required.

12.4.5 Acquisition

Acquisition is executed by the *acquisitionSnp.m* file. It is based on a FFT implementation and populates structure *acqResults*.[1]

12.4.6 Navigation Solution

The navigation solution without an accurate initial time or position reference is calculated as per Section 9.3. The receiver calculates a solution every 3 hours during the time uncertainty period, starting from the initial time, and adding/subtracting a 3-hour multiple until the uncertainty period is covered, or a solution is found. For each reference time, the receiver calculates a coarse-time Doppler solution and, if plausible (residuals vector module below a threshold defined in the initial settings), a coarse-time pseudorange solution. If the coarse-time pseudorange solution is considered plausible (i.e. residuals are below a threshold), then the receiver outputs the solution and stops the loop. The receiver performs this process for every acquisition. The receiver reports that no plausible position solution is found in the time uncertainty period, if this is the case. The acquisition phase may take a few seconds. Time uncertainties in the order of one day are solved instantaneously. Higher time uncertainties may take some seconds to be resolved.

---

[1] Note that the SAM and double FFT methods described in the theory are not implemented.

12.5   Other MATLAB Functions

These functions can be found under folder "/otherfunctions/" and generate some of the figures of the book, mostly related to BOC modulations.

– Fct_genericBOC: Function creating a BOC modulation signal in M-stages, M is given by the dimension of input N_BOC_vec.
– Fct_idealACF_AltBOC: Theoretical model for the computation of ACF of an AltBOC signal.
– Fct_idealACF_generic_BOC: compute the ideal ACF of a generic BOC modulation, as the superposition of triangular pulses.
– Fct_idealACF_MBOC: Theoretical model for the computation of ACF of an MBOC signal.
– Fct_idealACF_generic_BOC: compute the ideal ACF of a generic BOC modulation, as the superposition of triangular pulses.
– Fct_idealACF_MBOC: Theoretical model for the computation of ACF of an MBOC signal.
– Fct_PSD_AltBOC: Theoretical model for the computation of PSD of AltBOC modulation with nonconstant envelope.
– Fct_PSD_MBOC: Theoretical model for the computation of ACF of an MBOC signal.
– Main_AltBOC: Theoretical ACF and PSD of AltBOC signals.

12.6   Sample Files

This section includes the filenames of the sample files accompanying the MATLAB receiver. Note that by *sample* we mean the exemplary files used in the receiver, including configuration files, *.mat* files with the acquisition results and binary sample files. Some of the files are used to obtain the results presented in the previous chapters. Table 12.7 lists the files used to generate figures in different chapters using FGI-GSRx multi-GNSS software receiver. The associated chapter number is also offered for ease of understanding. Table 12.8 lists the files for the dual-frequency receiver and Table 12.9 for the snapshot receiver.

12.7   GNSS SDR Metadata Standard

As seen in the previous chapter, there is a wide range of different front ends available out there in the market. The list shown in Table 11.2 is just an example of devices relying on the SDR paradigm, but plenty more devices are also available using proprietary configurations. Such a diversity of hardware equipment poses many difficulties from the software processing point of view because different sampling rates, intermediate frequencies, quantization levels and encoding schemes are involved. All these parameters must be clearly indicated beforehand so that the samples become intelligible to our software receiver. This has been done in the past by taking note of all

Table 12.7 FGI-GSRx sample files names.

File name	File type	Reference chapter	Additional remarks
default_param_GPSL1_Chapter2.txt	Text	2	Default parameter file for GPS L1 constellation
default_param_GLONASSL1_Chapter3.txt	Text	3	Default parameter file for GLONASS L1 constellation
default_param_GalileoE1_Chapter4.txt	Text	4	Default parameter file for Galileo E1B constellation
default_param_BeiDouB1_Chapter5.txt	Text	5	Default parameter file for BeiDou B1 constellation
default_param_NavICL5_Chapter6.txt	Text	6	Default parameter file for NavIC L5 constellation
default_param_MultiGNSS_Chapter7.txt	Text	7	Default parameter file for Multi-GNSS constellation
defaultReceiverConfiguration.txt	Text	2–7	Default parameter file for all the constellations
trackData_GPSL1_Chapter2.mat	Mat	2	Saved tracked data file for GPS L1 constellation
trackData_GLONASSL1_Chapter3.mat	Mat	3	Saved tracked data file for GLONASS L1 constellation
trackData_GalileoE1_Chapter4.mat	Mat	4	Saved tracked data file for Galileo E1B constellation
trackData_BeiDouB1_Chapter5.mat	Mat	5	Saved tracked data file for BeiDou B1 constellation
trackData_NavICL5_Chapter6.mat	Mat	6	Saved tracked data file for NavIC L5 constellation
trackData_MultiGNSS_GPSL1_GalileoE1_BeiDouB1_Chapter7.mat	Mat	7	Saved tracked data file for Mult-GNSS constellation
rawData_GPSL1GalileoE1B_Chapters2_4.dat	Binary	2, 4	Saved binary raw data file for GPS L1 and Galileo E1B constellations
rawData_GLONASSL1_Chapter3.dat	Binary	3	Saved binary raw data file for GLONASS L1 constellation
rawData_BeiDouB1_Chapters5_7.dat	Binary	5, 7	Saved binary raw data file for BeiDou B1 constellation
rawData_GPSL1GalileoE1_Chapter7.dat	Binary	7	Saved binary raw data file for GPS L1 and Galileo E1B constellation
rawData_GPSL1_Chapter6.dat	Binary	6	Saved binary raw data file for GPS L1 constellation
rawData_NavICL5_Chapter6.dat	Binary	6	Saved binary raw data file for NavIC L5 constellation

Table 12.8 Dual-frequency receiver sample file names.

File name	File type	Reference chapter	Additional remarks
gps_L1_06_07_16.dat	Binary	8	Binary raw data file for GPS L1 signal
gps_L5_06_07_16.dat	Binary	8	Binary raw data file for GPS L5 signal
acqResults_L1L5.mat	mat	8	Saved acquisition results for GPS L1 and L5 signals
trackResults_L1L5_1ms.mat	mat	8	Saved track results at 1-ms integration time for GPS L1 and L5 signals
trackResults_L1L5_20ms.mat	mat	8	Saved track results at 20-ms integration time for GPS L1 and L5 signals

Table 12.9 Snapshot receiver sample file names.

File name	File type	Reference chapter	Additional remarks
02.03.10-1s.snp	Samples	9	Binary raw data file for GPS L1, 1 second
27.04.10-1s.snp	Samples	–	Binary raw data file for GPS L1, 1 second
modo1_30s.bin	Samples	–	Binary raw data file for GPS L1, 30 seconds
20100302-dgc-1s.cfg	Text	9	Configuration file for 02.03.10-1s.snp
20100427-dgc-1s.cfg	Text	–	Configuration file for 27.04.10-1s.snp
20140212-UAB-s1m1-30s.cfg	Text	–	Configuration file for modo1_30s.bin
20100302-dgc-1s.cfg_AcqResults .bak.mat	MATLAB	9	Acquisition results file for 02.03.10-1s.snp
20100427-dgc-1s.cfg_AcqResults .bak.mat	MATLAB	–	Acquisition results file for 27.04.10-1s.snp
20140212-UAB-s1m1-30s.cfg_ AcqResults .bak.mat	MATLAB	–	Acquisition results file for modo1_30s.bin

these parameters in a separate txt or csv file that accompanied the file with the raw GNSS samples. The information was introduced manually, and thus, it could be prone to errors and misunderstandings. Sometimes additional information was also missing, such as whether the samples stored in the file did correspond to a big- or little-endian scheme, an information that was often taken for granted.

With the widespread adoption of SDR devices, it became clear that some common agreement was needed in order to standardize the information accompanying the raw GNSS samples files. This need was identified by the Institute of Navigation (ION), which setup a specific working group on this matter in 2014. The goal was to promote the interoperability between GNSS SDR data collection systems and GNSS software receivers, by means of a free and open standard for the exchange of metadata [4, 5]. The results of the working group led to the definition of an xml file describing in an unique way, how the samples of a given GNSS SDR data collection system were recorded and what the collection topology was (i.e. either single-band or multiband, single-antenna or multiantenna, single-stream or multistream). The GNSS SDR Metadata Standard was officially approved by ION in January 2020. The interested reader may find full information about this standard as well as sample data files in the official website [6].

The FGI-GSRx offers a MATLAB function called "xmlParserIONMetaData.m." This function reads some mandatory front-end configuration parameters from the "xml" metadata file accompanying the raw GNSS data samples. The FGI-GSRx user can enable the metadata reading by enabling "loadIONMetaDataReading" as true in the default parameter settings. In case the user enables loadIONMetaDataReading as true, the receiver will then expect the metadata file name to be provided with a complete path in the variable "metaDataFileIn." In such a case, the receiver over-writes the existing front-end configuration parameters given manually in the parameter text file (i.e. "default_param_MultiGNSS_Chapter7.txt"). The FGI-GSRx software receiver was tested to work with the following set of front ends: (i) SiGe, (ii) BladeRF, (iii) LimeSDR, (iv) NSL's Stereo v1 and v2, and (v) Teleorbit's GTEC. Some of the example raw GNSS data samples from different front ends can be downloaded from the ION SDR website [6].

References

[1] Mathworks, "MATLAB Installation Guide." https://se.mathworks.com/help/install/ug/install-mathworks-software.html.

[2] J. F. Kaiser, "On a Simple Algorithm to Calculate the Energy of a Signal," in *Proceedings of IEEE International Conference on Acoustics, Speech, and Signal Processing (ICASSP), Alburquerque, NM*, pp. 381–384, 1990.

[3] K. Borre, D. Akos, N. Bertelsen, P. Rinder, and S. H. Jensen, *A Software-Defined GPS and Galileo Receiver, Single Frequency Approach.* Birkhäuser, 2007.

[4] S. Gunawardena and T. Pany, "Initial Report of Activities of the GNSS SDR Metadata Standard Working Group," in *Proceedings of the 27th International Technical Meeting of the Satellite Division of the Institute of Navigation (ION GNSS+ 2014), Tampa, FL*, pp. 1426–1432, 2014.

[5] S. Gunawardena and T. Pany, "GNSS SDR Metadata Standard Working Group Report," in *Proceedings of the 28th International Technical Meeting of the Satellite Division of the Institute of Navigation (ION GNSS+ 2015), Tampa, FL*, pp. 3218–3221, 2015.

[6] Institute of Navigation (ION), "GNSS Software Defined Receiver Metadata Standard." https://sdr.ion.org/. Accessed: April 2, 2020.

Index